Chemistry for GCSE

Chemistry for GCSE

E. N. RAMSDEN
BSc, PhD, DPhil

Basil Blackwell

© 1986 E. N. Ramsden

First published 1986

Published by Basil Blackwell Ltd
108 Cowley Road
Oxford OX4 1JF
England

All rights reserved. No part of this publication
may be reproduced, stored in a retrieval system,
or transmitted, in any form or by any means,
electronic, mechanical, photocopy, recording
or otherwise, without the prior permission of
the publisher.

British Library Cataloguing in Publication Data

Ramsden, E. N.
   Chemistry for GCSE.
   1. Chemistry
   I. Title
   540   QD33

   ISBN 0-631-90047-0

Illustrated by Angela Lumley, Anne Langford,
Lorraine White and Jane Bottomley

Typeset in 11 on 13pt Melior
by Katerprint Typesetting Services, Oxford

Printed and bound in Great Britain by
Butler and Tanner, Frome

Contents

Acknowledgements viii

Preface ix

1 Solids, liquids and gases 1
1.1 Liquid crystals 1
1.2 The states of matter 2
1.3 Change of state 2
1.4 Particles 4
1.5 Pure substances and mixtures 7
Exercise 1 12

2 Pure substances from mixtures 14
2.1 The gold rush 14
2.2 Separating solids 15
2.3 Separating a solid from a liquid 15
2.4 Separating a solute and a solvent from a solution 17
2.5 Separating miscible liquids 18
2.6 Separating immiscible liquids 21
2.7 Separating the solutes in a solution: chromatography 21
2.8 Testing for purity 23
Exercise 2 26

3 Elements and compounds 28
3.1 Diamond 28
3.2 Graphite 30
3.3 Charcoal 31
3.4 Silicon 31
3.5 Gold 32
3.6 Iron 34
3.7 Elements 34
3.8 Compounds 35
3.9 Symbols 37
3.10 Formulas 38
3.11 Equations 40
Exercise 3 41

4 The atom 42
4.1 A momentous discovery 42
4.2 Protons, neutrons and electrons 43
4.3 The arrangement of particles in the atom 44
4.4 Electrons in orbit 45

*4.5 How are the electrons arranged? 46
4.6 The repeating pattern of the elements 47
4.7 Isotopes 48
4.8 Radioactivity 49
4.9 The oldest murder in history? 50
4.10 Making use of radioactivity 50
4.11 The dangers of radioactivity 51
4.12 The missing yellowcake 54
4.13 The atomic bomb 55
4.14 Nuclear reactors 57
4.15 Nuclear fusion 57
*4.16 The hydrogen bomb 58
Exercise 4 58

5 Electrochemical reactions 60
5.1 Electroplating 60
5.2 Which substances conduct electricity? 61
5.3 What happens when solutions conduct electricity? 61
5.4 Molten salts 63
5.5 What is the difference between an ion and an atom? 63
5.6 What happens at the electrodes? 65
*5.7 More complicated examples of electrolysis 66
5.8 Which ions are discharged? 67
5.9 Electrolysis with reactive electrodes 68
5.10 Applications of electrolysis 69
5.11 Chemical cells 72
Exercise 5 73

6 The chemical bond 74
6.1 Getting in the swim with ions 74
6.2 The ionic bond 74
6.3 The covalent bond 78
6.4 Shapes of covalent molecules 79
6.5 Ionic and covalent compounds 80
*6.6 Formulas of ionic compounds 82
*6.7 Formulas of covalent compounds 84
*6.8 Equations: the balancing act 84
Exercise 6 86

7 Acids, bases and salts 87
7.1 The acid taste 87
7.2 Acids 87
7.3 Bases 88
7.4 Neutralisation 90
7.5 Strengths of acids and bases 91
*7.6 Acids and bases redefined by Brönsted and Lowry 93

---
*More advanced material required for higher grades.

*7.7 Weak acids and bases 94
*7.8 Amphoteric oxides and hydroxides 95
7.9 Salts 96
7.10 Uses of salts 97
7.11 Methods of making salts 98
Exercise 7 103

8 Air 105

8.1 Saving lives with heroism and oxygen 105
8.2 Air 105
8.3 How to obtain oxygen and nitrogen from air 106
8.4 Uses of oxygen 106
8.5 Nitrogen 109
8.6 Carbon dioxide and water vapour 109
8.7 The noble gases 110
8.8 How can oxygen be prepared in the laboratory? 111
8.9 The reactions of oxygen with some elements 112
8.10 Combustion 114
8.11 Breathing and respiration 115
8.12 Rusting 116
8.13 Pollution 116
Exercise 8 127

9 Water 129

9.1 We need it 129
9.2 Why is water so important? 129
9.3 The water cycle 130
9.4 Dissolved oxygen 131
9.5 Water treatment 132
9.6 Sewage works 132
9.7 What do we use it for? 134
9.8 Chemically speaking 134
9.9 Water as a solvent 135
9.10 Water in limestone regions 135
9.11 Soaps and detergents 136
9.12 Hard and soft water 137
9.13 Methods of softening hard water 138
9.14 Advantages of hard water 139
9.15 Pure water 139
9.16 Pollution of water 139
9.17 Hydrogen 143
Exercise 9 145

10 Metals 147

10.1 Metals and alloys 147
10.2 The metallic bond 148
10.3 The chemical reactions of metals 149
10.4 The reactivity series 153
10.5 Metals in the Periodic Table 154
10.6 What predictions can be made from the reactivity series? 155
10.7 Uses of metals and alloys 157
10.8 Compounds and the reactivity series 158
10.9 Extraction of metals from their ores 159
10.10 Focus on iron and steel 164
10.11 Rusting of iron and steel 167
10.12 Focus on aluminium 170
10.13 Some problems which metallurgists have solved 174
Exercise 10 175

11 Chemical calculations 179

11.1 Relative atomic mass 179
11.2 Relative molecular mass 179
11.3 Relative formula mass 180
11.4 Percentage composition 180
*11.5 The mole 181
*11.6 Calculating the mass of reactant or mass of product 184
*11.7 Finding the equation for a reaction 185
*11.8 Finding formulas 186
*11.9 Molecular formula and empirical formula 188
*11.10 The volumes of reacting gases 189
*11.11 Reactions in solution 191
Exercise 11 195

12 Limestone, chalk and sand 197

12.1 Concrete 197
12.2 Limestone and lime 199
12.3 Carbonates 201
12.4 Hydrogencarbonates 201
12.5 Carbon dioxide 202
12.6 The carbon cycle 203
12.7 The greenhouse effect 204
12.8 Fighting fire with carbon dioxide 206
12.9 Sand and limestone: a four thousand year old combination 208
12.10 Silicon 209
12.11 Some problems and their solutions 209
Exercise 12 211

13 Sulphur 213

13.1 Sulphur 213
13.2 Allotropes of sulphur 214
13.3 Sulphur dioxide 216
*13.4 Sulphites 217
13.5 Sulphuric acid 218
13.6 Manufacture of sulphuric acid: the Contact Process 219

---

*More advanced material required for higher grades.

13.7 Reactions of sulphuric acid 221
13.8 Sulphates 223
13.9 Methods of preparing sulphates 223
Exercise 13 224

14 Nitrogen and agriculture 226

14.1 The nitrogen cycle 226
14.2 Nitrogen 227
14.3 Fertilisers 227
14.4 Ammonia 230
14.5 Nitric acid 233
14.6 Nitrates 236
14.7 Agricultural chemicals 236
Exercise 14 239

15 Chemicals from salt 242

15.1 Sodium chloride 242
15.2 Hydrogen chloride 244
15.3 Reactions of hydrochloric acid 245
*15.4 Solutions of hydrogen chloride in organic solvents 246
15.5 Preparation of chlorides 246
15.6 Chlorine in war 247
15.7 Chlorine in peace 248
15.8 Properties and reactions of chlorine 248
*15.9 Chlorine as an oxidising agent 250
15.10 The halogens 253
15.11 Fluoride: friend or foe? 254
15.12 Iodides can be useful too 255
Exercise 15 256
Crossword on Chapter 15 257

16 Hydrocarbon fuels 258

16.1 Biogas 258
16.2 Alkanes 259
16.3 Halogenoalkanes 261
16.4 Hydrogen 262
16.5 Combustion 262
16.6 Petroleum oil and natural gas 263
16.7 Coal 269
Exercise 16 269

17 Alkenes, alcohols and acids

17.1 Plastic sand: what next? 270
17.2 Alkenes 270
17.3 Reactions of alkenes 271
17.4 Polymers 273
17.5 Alcohols 278
17.6 Ethanoic acid 282
17.7 Esters 283

17.8 The chemical industry 283
17.9 Organic compounds in medicine 284
Exercise 17 288

18 Energy 290

18.1 The energy crisis 290
18.2 Some solutions 290
18.3 Exothermic reactions 293
18.4 Endothermic reactions 295
18.5 Heat of reaction 296
18.6 Getting over the barrier 298
*18.7 Calculations on heat of reaction 298
18.8 The nuclear debate 300
Exercise 18 303

19 The speeds of chemical reactions 305

19.1 Why reaction speeds are important 305
19.2 How does the size of solid particles affect the speed at which they react? 306
19.3 Concentration and speed of reaction 307
19.4 Pressure and speed of reaction 309
19.5 Temperature and speed of reaction 309
19.6 Light affects the speeds of some reactions 309
19.7 Catalysis 310
Exercise 19 312

Table of symbols, atomic numbers and relative atomic masses 316

The Periodic Table 318

Numerical answers 320

Index 321

---

*More advanced material required for higher grades.

Acknowledgements

I am grateful to Mr G. H. Davies, Mr D. Haslam and Mr M. Vokins for their helpful comments and suggestions. I thank Ms Laura Hale for the original drawings on which the cartoons are based and Mr Keith Waters for taking many of the photographs. Organisations which have kindly supplied photographs and given permission for their use are:

Alyeska Pipeline Service Company 4.14(c); Anglian Water 16.3; Associated Press 9.4, 9.18; Austin Rover 10.30, 13.12; Australian Information Service 10.18; Automobile Association 10.1; Barnaby's Picture Library 15.2; BBC Hulton Picture Library 2.1, 7.8, 8.24, 8.26, 10.29, 13.9; BOC Ltd. 2.7, 8.4, 8.5; Brazilian Institute of Sugar and Alcohols 17.17; British Airports Authority 10.3; British Alcan Aluminium Ltd. 10.32; British Museum of Natural History 1.9; The British Petroleum Company Ltd. 2.11(b), 3.2, 3.3, 16.8; British Steel Corporation 10.21; Cambridge University Collection of Air Photographs 9.7; W. Canning Materials Ltd. 5.1, 5.2; Capper Pass 10.22; Judy Cass 16.2; Central Electricity Generating Board 14.13; Chubb Fire Security Ltd. 12.20; Cookson Industrial Materials Ltd. 4.10, 4.15(b); Peter Crawshaw 8.12; DJB Engineering Ltd. 10.31, 13.3; Duracell 5.13; Esso Petroleum Company Ltd. 16.9; Farmers Weekly 12.7; Ferranti 3.6; Fisons plc 13.13; Gold Information Office 3.7(a), 3.7(b); The Goodyear Tyre & Rubber Company (Great Britain) Ltd. 13.2; Professor G.W. Gray FRS 1.1; Greenpeace 8.22; J. Hazzard 8.6; HMSO cover, 8.23; Humber Bridge Board 12.1; ICI plc 2.5, 5.11, 14.4, 14.5, 14.7, 15.4; Imperial War Museum 4.17; Institute of Geological Sciences 9.10; Jeyes Ltd. 14.9; Professor D.A. Jones, University of Hull 14.1; London Fire Brigade 12.17; The Mary Rose Trust Ltd. 10.25; Metropolitan Police 17.15; Miller Homes 18.3; John Mills Photography Ltd. 4.3; NASA 3.8, 8.7; New Scientist 8.11; New Zealand Forest Products Ltd. 13.10; Oxfam 9.1, 9.2; Oxford Scientific Films 9.17; Pacemaker Press International Ltd. 10.24; Pepsi-Cola 12.11; Pilkington Group 12.22; Popperfoto 8.1; Shell Photographs 9.19, 16.7; The Sport and General Press Agency Ltd. 18.6; Thames Water 8.8, 9.3; Tyrrell 18.5; UKAEA 4.14(a), 4.14(b), 4.15(a), 4.15(c); Van den Burghs & Jurgens Ltd. 9.20; K. Waters 1.12(a–d), 2.11(a), 4.11, 7.1, 7.3, 7.4, 8.9, 9.11, 10.17, 13.8, 14.17, 15.1, 15.13, 16.5, 17.2, 17.4, 17.5, 17.13, 17.18; The Wind Energy Group 18.4; Yachting Monthly 17.11; G.H. Zeal 1.2.

In addition, the following illustrations have been redrawn with permission:

Figure 10.1 is taken from *AA Book of the Car*, The Reader's Digest Association Limited, London. Used with permission.
Figure 12.5 is redrawn by permission of Blue Circle Industries PLC.
Figure 12.5 is taken from *New Scientist*, 15 March 1984.
Figure 16.13 is adapted from *Energy* by J. Ramage published by Oxford University Press 1983.
The graphs on p. 313 are adapted from B. Selinger, *Chemistry in the Marketplace*, by permission of Brace, Harcourt, Jovanovich.

I thank the staff of Basil Blackwell for the care which has gone into the editing, illustration and production of the book, and my family for their encouragement.

E. N. Ramsden, 1986

Preface

This text covers the GCSE Chemistry syllabuses of the London and East Anglian Group, the Midlands Examining Group, the Northern Examining Association, the Northern Ireland Schools Examinations Council, the Southern Examining Group and the Welsh Joint Education Committee. Most of the examining boards indicate the distinction between core material and material which is required for the higher grades. In this text, the more advanced material is denoted by asterisks. It is assumed that readers can find out what they need for the syllabus they are following.

The text emphasises the social, economic, technological and environmental aspects of chemistry. Each topic is introduced in a social context. Throughout, the practical applications and social implications of the subject are interwoven with the theory.

Some of the topics of economic importance which are included are the contribution of fertilisers to the world's food supply, the energy crisis and the need for careful use of the Earth's resources. The importance of chemical technology is apparent in the coverage of metals, crystals, liquid crystals, plastics and other topics. The text spotlights some of the medical benefits of chemistry, such as anaesthetics, antiseptics and painkillers. Some instances of the impact of the chemical industry on society are described in the accounts of the pollution and purification of water, the pollution and purification of air and the effects on the landscape of mining and quarrying. The idea that science is a rewarding human activity is fostered by giving credit for their discoveries to people like Alfred Nobel, Marie Curie, Roy Plunkett and other present-day scientists. Scientific method is illustrated in the accounts of the development of the fertiliser industry, the analysis of the effect of fluoride ion on tooth decay and the discovery of radioactivity. Some case histories of the solving of scientific and technical problems, such as the manufacture of plate glass and 'gasohol', are included.

Where possible, information is presented in the form of charts and diagrams. At intervals in the text, margin summaries focus the readers' attention on salient points. At the end of each short topic, there is a brief 'Just Testing' section to allow readers to reinforce their grasp of that topic before they proceed to the next. The exercises at the ends of the chapters give readers a chance to apply the knowledge they have gained to different situations and to problem solving.

The author assumes that readers will be following a programme of practical work. The reason why the text does not include practical work is that it is difficult for experiments to be genuinely investigative if they are incorporated in the main text, close to a description of the results of those experiments. The experiments to which the text refers are to be found in the companion book, *Practical Chemistry for GCSE*, which contains pages for photocopying.

Thumbnail sketches of careers in chemistry have been included. Pupils who are interested in planning such careers will find them a useful starting point.

E. N. Ramsden, 1986

CHAPTER 1 Solids, liquids and gases

Fig. 1.1 Liquid crystals seen through a microscope

SUMMARY NOTE

Liquid crystals are an unusual state of matter. They change colour with temperature. They are used in thermometers and in LCDs.

1.1 Liquid crystals

Liquid crystals are an unusual kind of matter. These substances look like liquids, but seen through a microscope they show strange patterns (Fig. 1.1). They flow like liquids, but in other ways they behave like crystals.

You will have seen how widely liquid crystals are used in digital watches and calculators. The letters 'LCD' stand for 'liquid crystal display'. You have probably seen them in use as room thermometers. This is because liquid crystals change colour as their temperature changes. Doctors use them as thermometers to help in cancer diagnosis. The temperature of a cancer is higher than that of the surrounding tissue because more blood circulates in the tumour area. Liquid crystals can show up the cancer as a 'hot spot'. The doctor paints the patient with a film of liquid crystal. Different liquid crystals have different colour ranges. For this use, the doctor might choose one that is red at normal body temperature and blue at a higher temperature. Then a cancer would show up as a blue 'hot spot' on a red background. This technique is used to detect cancer of the face, throat or breast, where the suspect area is easy to paint.

Liquid crystals are an unusual **state of matter**. You can read about the chief states of matter in Section 1.2.

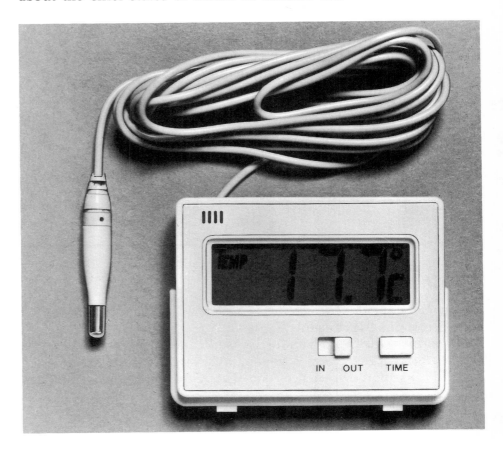

Fig. 1.2 An LCD display on a thermometer

1.2 The states of matter

Water, air, minerals, trees and animals are all **matter**. Anything that occupies space and has mass is matter. There are different kinds of matter: **solid**, **liquid** and **gas**. These are referred to as the **states of matter**. You are familiar with water in all three states: solid **ice**, liquid **water** and gaseous **water vapour**. Table 1.1 summarises the differences between the three states of matter. They are given **state symbols**: solid (s), liquid (l) and gas (g).

Table 1.1 States of matter

State	Description
Solid (s)	Has a definite mass. Has a definite volume. The shape is usually difficult to change.
Liquid (l)	Has a definite mass. Has a definite volume. Changes its shape to fit the container.
Gas (g)	Has a definite mass. Changes its size and shape to fit the size and shape of its container.

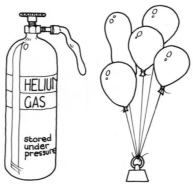

> **SUMMARY NOTE**
>
> The three chief states of matter are:
> - solid (s): fixed volume and shape
> - liquid (l): fixed volume; shape changes
> - gas (g): neither volume nor shape is fixed.

1.3 Change of state

Matter can change state (see Experiment 1.1). Some changes of state are summarised below.

Sometimes, the word **vapour** is used to describe a gas. A liquid evaporates to form a vapour. All vapours are gases. A gas is called a vapour when it is cool enough to be liquefied by compression.

$$\text{Vapour} \xrightarrow{\text{Either cool or compress without cooling}} \text{Liquid}$$

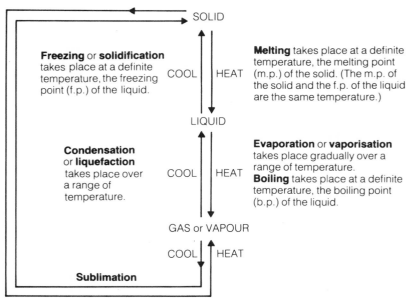

Some solids do not melt: they change directly into a vapour when heated. The vapour solidifies on cooling. This behaviour is **sublimation**.

Solid carbon dioxide and iodine sublime.

SUMMARY NOTE

Matter can change from one state to another. The changes in state are:
- melting
- freezing or solidification
- evaporation or vaporisation
- condensation or liquefaction
- sublimation.

A vapour is a gas which can be liquefied by compression.

The hotter a liquid is, the more quickly it evaporates. Eventually, it becomes hot enough for bubbles of vapour to appear inside the liquid. When this happens, the liquid is boiling. The temperature at which this happens is the **boiling point** of the liquid (see Chapter 2).

Physical and chemical changes

All the changes discussed so far are **physical changes**. The substance heated or cooled changes its state only. It does not change into a different substance. Physical changes are easily reversed. Many of the changes we shall be studying are **chemical changes**. When a substance changes into a new substance, a chemical change or **chemical reaction** has happened (see Experiments 3.1–3.4). Chemical changes are difficult to reverse.

SUMMARY NOTE

Changes of state are **physical changes**. They are different from chemical changes. In a **chemical change**, a new substance is formed.

JUST TESTING 1

1 Copy this scheme. Fill in the names of the processes represented by arrows. Some of the processes have two names.

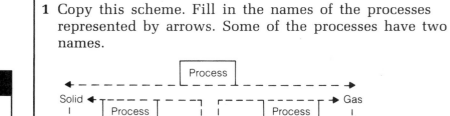

1.4 Particles

Scientists believe that matter is made up of tiny **particles**. The way in which solids, liquids and gases behave can be explained with the help of this theory. It is called the **particle theory of matter**.

Change of state explained by the particle theory

Vaporisation. The vaporisation of liquids can be explained. Figure 1.3 shows what happens when a drop of the brown liquid, bromine, is put into a gas jar. You can explain what happens if you assume that bromine consists of tiny particles. Given all the space in the gas jar to occupy, particles leave the liquid state and move about until they are spread throughout the gas jar.

Solid, liquid and gaseous states. We can explain the relationship between the solid, liquid and gaseous states of matter.

In a solid, the particles are arranged in a regular three-dimensional pattern. They cannot move out of position. The only movement they can make is to vibrate.

If a solid is heated, the particles gain energy. They vibrate more and more until they break away from their regular arrangement. When this happens, the solid has turned into a liquid: it has melted.

A liquid has no fixed shape because the particles are free to move. If a liquid is heated further, the particles gain still more energy. Some of them will break free from the body of the liquid: they will become a gas.

A gas occupies a very much larger volume than the liquid which vaporised to make it. One litre of water forms 1200 litres of water vapour. The particles are very much further apart in a gas than in a liquid, and are free to move.

Particles in motion: the kinetic theory

The **kinetic theory of gases** tells us that gases consist of particles which are always moving. (*Kinetic* comes from the Greek word for *moving*.) Most of a gas is space, and the particles shoot through the space at high speed. Occasionally they collide with other particles or with the walls of the container.

The kinetic theory can explain the connection between the pressure and the volume of a fixed mass of gas:

- Increase the pressure → the volume decreases (Fig. 1.5a)
- Decrease the pressure → the volume increases (Fig. 1.5b)

Gases are much more compressible than other forms of matter. Since a gas consists largely of space, the particles occupying the space can move closer together. Gases exert pressure because their particles are colliding with the walls of the container. If the volume of the gas is decreased, the particles will hit the walls more often, and the pressure will increase.

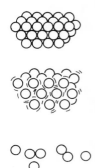

Fig. 1.3 Liquid bromine vaporising

Fig. 1.4 The arrangement of particles in a solid, a liquid and a gas

SUMMARY NOTE

The particle theory of matter is the theory that matter is made up of tiny particles. The theory explains the differences between the solid, liquid and gaseous states, and how matter can change state.

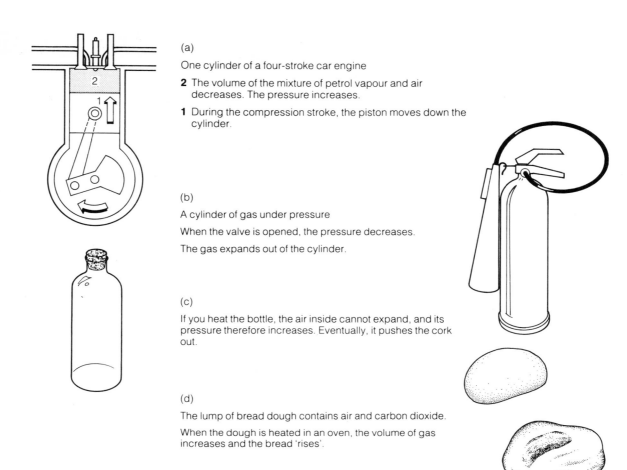

(a) One cylinder of a four-stroke car engine

2 The volume of the mixture of petrol vapour and air decreases. The pressure increases.

1 During the compression stroke, the piston moves down the cylinder.

(b) A cylinder of gas under pressure

When the valve is opened, the pressure decreases.

The gas expands out of the cylinder.

(c) If you heat the bottle, the air inside cannot expand, and its pressure therefore increases. Eventually, it pushes the cork out.

(d) The lump of bread dough contains air and carbon dioxide.

When the dough is heated in an oven, the volume of gas increases and the bread 'rises'.

Fig. 1.5 The connection between pressure, volume and temperature

The kinetic theory can also explain how the pressure and the volume of a fixed mass of gas change with temperature:

- Increase the temperature → the pressure increases (Fig. 1.5c) (while keeping the volume constant)
- Increase the temperature → the volume increases (while keeping the pressure constant): the gas expands (Fig. 1.5d)

If the temperature of a gas increases, the particles will have more energy, move faster and collide more often with the walls of the container. If the gas cannot expand, the pressure will increase. If the pressure is kept constant while the temperature is raised, the gas will expand.

Evidence for the kinetic theory

Diffusion. What evidence have we that the particles of a gas are moving? The **diffusion** of gases can be explained. Diffusion is the way gases spread out to occupy all the space available to them. Figure 1.6 shows what happens when a jar of the dense green gas, chlorine, is put underneath a jar of air. If gases consist of small particles which are always on the move, it is easy to see how diffusion can occur. Moving particles of air and chlorine can spread themselves between the two gas jars.

Dense gases, such as carbon dioxide, diffuse more slowly than

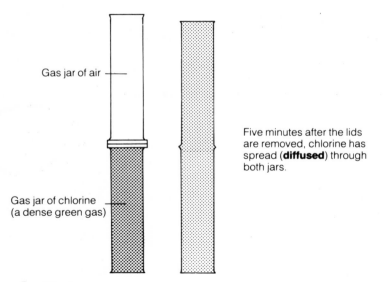

Fig. 1.6 Gaseous diffusion

> **SUMMARY NOTE**
>
> The kinetic theory of gases says that the particles in a gas are always in motion. The theory explains the compressibility of gases. It explains gas pressure and why gas pressure increases with temperature. It can also explain gaseous diffusion.
>
> The particles in a liquid move too. Brownian motion is evidence of this.

gases with a low density, such as hydrogen. This is because dense gases have heavy particles, which move more slowly than light particles. Figure 1.7 shows an experiment to compare the speeds of diffusion of the gases ammonia and hydrogen chloride.

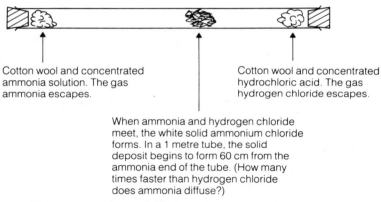

Fig. 1.7 An experiment to compare the rates of diffusion of hydrogen chloride and ammonia

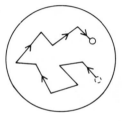

(a) A view through a microscope of the path taken by a grain of pollen floating in water

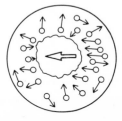

(b) A magnification of (a) (not to scale). Water particles, which are invisible under a microscope, strike the pollen grain from every direction. If at any instant more water particles collide with the right-hand side than with the left-hand side, as shown here, the pollen grain will move from right to left.

Fig. 1.8 Brownian motion

Brownian motion. What evidence have we that the particles of a liquid are moving? A botanist called Robert Brown saw something mysterious in 1785. He was looking through a microscope at grains of pollen floating on water. He saw that they were constantly moving about (Fig. 1.8a). We now call this movement **Brownian motion**. We can explain it by the particle theory (Fig. 1.8b).

Career sketch: Laboratory assistant 1

Bob works in industry. His job is to make a complete analysis of the firm's products. He must make sure that the analysis of each product matches that on the label. His laboratory is well equipped with instruments to use for titration, chromatography, radiochemistry and other methods of analysis. Bob left school after GCSE.

JUST TESTING 2

1. Mother is out and two hungry sisters approach the kitchen.
   Alice That soup smells good, Dad.
   Dad It shows that Mum isn't the only cook round here.
   Monica (just back from school) I *think* it shows that the particle theory of matter might be true.
   Alice What are you talking about?
   Dad Have you been doing chemistry at school?
   Explain why the fact that Monica could smell the soup outside the kitchen is evidence for the particle theory of matter.
2. Explain how a teaspoon of salt can flavour a whole pan of soup.
3. A girl enters a room wearing perfume. Soon a boy standing at the other side of the room can smell it. How does the perfume reach his nose?
4. The smoke from a bonfire rises lazily. Hot air rises and takes the smoke with it, but why does smoke sometimes waft from side to side?

1.5 Pure substances and mixtures

You have seen how matter can be subdivided (**classified**) into different states. Another way of classifying matter is to divide it into pure substances and mixtures.

Pure substances

A **pure** substance is a **single** substance. No other substance is present in a pure substance. Pure X is 100% X. Impure X contains another substance or substances as well as X.

We sometimes use the word *pure* in a different sense. We talk about 'pure' water, 'pure' air or 'pure' food. If a health official describes water as 'pure', he or she means that the water is safe to drink. In this sense, *pure* means *containing no substance which is harmful to health*. 'Pure' drinking water does in fact contain many substances dissolved in it (see Chapter 11). Chemists use the word 'pure' to mean 'consisting of one substance only'. If a chemist describes water as *pure*, he or she means that the water *contains no other substance*.

Crystals. Many pure solids are **crystalline**. A **crystal** is a piece of solid which has a regular shape and smooth faces which reflect the light (Fig. 1.9). Most crystalline material is made up of many small crystals. It may therefore not appear to be crystalline until you look at it under a microscope.

Solids which do not consist of crystals are **amorphous** (without shape).

Speaking of crystals. These beautiful forms of matter are useful too. Quartz crystals vibrate when an electric current is applied.

Fig. 1.9 Crystals of quartz

SUMMARY NOTE

A pure substance is a single substance.

The rate of vibration is absolutely constant. A quartz crystal (silicon(IV) oxide) is used to keep a radio transmitter on a fixed frequency. This is why you always find your favourite programme at the same spot on the tuner. A quartz crystal controls the timing of the liquid crystal display in your quartz watch.

When light falls on a crystal of a substance called selenium, electricity is generated. In a photographer's light meter, the amount of light falling on a crystal of selenium determines the amount of electric current generated. This shows as a deflection of the needle in the light meter.

In satellites there are solar cells consisting of thousands of silicon crystals. These crystals also convert light energy into

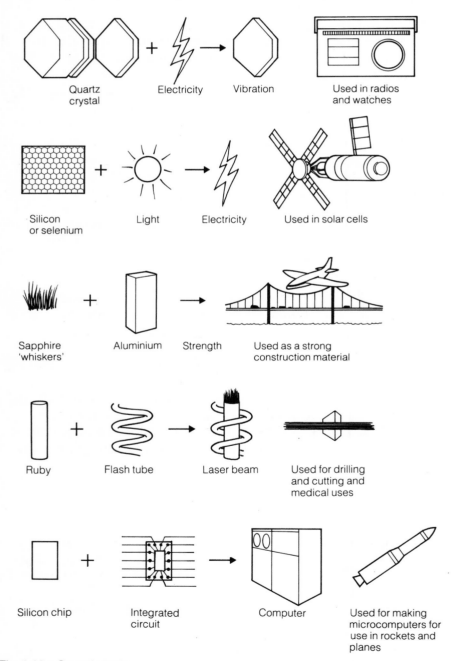

Fig. 1.10 Crystal magic

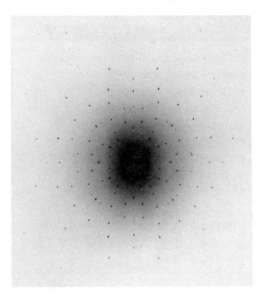

Fig. 1.11 X-ray pattern from crystals of a metallic alloy

electrical energy. They provide the current that powers the satellite.

Sapphires (aluminium oxide with impurities) can be grown as whisker-shaped crystals. They have amazing strength. When they are combined with metals, they give materials which are strong even at high temperatures. They are used in the construction of supersonic aircraft. The wings and fuselage reach high temperatures because of the heat generated by friction at such high speeds.

Lasers are intense beams of light. They can drill holes, cut through metals and destroy cancers. Laser beams shoot from ruby crystals (aluminium oxide with impurities). A flash tube coiled around the crystal gives a sudden powerful burst of light. This interacts with the ruby to give rise to a laser beam.

Silicon crystals are the basis of the computer industry. The crystals are sliced into 'chips'. Miniature electronic circuits are built on to the chips to make **integrated circuits** (ICs). These ICs are the brainpower of the microcomputer. (See Section 3.4.)

Why are substances crystalline? The reason for the regular arrangement of faces is that the particles of which the crystal is made are arranged in a regular manner. Figure 1.11 shows an X-ray pattern produced by passing a beam of X-rays through a crystal on to a photographic plate. The dark area in the centre is produced by X-rays which have passed straight through the crystal. The pattern of dots is produced by X-rays which had to change direction because they met particles. Since the pattern of dots is regular, the arrangement of particles in the crystal must be regular. In this way, X-ray photographs give evidence that matter consists of particles.

Elements and compounds. There are two classes of pure substances: **elements** and **compounds**. **Elements** are substances which cannot be split up into simpler substances. Copper is an element; whatever you do with it, you cannot split copper into simpler substances.

Compounds are pure substances which *can* be split into simpler substances. Often, they can be split into elements. Usually, energy must be supplied in the form of heat or light or electricity to split up a compound. A chemical reaction takes place when a compound splits up. A compound contains two or more elements chemically combined. You will learn more about compounds in Chapter 3.

SUMMARY NOTE

Many pure solids are crystalline. Some crystals have important uses. Radio transmitters, light meters, solar cells, construction materials, lasers and microcomputers all depend on crystals.

X-ray photographs of crystals provide evidence for the particle theory of matter.

'Elementary, my dear Watson!'

'By **elementary**, do you mean **simple**, Holmes?'

'It's the same in chemistry, Watson. An **elementary** substance, an **element**, is a **simple** substance: whatever you do, you can't split it up into simpler substances.'

The atom. Elements are made up of minute particles. The smallest particle of an element is called an **atom**. All the atoms in an element behave in the same way. Atoms are so small that it is difficult to imagine just how small they are. It takes 4 million hydrogen atoms to stretch across 1 millimetre.

In some elements, atoms join together by chemical bonds to form larger particles called **molecules** (Fig. 1.12).

SUMMARY NOTE

Pure substances are either elements or compounds. Compounds are pure substances which are composed of two or more elements. Elements are pure substances which cannot be split up into simpler substances.

The smallest particle of an element is an atom. Molecules are groups of atoms joined by chemical bonds.

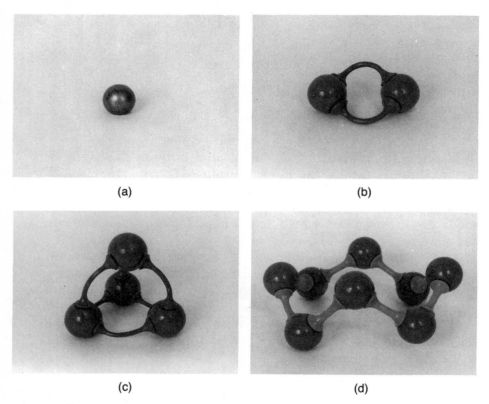

Fig. 1.12 Molecules of (a) helium, (b) oxygen, (c) phosphorus, (d) sulphur

Just think! One million hydrogen atoms in a row would only stretch across one grain of sand!

There's room for 5 million million hydrogen atoms on a pinhead!

Mixtures

There are two kinds of mixtures:
- **Solutions**: Every part of a solution has the same composition. A solution is a **homogeneous mixture**; it is the same all through.
- **Heterogeneous mixtures**: When a mixture is not the same all through (that is, when different parts of the mixture have different compositions), the mixture is described as **heterogeneous**.

Solutions. A solution consists of a **solvent** and a **solute**. The solid or liquid or gaseous solute **dissolves** in the solvent, which is usually a liquid. Water is the most common solvent, but there are many others, e.g. alcohols such as ethanol (Chapter 17) and chloro-compounds such as trichloroethane (Chapter 16).

Dissolving is another change that can be explained in terms of particles. Figure 1.13 shows a coloured substance dissolving. Coloured particles split off from the crystal and move through the solution until they are evenly spread out.

A **concentrated** solution is one which contains a high

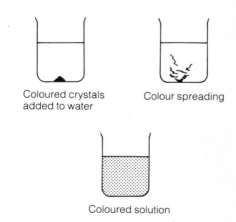

Fig. 1.13 Crystals dissolving

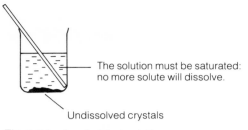

Fig. 1.14 A saturated solution

proportion of solute. A **dilute** solution is one which contains only a small proportion of solute.

A **saturated** solution is a solution which has as much solute dissolved in it as is possible at that temperature. If there is undissolved solute present in the bottom of the container, you know that the solution must be saturated (Fig. 1.14).

The concentration of solute in a saturated solution is the **solubility** of the solute. The solubility of a solute is the mass (in grams) of the solute that will saturate 100 grams of solvent at a certain temperature. You must state the temperature because solubility changes with temperature. A graph of solubility against temperature is called a **solubility curve** (see Fig. 1.15).

For most solutes, the solubility increases with temperature. It follows that when a saturated solution is cooled, the solution can hold less solute at the lower temperature. Some solute comes out of solution: it **crystallises** (see Experiments 1.2, 1.3). In contrast, gases are less soluble at higher temperatures (see Chapter 8).

SUMMARY NOTE

Mixtures can be homogeneous or heterogeneous.

Solutions are homogeneous mixtures. They may be dilute, concentrated or saturated. The concentration of solute in a saturated solution is the **solubility** of the solute. Solubility is the maximum mass of solute that will dissolve in 100 g of solvent at a stated temperature.

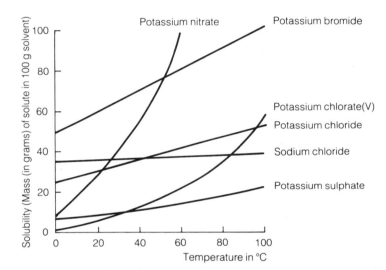

Fig. 1.15 Solubility curves for some solutes

JUST TESTING 3

1 Figure 1.15 shows some solubility curves.
   a 1 kg of a saturated solution of sodium chloride is cooled from 100 °C. What mass of sodium chloride crystallises out?
   b A solution is saturated with both potassium chloride and potassium chlorate(V). If 1 kg of the solution is cooled to 60 °C, what mass of (i) potassium chloride and (ii) potassium chlorate(V) crystallises out?
   c A solution containing enough potassium nitrate and potassium bromide to saturate it at 60 °C is cooled to 20 °C. What masses of the two solutes will separate from 1 kg of the solution?

2 A concentrated solution of copper(II) sulphate is deep blue. When you dilute it with water, it becomes a paler blue. How can you explain this in terms of the particle theory of matter?

Heterogeneous mixtures. In a heterogeneous mixture, particles of one substance are dispersed (spread) through another substance (see Table 1.2).

Table 1.2 Some heterogeneous mixtures

Type	Description	Example
Suspension	Solid particles dispersed in liquid	Clay
Emulsion	Liquid droplets dispersed in liquid	Milk, mayonnaise
Foam	Gas bubbles dispersed in liquid	Soap suds, whipped cream
Mist	Liquid droplets dispersed in gas	Fog, aerosol products
Smoke	Solid particles dispersed in gas	Dust-laden air, smoke

SUMMARY NOTE

There are various types of heterogeneous mixtures: suspensions, emulsions, foams, mists and smokes.

JUST TESTING 4

1 Copy this scheme and fill in the blanks.

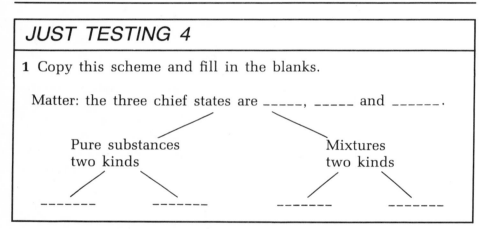

Matter: the three chief states are _____, _____ and _____.

Pure substances — two kinds

Mixtures — two kinds

Exercise 1

1 Name some common substances which are **a** solids, **b** liquids and **c** gases.

2 At room temperature, oxygen is a gas, water is a liquid and sugar is a solid. Which of these substances has the strongest forces of attraction between its particles? Which has the weakest?

3 When mothballs are put into a cupboard, they slowly disappear. Do they **a** melt, **b** condense, **c** sublime or **d** distil?

4 If the sun comes out after a rainshower, the puddles dry up faster than if it stays cloudy. Explain why this happens.

5 Of the processes:
Evaporation, Condensation, Distillation, Filtration, Sublimation,
select the process which

  a makes water collect on window panes in cold weather
  b makes water disappear from puddles
  c can be used to separate water from sand
  d includes the changes:
    water → steam → water
  e describes the change:
    ice → water vapour.

6 A weather balloon is being inflated. It will rise 11 km into the atmosphere. It is only partially filled when it is inflated on the ground. Why should it not be filled completely?

7 Copy this passage and fill in the blanks.
Ice is solid _____. Dry ice is solid _____ _____. Ice feels wet because _____. Dry ice feels dry because _____. When ice takes

heat from the surroundings, it _____. When dry ice takes heat from the surroundings, it _____. As dry ice cools the surrounding air, water vapour condenses out of the air. This is why when pop groups use dry ice on stage you see _____.

8 The air pressure in the tyres on a car was measured on a cool morning. The car was driven all through the day, which turned out to be sunny and hot. What happened to the pressure in the tyres? Give an explanation according to the kinetic theory of gases.

9 Which goes 'flat' faster, an open bottle of Coke on the kitchen table or an open bottle in the fridge? Explain your answer.

10 One 2 litre balloon is filled with helium. A second 2 litre balloon is filled with air at the same temperature and pressure. It takes 0.32 g of helium to fill the first balloon and 2.4 g of air to fill the second. If the two balloons are punctured (with the same size holes), which balloon will deflate faster? Give a reason for your answer.

11 Name some common substances which are **a** homogeneous solids, **b** heterogeneous solids, **c** homogeneous liquids and **d** heterogeneous liquids.

12 Name some common substances which are **a** solutions, **b** solutes present in solutions, **c** solvents.

Study Fig. 1.16 before answering questions **13–15**.

13 What is the solubility of potassium sulphate at 80 °C?

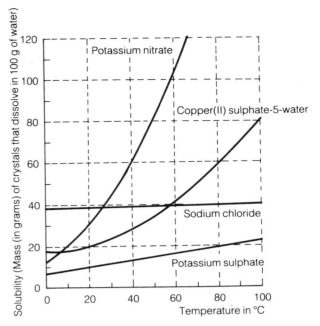

Fig. 1.16   Some solubility curves

14 a What mass of copper(II) sulphate-5-water will dissolve in 100 g of water at 100 °C?
  b What happens when the solution is cooled to 50 °C?

15 a What is the solubility at 60 °C of (i) potassium nitrate, (ii) sodium chloride?
  b What is the solubility at 20 °C of (i) potassium nitrate, (ii) sodium chloride?
  c A solution is saturated at 60 °C with potassium nitrate and with sodium chloride. A 100 g sample of solution is cooled from 60 °C to 20 °C. What mass of crystals forms?

CHAPTER 2 Pure substances from mixtures

2.1 The gold rush

In the Gold Rush days, some prospectors found nuggets of pure gold. They were the lucky ones. Most gold miners had to separate particles of gold from a large mass of rock or sand. Figure 2.1 shows an old-timer separating gold dust from sand. He is swirling the mixture with water in his miner's pan. Sand is less dense than gold dust. Therefore it swishes out of the pan with the wash water, while gold dust settles in the bottom of the pan. This is not an efficient method, but many people made a living using it.

Very few substances are found pure in nature. It is more usual to find the materials we want to use mixed with other materials. Chemists have to find ways of separating useful substances from the materials with which they are mixed.

> **SUMMARY NOTE**
>
> We need to find ways of separating useful substances from natural sources.

Fig. 2.1   Panning for gold

2.2 Separating solids

Dissolving one of the substances

Salt is mined in Cheshire. There are two ways of getting it out of the ground (see Section 15.1). One way is to dissolve it out: salt is soluble, whereas rock is not. A hole is bored into the salt deposit and two pipes are lowered into it. Water is pumped down one pipe and allowed to dissolve the salt. Brine (a solution of salt) is pumped up to the surface through the second pipe (Fig. 2.2). The brine obtained in this way is used by industry in manufacturing processes. In Experiment 2.1, you can find out how to obtain pure salt from rock salt.

> **SUMMARY NOTE**
>
> Sometimes it is possible to separate one solid from others by dissolving it and filtering the solution to leave undissolved material behind.

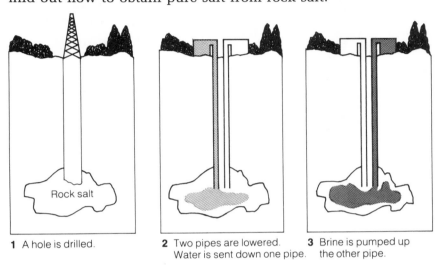

Fig. 2.2 Solution mining: dissolving salt in an underground mine

1 A hole is drilled.
2 Two pipes are lowered. Water is sent down one pipe.
3 Brine is pumped up the other pipe.

2.3 Separating a solid from a liquid

From a heterogeneous mixture: by filtration

Filtration separates a solid from a liquid by using a sieve which allows liquid to pass through but does not allow solid particles to pass through. In the laboratory, the sieve used is often **filter paper**. A liquid can pass through pores in the filter paper; a solid cannot. Figure 2.3 shows laboratory apparatus for filtration. Figure 2.4 shows a faster method: filtering under reduced pressure.

> **SUMMARY NOTE**
>
> Filtration will separate a solid from a liquid.

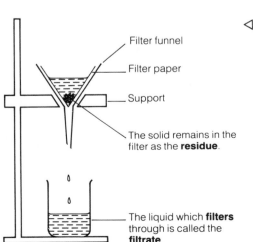

◁ Fig. 2.3 Filtration

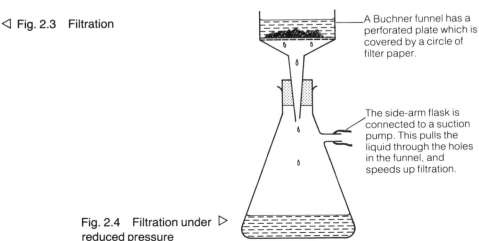

Fig. 2.4 Filtration under reduced pressure ▷

A Buchner funnel has a perforated plate which is covered by a circle of filter paper.

The side-arm flask is connected to a suction pump. This pulls the liquid through the holes in the funnel, and speeds up filtration.

Filter funnel
Filter paper
Support
The solid remains in the filter as the **residue**.
The liquid which **filters** through is called the **filtrate**.

Fig. 2.5 Salt pans in Australia

From a solution: by crystallisation

In Fig. 2.5, you see how sea water has been allowed to flow into large **salt pans**. They are in Australia where the sun is hot enough to evaporate much of the water. When the brine has become a saturated solution, the solute (salt) begins to crystallise. Scoops are used to remove salt crystals from the sea water.

Experiment 1.2 enables you to grow crystals. A laboratory method of evaporating a solution until it crystallises is shown in Fig. 2.6.

SUMMARY NOTE

A solute separates from a saturated solution by crystallisation. If a solution is unsaturated, some of the solvent has to be removed by evaporation before crystals will form.

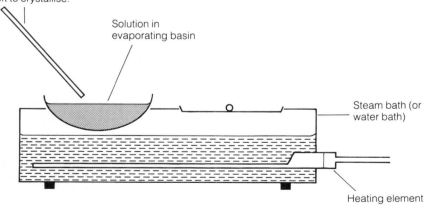

Fig. 2.6 Evaporating a solution to obtain crystals of solute

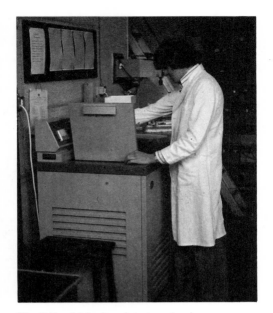

Fig. 2.7 A biochemist at work

From a suspension: by centrifuging

Suppose the biochemist in Fig. 2.7 wants to separate blood **cells** (which are solid) from blood plasma (which is liquid). Blood cells are very small solid particles. Because they are small they **disperse** (spread out) through the plasma and remain **in suspension**. They do not settle to the bottom of a container as heavy particles would. How can small solid particles be separated from the liquid around them? The method of **centrifuging** (or centrifugation) provides the answer. A **suspension** (such as blood) is put into glass tubes inside a centrifuge. This instrument spins round at high speed. The solid particles (in this case, blood cells) are flung to the bottom of the centrifuge tubes. After the machine has been stopped, the liquid (plasma) can be **decanted** (poured off) from the mass of solid at the bottom of the tube. This method is used for many different types of suspension, not just blood.

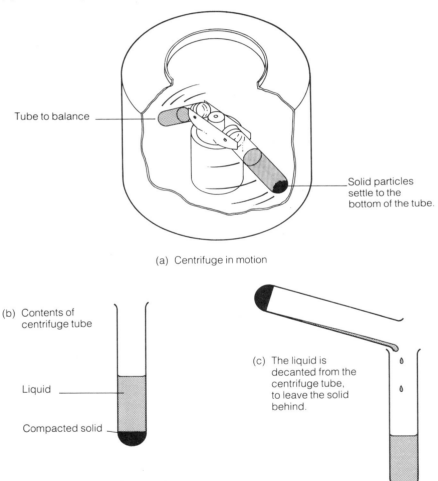

Fig. 2.8 How a centrifuge works

SUMMARY NOTE

Centrifuging will separate a suspended solid from solution.

2.4 Separating a solute and a solvent from a solution

In some parts of the world, drinking water is obtained from sea water. The solvent (water) is separated from the solute (salt) by **distillation**. This process uses a great deal of heat. It is only in

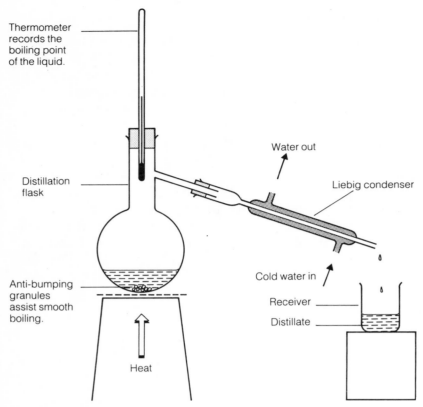

Fig. 2.9 Distillation

the Persian Gulf that the need for water is great enough and the price of oil is low enough to make this method of obtaining drinking water a practical proposition. Figure 2.9 shows a laboratory-scale distillation apparatus. The processes that take place are

- in the distillation flask, **vaporisation** (liquid → vapour)
- in the condenser, **condensation** (vapour → liquid).

Vaporisation followed by condensation is called **distillation**.

2.5 Separating miscible liquids

Distillation

A simple distillation apparatus like that in Fig. 2.9 will separate a liquid solvent from a solid solute. Often, people want to separate **miscible** liquids, that is, liquids that have dissolved in one another to form a solution. For example, whisky manufacturers need to separate alcohol from water in a solution of the two liquids. Alcohol is the substance called **ethanol**, which boils at 78 °C. Water boils at 100 °C. Figure 2.10 shows laboratory apparatus which will separate a mixture of ethanol and water into its parts or **fractions**. The process is called **fractional distillation**.

The **fractionating column** has a large surface area. Three designs of fractionating column are shown in Fig. 2.11a. Vaporisation of the liquid followed by condensation of the vapour takes place many times on the surface of the fractionating

SUMMARY NOTE

Distillation is used to separate a solvent from a solution.

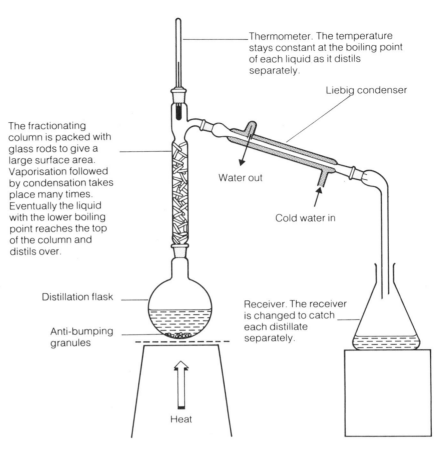

Fig. 2.10 Fractional distillation

Fig. 2.11 (a) Three designs of fractionating column; (b) a distillation column at B.P. Chemicals, Hull

column. The liquid with the lower boiling point (ethanol) reaches the top of the column first and distils over. The temperature stays constant at 78 °C while all the ethanol distils into the receiver. Then the temperature starts to rise, and the receiver is changed. At 100 °C, water starts to distil, and a fresh receiver is put into position to collect it.

Fractional distillation

Fractional distillation can be made to run continuously. This is done in the petroleum industry. Crude oil is not very useful in its raw state. By fractional distillation, it can be separated into a number of fractions which are very useful indeed (Fig. 2.12). These fractions are not pure substances. Each is a mixture of substances which boil at similar temperatures. The fractions with low boiling points are collected from the top of the column. Fractions with high boiling points are collected from the bottom of the fractionating column.

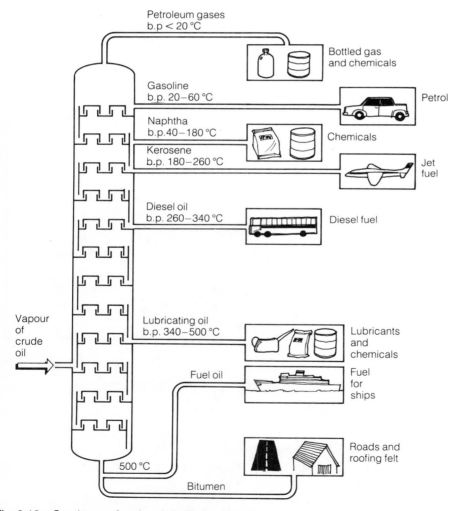

Fig. 2.12 Continuous fractional distillation in the oil industry

SUMMARY NOTE

Fractional distillation is used to separate a mixture of miscible liquids. It can be made to run continuously, e.g. in the fractionation of crude petroleum oil.

2.6 Separating immiscible liquids

A cook who wants to separate oil from gravy skims off the oil as it floats on top of the gravy (Fig. 2.13). Chemists have a better way of separating **immiscible** liquids (liquids which do not mix). They use a separating funnel (Fig. 2.14).

◁ Fig. 2.13 A cook separating immiscible liquids

1 Pour in the mixture of liquids. It will settle into two layers.
2 Open the tap. Let the bottom layer run into a receiver.
3 Close the tap. Change the receiver. Let the top layer run into a fresh receiver.

Fig. 2.14 A separating funnel ▷

SUMMARY NOTE

Immiscible liquids can be separated by means of a separating funnel.

JUST TESTING 5

1 A crew has to abandon ship. They find themselves in a lifeboat which has tins of food, a tin opener, a packet of plastic straws, a camping stove, matches, but no drinking water. They know that it is fatal to drink sea water. How can they obtain drinking water?
2 Describe how you could separate the oil and the vinegar in a salad dressing.
3 Somehow the car wax in your house has become mixed with sand. Describe how you could remove the sand from it.

2.7 Separating the solutes in a solution: chromatography

You can start this topic by doing Experiments 2.7–2.9. These experiments all separate the solutes present in a solution. They all work with a small volume of solution. They use the method called **chromatography**. When a spot of solution is applied to the chromatography paper, the paper **adsorbs** the solutes, that is, binds them to its surface. As solvent rises through the paper, a competition takes place. The paper can adsorb the solutes; the solvent can dissolve them. Which one wins depends on the individual solute. Some solutes stay put; others dissolve in the solvent and travel in it up the paper. A

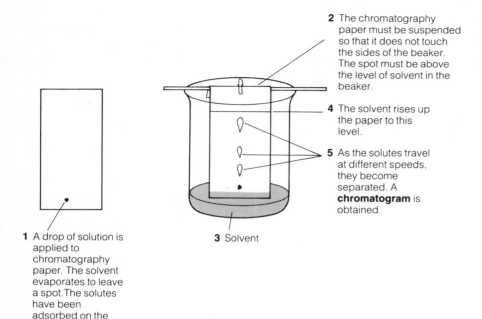

Fig. 2.15 Chromatography

solute which is very soluble in the solvent travels through the paper faster than a solute which is only slightly soluble. When the solvent reaches the top of the paper, the process is stopped. Different solutes will have travelled different distances. The result is a **chromatogram** (Fig. 2.15).

Water is not the only solvent that can be used (see Experiment 2.9). It is necessary to experiment to find out which solvent gives a good separation of the solutes. When you are using a solvent other than water, you should use a closed container so that the chromatography paper is surrounded by the vapour of the solvent (Fig. 2.16).

Frequently, chemists use chromatography to **analyse** a mixture. **Analysis** of a mixture means finding out which substances are present in it. What they do is to make a chromatogram of the unknown mixture together with known substances which they suspect may be present. For example, a food chemist might use known food additives as well as the unknown mixture. Figure 2.17 shows how results can be obtained by this method.

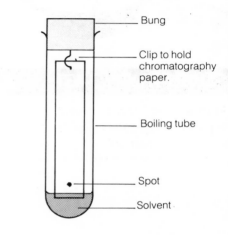

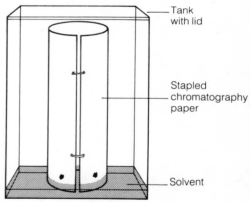

Fig. 2.16 Chromatography with a solvent other than water

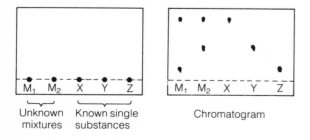

The results show that M_1 is a mixture of X and Z and M_2 is a mixture of X and Y.

Fig. 2.17 How to obtain results from the chromatograms

SUMMARY NOTE

Chromatography is a method of separating the solutes in a solution. It is used in analysis, that is, in finding out what substances are present in a mixture.

JUST TESTING 6

1.

   Mr Dunn writes a cheque for £7 to Mr Fixit. Mr Fixit alters the amount. Mr Dunn reports the matter to the police. How can the police chemist prove that the 'ty' and the '0' have been added by a different person?

2. A biochemist treats a protein food with an enzyme. She uses chromatography to separate the mixture that she obtains. Her chromatogram is shown in Fig. 2.18a. She knows that the mixture is a mixture of **amino acids**. She suspects that glycine, arginine, aspartic acid, serine, histidine and phenylalanine may be present. When she runs chromatograms on these amino acids one at a time, she obtains the result shown in Fig. 2.18b. What conclusion can you draw from her experimental results?

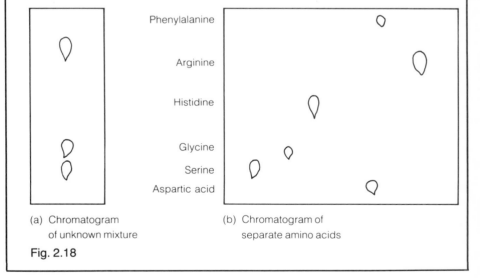

(a) Chromatogram of unknown mixture

(b) Chromatogram of separate amino acids

Fig. 2.18

2.8 Testing for purity

Solids

If a solid consists of a single substance, it is said to be a **pure** substance. If the solid is a mixture of substances, it melts gradually over a range of temperature. If the solid is a pure

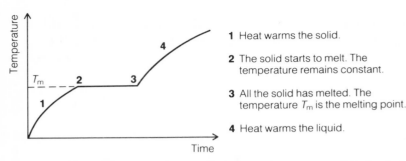

Fig. 2.19 The temperature changes that occur when a pure solid is heated

1 Heat warms the solid.
2 The solid starts to melt. The temperature remains constant.
3 All the solid has melted. The temperature T_m is the melting point.
4 Heat warms the liquid.

substance, it melts suddenly at a definite temperature, the **melting point** of the solid. While a solid is melting, its temperature stays constant as long as there is some solid left. This is because all the heat entering the solid is used to change the solid into a liquid, and not to raise its temperature (see Fig. 2.19).

To find out whether a solid is pure, you need to find its melting point (see Fig. 2.20). If an impurity is present, it lowers the melting point and also makes the impure substance melt over a range of temperatures.

People have listed the melting points of solids at standard pressure, that is, atmospheric pressure at sea level. When you know the melting point of a solid, you can find out what the solid is. Look through the list of melting points until you find the name of a solid with that melting point.

> **SUMMARY NOTE**
>
> The melting point of a solid can be used to test its purity. When a pure solid melts, the temperature stays constant as long as there is some solid left. The solid can be identified from tables of melting points. The presence of an impurity lowers the melting point.

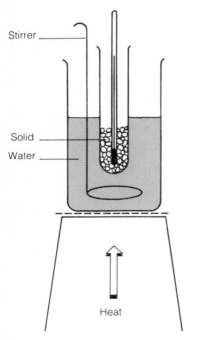

1 Heat the water in the beaker until the solid in the tube melts. Switch off the Bunsen.
2 Let the molten solid cool. Stir. Watch the thermometer.
3 While the liquid begins to solidify, the temperature stops falling. It stays constant until no liquid remains. This temperature is the melting point of the solid (and also the freezing point of the liquid).

Fig. 2.20 Finding the melting point of a solid. (When water is used in the beaker, the solid must melt below 100 °C. If the solid has a higher melting point, a different liquid must be used. See Figure 2.19 for a graph of the way the temperature falls.)

Liquids

A mixture of liquids boils over a range of temperatures. While a pure liquid is boiling, its temperature remains constant at its

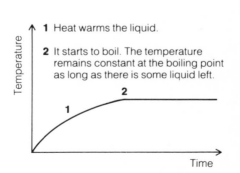

1 Heat warms the liquid.
2 It starts to boil. The temperature remains constant at the boiling point as long as there is some liquid left.

Fig. 2.21 The temperature changes that occur when a pure liquid boils

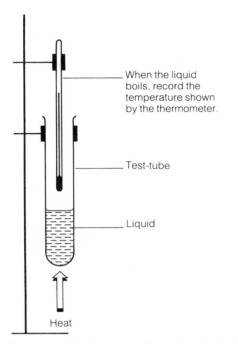

Fig. 2.22 Finding the boiling point of a liquid

boiling point. If you heat the liquid more strongly, it will boil faster but the temperature will not go up. Heat going into the liquid is used to convert the liquid into vapour and not to raise its temperature (see Fig. 2.21).

To find out whether a liquid is pure, you find its boiling point (Fig. 2.22). A pure liquid boils at a certain temperature. A mixture of liquids boils over a range of temperatures. If a solid is dissolved in the liquid, it raises the boiling point.

The boiling point of a liquid depends on the surrounding pressure. A drop in the surrounding pressure makes the liquid boil at a lower temperature. An increase in the surrounding pressure makes the liquid boil at a higher temperature.

JUST TESTING 7

1 On high mountains, the atmospheric pressure is low.
   Does water boil above or below 100 °C on a high mountain?
   Do mountaineers find it easy or difficult to make a pot of strong tea?
   Explain your answers.

2 In a pressure cooker, food cooks at a pressure greater than atmospheric pressure.
   Is the temperature of water boiling in a pressure cooker above or below 100 °C?
   Does food cook more quickly or more slowly in a pressure cooker than in a saucepan?
   Explain your answers.

SUMMARY NOTE

The boiling point of a liquid can be used to test its purity. A pure liquid boils at a constant temperature. Impurities raise the boiling point of a liquid. The boiling point of a liquid depends on the surrounding pressure.

Tables of boiling points at standard pressure (that is, atmospheric pressure at sea level) have been drawn up. By checking through these lists, you can find the name of the liquid which has the boiling point you have just measured.

Alternatively, you can find the freezing point of a liquid (Fig. 2.23).

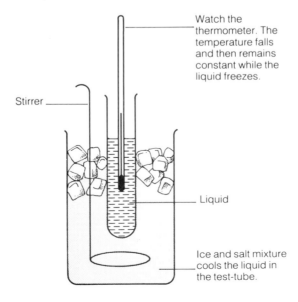

Fig. 2.23 Finding the freezing point of a liquid

JUST TESTING 8

1. A solid melts at 43 °C. Which of the solids listed below could it be?
   a phenol (m.p., 41 °C), **b** benzoic anhydride (m.p., 42 °C), **c** chlorophenol (m.p., 43 °C), **d** nitrophenol (m.p., 45 °C).

2. An impure sample of a solid X melts at 83 °C. Which of the solids in this list could be X?

Substance	Melting point in °C
Dibromobenzene	87
Naphthalene	80
Di-iodoethane	81
Glycollic acid	79

3. An impure sample of a liquid Y boils at 66 °C. Which of the liquids in this list could be Y?

Liquid	Boiling point in °C
Chloroform	61
Hexene	64
Hexane	69
Benzene	80

Exercise 2

1. Say which of the pure substances listed below are **a** solids, **b** liquids and **c** gases at room temperature and standard pressure. Which substance cannot be a metal? In which substance are the forces of attraction between particles the greatest?

Pure substance	Melting point in °C	Boiling point in °C
A	29	690
B	−101	−35
C	−7	59
D	114	184
E	−39	357
F	820	1100

2. The magnetic separator shown in Fig. 2.24 is in use in a metal recycling plant. It is separating iron and steel (which are magnetic)

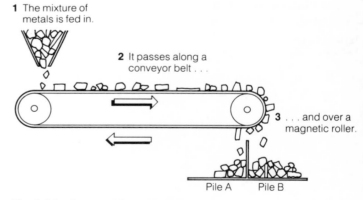

Fig. 2.24 A magnetic separator

from aluminium (which is non-magnetic). Which pile, A or B, is aluminium? Explain how the machine works.

3. Select the correct ending to this sentence: A solid can be purified by crystallisation from water because

A it dissolves in cold water but not in hot water.
B it is insoluble in hot and cold water.
C it is very soluble in hot and cold water.
D it is more soluble in hot water than in cold water.
E it is more soluble in cold water than in hot water.

4 Figure 2.25 shows the changes that occur when a pure liquid is cooled. Explain what changes are occurring **a** in region 1, **b** between point 2 and point 3, **c** in region 4. What do you call the temperature shown as T?

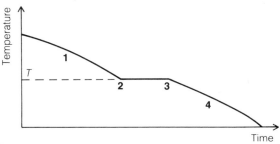

Fig. 2.25 A liquid cools

5 The melting point of a sample of a solid compound A is 133 °C. Which of the solids in this list could be A?

Solid	Melting point in °C
Benzoic acid	122
Maleic acid	130
Urea	136
Oxalic acid	190

6 If you sprinkle salt on to ice, some of the salt will dissolve. Can you explain why this leads to the ice melting?

7 A mixture contains two liquids, A and B. A boils at 70 °C, and B boils at 110 °C. Draw the apparatus you would use to separate A and B. Label the drawing.

8 A chemist extracts an acidic solid from rancid butter. What should he do to find out whether it is a pure substance or not?

9 The police are investigating a case of forged £5 notes. When they search a suspect's premises, they find paper and ink but no counterfeit money. How can they compare the ink they have found with the ink in the forged £5 notes?

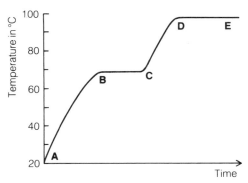

Fig. 2.26 Distillation of a mixture of hexane and heptane

10 Figure 2.26 shows how the temperature changes during the distillation of a mixture of hexane and heptane. The boiling points of the two liquids are: hexane, 69 °C; heptane, 98 °C. Explain what is happening over the regions A–B, B–C, C–D and D–E.

11 In Fig. 2.27, point A shows ice at −25 °C. The graph shows how the temperature changes when ice is heated at a constant rate. What changes take place in regions A–B, B–C, D–E and E–F?

*12 Ethanol (alcohol) melts at −117 °C and boils at 78 °C. Draw a heating curve for ethanol similar to Fig. 2.27, and label
a the melting point
b the boiling point
c the region where ethanol is a gas
d the region where ethanol is a liquid
e the region where ethanol exists as both solid and liquid
f the region where ethanol exists as both liquid and gas.

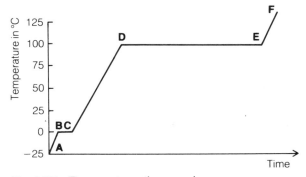

Fig. 2.27 Temperature–time graph

CHAPTER 3 Elements and compounds

3.1 Diamond

Probably the most famous of diamonds is the Hope diamond. This is a huge (20 g) blazing diamond with a bluish tint. It once formed the single eye of a statue of Sita, a Hindu goddess. The diamond was stolen from the statue, and disappeared for many years. It eventually turned up in France in 1668, by which time a legend had grown up around it. The 'French blue', as it was called, was said to carry the curse of the goddess Sita, who resented the theft. Certainly, the diamond brought no luck to one wearer, Marie Antoinette, who was guillotined during the French Revolution in 1792. Part of the French crown jewels, the diamond was stolen again at the time of the Revolution. In 1830, it turned up in London where it was bought by the banker, Henry Hope. The Hope diamond, as it has been called ever since, brought no luck to his family. They lost all their father's wealth and died bankrupt. The diamond passed to an Eastern European prince. He gave it to an actress in the Folies Bergères, but afterwards shot her in a fit of jealous rage. Later, a Greek owner drove over a precipice with his family in a car accident. Sultan Abdul Hamid II of Turkey had owned the diamond for only a few months when he was overthrown by a military revolt in 1909. An American woman, Mrs Evalyn Walsh, then bought the Hope diamond. She did not believe the story of Sita's curse, and wore the diamond as a necklace. Her children died in accidents, and mental illness afflicted her husband. When she died, a dealer bought her jewels, and donated the Hope diamond to the American people. It is now on view at the Smithsonian Institute in Washington, DC.

Why have people always been fascinated by diamonds? The brilliance of diamonds is due to their ability to reflect light. A skilful diamond cutter can increase this brilliance (Fig. 3.1). The 'fire' of a diamond is a result of its ability to split light into flashes of colour. Diamond is the hardest of natural materials. The only thing that can scratch a diamond is another diamond, so 'diamonds are for ever'. This is why diamonds are so often used in engagement rings.

Eye surgeons now use diamond knives to remove cataracts. The edges of these knives are so sharp and so even that a clean cut is made with no tearing.

In a well-cut diamond the rays of light are reflected.

This diamond has been cut too deep and 'leaks' light.

This diamond has been cut too shallow. It also 'leaks' light.

Fig. 3.1 How diamonds are cut

That space capsule the Americans sent to Venus — it had a window in it! They had to have one to let in infrared radiation for their scientific instruments to record. What could the window have been made of?

Well it's hot on Venus (about 500 °C) and there's about 100 atmospheres pressure. But in space it's cold and there's a vacuum. The only transparent material I can think of which could stand up to all those conditions is diamond.

Fig. 3.2 A diamond-edged saw cutting concrete

Fig. 3.3 Diamond-edged drill bits

Diamond is the hardest naturally occurring substance. In industry there are thousands of applications for diamonds. Diamonds are used for grinding, cutting, grooving, sharpening, etching and polishing. Stone, metals, ceramics, glass and concrete are all cut with diamond-edged tools (Fig. 3.2). The bits of oil-rig drills are studded with small diamonds. Often they have to drill 7000 metres through hard rock. This was almost impossible before the diamond bit (Fig. 3.3).

What is diamond, this beautiful and useful material? What gives it such extraordinary properties? Diamond consists only of carbon atoms. It is a form of the element carbon. **An element is a pure substance which cannot be separated into simpler substances** (see Section 1.5).

Deep down in the Earth the temperature and pressure are both very high. There, millions of years ago, carbon atoms crystallised in a regular solid arrangement. You can see from Fig. 3.4 how

SUMMARY NOTE

Diamonds have always been valued for their sparkle and fire. Diamond is a very hard material. It is used in industry for cutting, grinding and drilling. Diamond is a form of the element carbon. The carbon atoms are joined by chemical bonds to form a giant molecular structure.

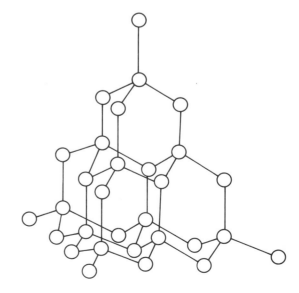

Fig. 3.4 The structure of diamond

each carbon atom is joined by chemical bonds to four other carbon atoms. A crystal of diamond contains millions of carbon atoms. It is very difficult to break them apart. This is the reason why diamonds are so hard. You will read in Chapter 6 that the structure of diamond is described as a **macromolecule** or **giant molecule**.

3.2 Graphite

Graphite is a shiny dark grey solid. It is soft and, when you rub it, comes off on to your fingers. Mixed with clay, graphite is used to make pencil 'leads'. It is also used as a lubricant, in cars for example. Graphite conducts electricity (as in Experiments 5.1 and 5.2).

Diamond is brilliant. It is one of the hardest substances and is a non-conductor of electricity. Graphite is only slightly shiny. It is so soft that it is used as a lubricant, and it is an electrical conductor. Yet these two substances are both forms of the element carbon. Both diamond and graphite consist of atoms of carbon and nothing else. The reason for the difference in properties lies in the arrangement of atoms in the two solids. They are both crystalline solids: the atoms are arranged in a regular, ordered manner. Graphite has the layer structure shown in Fig. 3.5. The carbon atoms in each layer are joined by chemical bonds. Between the layers, there are only weak forces of attraction. These are weak enough to allow one layer to slide over the next layer.

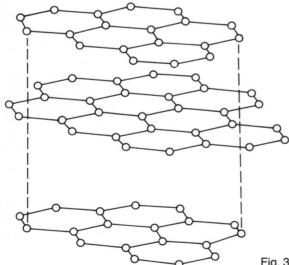

Fig. 3.5 The structure of graphite

Diamond and graphite are **allotropes** of carbon. **Allotropy** is the existence of different crystalline forms of the same element. It is possible to make diamonds from graphite by keeping graphite at a high temperature under high pressure.

SUMMARY NOTE

Graphite is a soft solid with a slight shine. It is used as a lubricant and as an electrical conductor. Graphite is another form of the element carbon. It differs from diamond because it has a different structure, a layer structure.

Diamond and graphite are **allotropes** of carbon.

3.3 Charcoal

There are also impure forms of the element carbon. Wood charcoal and animal charcoal are made by burning wood or bones in a limited supply of air. Animal charcoal is used as an **adsorbent**. It will adsorb the natural brown colour from sugar during the manufacture of white sugar. It is also used in the gas masks which industrial workers wear when they have to deal with poisonous gases.

> **SUMMARY NOTE**
>
> Charcoal is an impure form of carbon.

JUST TESTING 9

1. What are diamonds made of? What other forms of this element do you know? How does diamond differ from the other forms of the element in **a** behaviour and **b** structure?
2. What is the advantage of using diamond as a surgical knife?
   Why are diamonds often chosen as engagement rings?
   List the characteristics of diamond which make it useful in industry.
   What do the following people have in common: oil prospectors, glass carvers, stone cutters, eye surgeons?

3.4 Silicon

Radios and television sets contain electronic circuits. These circuits used to contain devices called **vacuum tubes**.
Their job was to allow current to pass in one direction only. In the 1950s, **transistors** made from silicon replaced vacuum tubes. A

Fig. 3.6   A silicon chip magnified 1000 times

transistor made from a speck of crystal the size of the letter '**o**' on this page can do the job of a vacuum tube. Transistors led to the miniaturisation of electronic circuits:

- Transistor radios could be made small enough to be held in the hand.
- Computers the size of a typewriter did the job of the older computers which needed a room full of vacuum tubes.
- Aeroplanes and spacecraft became able to use computers (the size and weight of the valve type of computer had made this impossible).

A second giant step was the packing of an entire electronic circuit (transistors, resistors etc.) on to a single chip of silicon. This saves weight, space and cost. The chip is little affected by moisture, ageing or vibration. This device is called an **integrated circuit** (Fig. 3.6).

Silicon is an element. It is a **semiconductor**: its electrical properties lie between those of an electrical insulator and an electrical conductor. (You can read in Section 12.9 how silicon is obtained.) The first step in making an integrated circuit is to slice large crystals of silicon into wafers. A thousand integrated circuits are then put on to each wafer. This is done by treating tiny areas of the silicon wafer with other elements which make it become a conductor.

The silicon wafer is cut into identical silicon **chips**. Each chip is sealed in plastic or metal. Fine gold wires (finer than human hair) are bonded to the chip to provide connections to an external circuit. This is done by operators looking at the chip through a microscope and using very delicate tools.

The uses of the integrated circuit are many and increase daily. Computers are used in business, commerce, education and medicine.

SUMMARY NOTE

The element silicon is a semiconductor of electricity. This property enables silicon to be used as a **transistor** (which allows current to flow in one direction only). Integrated circuits built on to silicon chips are the basis of the microcomputer industry.

3.5 Gold

People have always treasured gold. It never tarnishes, so the gold ornaments in the tombs of the pyramids are as bright today as when they were put there 3000 years ago.

Gold is a metallic element and like most metals it is shiny. Unlike most metals, however, it does not tarnish. It is not affected by air or water or any other chemicals in the environment. Like most metals, gold can be easily hammered into different shapes and it can be pulled into the form of a wire.

There are scientific uses for gold. Integrated circuits (see Section 3.4) operate at low current and low voltage. Contacts and switches must therefore have low resistance. At these low currents, even small areas of corrosion on the electrical contacts can increase the resistance. Extremely non-corrodable metals must therefore be used, and gold is the best choice. Not only is it an excellent electrical conductor, it is also one of the most unreactive of metals. It is expensive, of course, but it is used in small quantities to plate contacts made of less expensive metals.

SUMMARY NOTE

Gold is a beautiful metal which never tarnishes. It is an element. Its ability to be worked into different shapes is typical of metals.

Gold is used for jewellery and for electrical circuits which demand an electrical conductor which never tarnishes.

Fig. 3.7 (a) Gold jewellery (b) Gold ingots

Satellites are covered with gold foil protected by glass ceramic. I wonder why.

It's probably to reflect some of the heat from the sun. If a satellite got very hot, the scientific instruments inside it would be damaged.

Fig. 3.8 A satellite

3.6 Iron

Iron is another metal which has been used by the human race for thousands of years. It is an element. Between 2000 and 1000 B.C., people discovered how to obtain iron from iron-bearing rocks. They did not value it for its beauty, like gold, but for the useful tools they could make from it. Iron is hard and strong. It can be hammered into flat blades and then ground with a stone to give it a cutting edge. People were able to make hammers, chisels, axes and knives which were not brittle like their stone tools or easily bent like the bronze tools they were used to. The development of iron tools and weapons made a huge difference to the way people lived. The human race entered a new 'age', called by historians the Iron Age.

You can read about iron and steel, which is an alloy of iron, in Section 10.11.

> **SUMMARY NOTE**
>
> Iron is a metallic element. Being hard and strong, it has been used for thousands of years for making tools.

3.7 Elements

The properties of gold and iron are **typical** of metallic elements. They conduct electricity. Many of the metals we use are not elements. Bronze, brass, solder, steel and many others are **alloys**. They are combinations of two or more metallic elements and sometimes non-metallic elements also. The other elements you have read about in this chapter, carbon and silicon, are non-metallic elements. Some of their properties are typical of non-metallic elements, but some are **atypical** (not typical). The diamond allotrope of carbon is atypical in being shiny; most non-metallic elements are dull. The graphite allotrope of carbon is the only non-metallic element that conducts electricity. Silicon is atypical in being a semiconductor.

Table 3.1 lists some metallic elements and non-metallic elements with their state symbols. In all, there are 92 natural elements found on the Earth. Other elements have been made artificially (see Chapter 4). Table 3.2 summarises the differences in physical properties between metallic elements and non-metallic elements. You will discover the chemical differences later; in particular, their reactions with oxygen (Section 10.3).

Table 3.1 Some elements and their state symbols

Metallic elements	Non-metallic elements
Aluminium (s)	Carbon (s)
Copper (s)	Chlorine (g)
Gold (s)	Iodine (s)
Iron (s)	Hydrogen (g)
Lead (s)	Neon (g)
Mercury (l)	Nitrogen (g)
Silver (s)	Oxygen (g)
Sodium (s)	Silicon (s)
Zinc (s)	Sulphur (s)

> **SUMMARY NOTE**
>
> Gold and iron are metallic elements. Combinations of metallic elements are called **alloys**.
>
> Carbon and silicon are non-metallic elements. Other than graphite and semiconductors like silicon, non-metallic elements do not conduct electricity. This distinguishes them from metals, which are good electrical conductors. The differences between metallic and non-metallic elements are listed in Table 3.2.

Table 3.2 Physical properties of metallic and non-metallic elements

Metallic elements	Non-metallic elements
Solids (except mercury) Dense and hard	Gases, liquids and solids Most of the solid elements are softer than metals; diamond is exceptional.
A fresh surface is shiny; many metals tarnish in air, e.g. iron rusts.	Most are dull; diamond is exceptional.
Can be hammered into shape without breaking: are **malleable**. Can be pulled out into wire form: are **ductile**.	Solid non-metallic elements are easily fractured when attempts are made to change their shape. Diamond is exceptional.
Conduct heat, although highly polished surfaces reflect heat.	All are poor thermal conductors.
Are good electrical conductors.	Most are poor electrical conductors. Graphite is the exception. There are semiconductors, such as silicon.
Make a pleasing sound when struck: are **sonorous**.	Are not **sonorous**.

3.8 Compounds

This story from *The Times* shows what can happen when elements **combine**. Mr Martin had been using a mixture which is sold in the USA for unblocking drains. It reacts with water to form hydrogen. He had also been using a bleach which reacts with water to form chlorine. Hydrogen and chlorine are elements. In strong sunlight, an explosive chemical reaction takes place between the two elements. They combine to form the compound hydrogen chloride:

hydrogen + chlorine → hydrogen chloride
element + **element** → **compound**

A compound of chlorine with one other element is called a **chloride**.

What is a compound?

Compounds are pure substances which contain two or more elements chemically combined (see Section 1.5). Water is a compound. You can split it up by passing a direct electric current through acidified water (see Section 5.3). A chemical reaction takes place to give the elements hydrogen and oxygen. Water is a compound of these elements. You could call it hydrogen oxide. A compound of oxygen and one other element

---

Saved by the bell from the bathroom blast

Mr Hilton Martin was saved by the bell when a ringing telephone made him leave the bathroom. As he answered, the lavatory exploded at his home south of Cape Canaveral.

Mr Martin had just cleaned the lavatory using two different detergents. The water started bubbling, but he ran to answer the telephone instead of investigating.

'If I'd been in the bathroom at the time of the explosion, I would have been hurt. It blew the porcelain to pieces. It disintegrated the tank itself', Mr Martin said. 'It sounded like a hand grenade going off.'
(*The Times*, 21 September 1985, Satellite Beach, Florida)

is called an **oxide**. Water can be made by a chemical reaction between hydrogen and oxygen (see Section 9.8). Making a compound by combining elements is called **synthesis**.

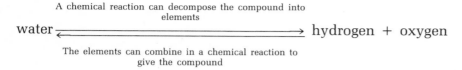

Some compounds can be split up by heat. A chemical reaction of this kind is called **thermal decomposition**. The compound mercury oxide decomposes when it is heated to give the elements mercury and oxygen. This is how oxygen was discovered by the British chemist, Joseph Priestley, in 1774 (see Fig. 3.9). Mercury oxide can be **synthesised** from mercury and oxygen.

To define the term compound:

A compound is a pure substance which consists of two or more elements chemically combined.

You can synthesise some compounds in Experiments 3.1–3.4.

> ### SUMMARY NOTE
> Elements combine to form compounds. Some compounds can be split up into their elements.
>
> A direct electric current splits water into the elements hydrogen and oxygen.
>
> Some compounds can be decomposed by heat. Mercury oxide is an example.

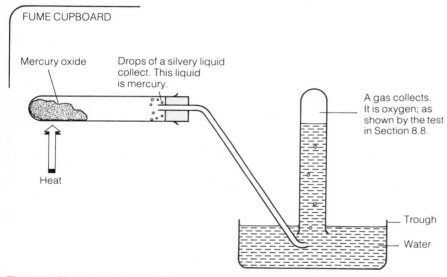

Fig. 3.9 Thermal decomposition of mercury oxide. (Note: This experiment must be done in a fume cupboard because mercury vapour is poisonous)

What is the difference between a compound and a mixture?

There are differences between a compound of elements and a mixture of elements (see Experiments 3.1–3.4). A mixture can contain its components in any proportions. A compound has a **fixed composition**. It always contains the same elements in the same percentages by mass. Calcium carbonate is a compound of calcium, carbon and oxygen. It is quarried as limestone, as chalk and as marble. Wherever it comes from, calcium carbonate always has the composition by mass:

calcium 40%; carbon 12%; oxygen 48%.

Table 3.3 summarises the differences between mixtures and compounds.

Table 3.3 Differences between mixtures and compounds

Mixtures	Compounds
1 A mixture can be separated into its parts by physical methods, such as those used in Chapter 2.	A chemical reaction is needed to split a compound into simpler substances.
2 No chemical change takes place when a mixture is made.	A chemical reaction takes place when a compound is made. Often heat is given out or taken in when a compound is formed.
3 A mixture behaves in the same way as the substances in it.	A compound does not have the properties of its elements. It has a new set of properties.
4 A mixture can be made by mixing elements in any proportions.	A compound always contains its elements in fixed proportions by mass; for example, aluminium iodide always contains 6.6% of aluminium and 93.4% of iodine by mass.

> **SUMMARY NOTE**
>
> A compound is a pure substance which consists of two or more elements chemically combined. The components of a mixture are not chemically combined. The differences between compounds and mixtures are listed in Table 3.3.

> **JUST TESTING 10**
>
> 1 How could you tell the difference between **a** a compound of iron and sulphur and **b** a mixture of iron and sulphur?
>
> 2 How can a mixture of oxygen and hydrogen be separated? How can the compound water be separated into hydrogen and oxygen?
>
> 3 Group the following into **a** mixtures and **b** compounds: sea water, air, common salt, rock salt, gold chloride, aluminium oxide, ink.
>
> 4 Name an element which can be used for each of the following uses:
> earrings, surgical knife, pencil 'lead', decolourising agent, integrated circuit, chisel, thermometer, plumbing, electrical wiring, coloured lights, saucepans, disinfecting swimming pools.
>
> 5 Explain the difference between a compound and a mixture.

3.9 Symbols

Chemists use **symbols** to represent atoms. Every element has a different symbol. For example, the symbol for carbon is C. The letter C stands for one atom of carbon. Sometimes, two letters are needed. The letters Ca stand for one atom of calcium: the symbol for calcium is Ca. The symbol for chlorine is Cl, and the symbol for chromium is Cr. **The symbol of an element is a letter or two letters which stand for one atom of the element.** Sometimes the letters are taken from the Latin name of the element, e.g. Cu from cuprum (copper) and Fe from ferrum (iron). A short list of symbols is given in Table 3.4. A complete list can be found on p. 316.

Table 3.4 Symbols of some common elements

Element	Symbol	Element	Symbol	Element	Symbol
Aluminium	Al	Gold	Au	Oxygen	O
Barium	Ba	Hydrogen	H	Phosphorus	P
Bromine	Br	Iodine	I	Potassium	K
Calcium	Ca	Iron	Fe	Silver	Ag
Carbon	C	Lead	Pb	Sodium	Na
Chlorine	Cl	Magnesium	Mg	Sulphur	S
Copper	Cu	Mercury	Hg	Tin	Sn
Fluorine	F	Nitrogen	N	Zinc	Zn

> **SUMMARY NOTE**
>
> The symbol of an element is a letter or two letters which stand for one atom of the element.

JUST TESTING 11

1 A well-educated crook sends this message to his partner in crime. If you have looked at the symbols on p. 316, you should be able to decipher it.

Au	Ba	In	V
Cu	Sn	Ru	Ni
Te	Sr	Kr	Ac

Write down the names of all the elements in the message. Can you see the message? (You have to use some of the letters in the names, and ignore others.)

3.10 Formulas

Formulas are written for compounds. The **formula of a compound** consists of the symbols of the elements present and some numbers. The numbers show the ratio in which the atoms are present. The compound carbon dioxide has the formula CO_2. The formula tells you that carbon dioxide contains two oxygen atoms for every carbon atom. The 2 below the line multiplies the O in front of it. Carbon dioxide consists of particles. Each particle contains one carbon atom joined by chemical bonds to two oxygen atoms (Fig. 3.10). The particle is called a **molecule** of carbon dioxide. To show three molecules, you write $3CO_2$. You cannot assume that every compound consists of molecules (see Section 6.2). The formulas of some of the compounds mentioned in Section 3.8 are given in Table 3.5.

The formula for sulphuric acid is H_2SO_4. This tells you that the compound contains two hydrogen atoms and four oxygen

Fig. 3.10 A molecule of carbon dioxide

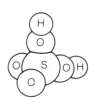

Fig. 3.11 A molecule of sulphuric acid

Table 3.5 Some formulas

Compound	Formula
Hydrogen chloride	HCl (one H atom and one Cl atom)
Water	H_2O (two H atoms and one O atom; the 2 multiplies the H in front of it)
Sodium chloride	NaCl (one Na, one Cl)
Mercury oxide	HgO (one Hg, one O)
Mercury iodide	HgI_2 (one Hg, two I)

atoms for every sulphur atom. The numbers below the line multiply the symbols immediately in front of them. Sulphuric acid consists of molecules containing two hydrogen atoms, one sulphur atom and four oxygen atoms (Fig. 3.11). To write three molecules, you write $3H_2SO_4$. The 3 in front of the formula multiplies everything after it. In $3H_2SO_4$, there are 6 H, 3 S and 12 O atoms, a total of 21 atoms.

The formula for calcium hydroxide is $Ca(OH)_2$. The 2 multiplies the symbols in the brackets. There are 2 oxygen atoms, 2 hydrogen atoms and 1 calcium atom. To write $4Ca(OH)_2$ means that the whole of the formula is multiplied by 4. It means 4 Ca, 8 O and 8 H atoms. Section 6.6 deals with how to work out the formula of a compound. Table 3.6 lists the formulas of some common compounds.

SUMMARY NOTE

Compounds have formulas. A formula is a set of symbols and numbers. The symbols tell you what elements are present in the compound. The numbers tell you the ratio in which the atoms of different elements are present.

Table 3.6 The formulas of some compounds

Water	H_2O	Aluminium chloride	$AlCl_3$
Sodium hydroxide	NaOH	Aluminium oxide	Al_2O_3
Sodium chloride	NaCl	Carbon monoxide	CO
Sodium sulphate	Na_2SO_4	Carbon dioxide	CO_2
Sodium nitrate	$NaNO_3$	Sulphur dioxide	SO_2
Sodium carbonate	Na_2CO_3	Ammonia	NH_3
Sodium hydrogencarbonate	$NaHCO_3$	Ammonium chloride	NH_4Cl
		Hydrogen chloride	HCl
Calcium oxide	CaO	Hydrochloric acid	HCl(aq)
Calcium hydroxide	$Ca(OH)_2$	Sulphuric acid	H_2SO_4(aq)
Calcium chloride	$CaCl_2$	Nitric acid	HNO_3(aq)
Calcium sulphate	$CaSO_4$	Copper(II) oxide	CuO
Calcium carbonate	$CaCO_3$	Copper(II) sulphate	$CuSO_4$

JUST TESTING 12

1 How many atoms are present in the following?
   a CCl_4, **b** SO_2Cl_2, **c** $COCl_2$, **d** $C_2H_3Cl_3$, **e** $CuSO_4$, **f** $4CuSO_4$, **g** $Pb(NO_3)_2$, **h** $Al_2(SO_4)_3$, **i** $3Fe(NO_3)_3$, **j** $2Al(OH)_3$.

2 What are the formulas of
   a water, **b** hydrochloric acid, **c** sodium hydroxide, **d** calcium oxide, **e** sodium chloride, **f** carbon dioxide, **g** calcium carbonate, **h** sulphuric acid?

3.11 Equations

You have studied symbols for elements and formulas for compounds. These enable you to write **equations** for chemical reactions. Zinc and sulphur combine to form zinc sulphide. You can write this information as a **word equation**:

zinc + sulphur → zinc sulphide

The arrow means 'form'. If you put symbols for the elements and a formula for the compound, you can write:

Zn + S → ZnS

This is a **chemical equation**. On the left-hand side, you have one atom of zinc and one atom of sulphur. On the right-hand side, you have one atom of zinc and one atom of sulphur combined as zinc sulphide. The two sides are **equal**, and this is why the expression is called an equation.

It gives more information if you include **state symbols** in the equation. These are (s) for solid, (l) for liquid, (g) for gas and (aq) for aqueous solution (in water).

Zinc carbonate decomposes to give zinc oxide and carbon dioxide. A word equation for this reaction is

zinc carbonate → zinc oxide + carbon dioxide

The chemical equation is

$ZnCO_3 \rightarrow ZnO + CO_2$

Putting in the state symbols

$ZnCO_3(s) \rightarrow ZnO(s) + CO_2(g)$

tells you that solid zinc carbonate decomposes to form solid zinc oxide and carbon dioxide gas.

Magnesium reacts with sulphuric acid to give hydrogen gas and a solution of magnesium sulphate. The word equation is

magnesium + sulphuric acid → hydrogen + magnesium sulphate

The chemical equation is

$Mg(s) + H_2SO_4(aq) \rightarrow H_2(g) + MgSO_4(aq)$

Hydrogen is written as H_2 because hydrogen gas consists of molecules containing two atoms.

There is more about equations in Section 6.8.

JUST TESTING 13

1 Write chemical equations for these reactions:
   a Iron and sulphur combine to form iron sulphide, FeS.
   b Mercury and sulphur combine to form mercury sulphide, HgS.
   c Copper carbonate decomposes to form copper oxide and carbon dioxide.
   d Silicon and oxygen combine to form silicon oxide, SiO_2.
   e Zinc and sulphuric acid react to form hydrogen and zinc sulphate.

Exercise 3

1. Write down from memory the symbols for: sulphur, sodium, potassium, magnesium, lead, copper, iron, zinc, aluminium.

2. In the word square are the names of 20 elements. They may read across, in reverse, upwards, downwards and diagonally. Write down the names of all the elements you can find.

   ```
   T I N E O N C T R Y M O
   P R R B U C A L C I U M
   H O R O T H R U O S I O
   O N D R N L B I P M N X
   S G R O P O O E P U I Y
   P O A N Q R N T E I M G
   H L D O Z I N C R S U E
   O D B I R N I C K E L N
   R E N O U E Z C L N A D
   U C U E W M O L E G T A
   S L T S I L V E R A I E
   F H Y D R O G E N M N L
   ```

3. Say whether each of the changes listed is a chemical reaction or a physical change. Give reasons for your answers.
   a Letting off a firework
   b Adding sugar to tea
   c Breaking an egg
   d Frying an egg
   e Baking a cake
   f Making an ice lolly
   g Writing on a piece of glass with a diamond 'pencil'.

4. Can you decipher this message? You will need the symbols on p. 000. Write down the name of each element. Using some of the letters and ignoring others, see if you can spot the message.

   Ar Kr Ru Ba At W
   Te Sb Kr Md Ru Al
   Ru Ti V Bi Br Sc
   Ru Ag K Bi Cf Ru

5. How could you prove that the change that takes place when salt is added to water is a physical change?
   How could you prove that the change that takes place when iron filings are added to water is a chemical change?

6. How could you find out whether the following substances are mixtures or compounds?
   a salt, b wine, c city air, d honey, e brass.

7. Can you recall the formulas for these important compounds?
   sodium chloride, sodium hydroxide, hydrochloric acid, sulphuric acid, calcium oxide, copper(II) sulphate, ammonia, water.

8. Write chemical equations for the following reactions:
   a Copper and sulphur form copper(II) sulphide, CuS.
   b Mercury and iodine combine to form mercury iodide, HgI_2.
   c Carbon burns in oxygen to form carbon dioxide.
   d Zinc carbonate decomposes to form zinc oxide and carbon dioxide.
   e Iron reacts with sulphuric acid to form hydrogen and iron(II) sulphate, $FeSO_4$.

CHAPTER 4 The atom

4.1 A momentous discovery

In Paris in 1898 a young physicist called Marie Curie began her research work. She had decided to look for an explanation of a peculiar property of uranium salts. They can 'fog' photographic film, even when the film is some distance away. Could the uranium salts be giving off some kind of rays that travel through air?

Marie Curie soon found that this strange effect happened with all uranium salts. It depended only on the *amount* of uranium present in the salt and *not* on the chemical bonds that were present. She realised that this ability to give off rays must be a property of the *atoms* of uranium. It was a completely new type of property, quite different from the chemical reactions of uranium salts. This was an adventurous new idea. Marie Curie called this property of the uranium atom **radioactivity**.

Marie's husband, Pierre, left his own research work to help her in this exciting new science of radioactivity. In 1898, they discovered two new radioactive elements. They called one polonium, after Madame Curie's native country, Poland. The second element they called radium, meaning 'giver of rays'.

The Curies worked in dismal conditions. A scientist called Wilhelm Ostwald asked to be shown the laboratory in which they had discovered radium. He said it was 'a cross between a horse stable and a potato cellar' and that 'if I had not seen the work table with the chemical apparatus, I would have thought it was a practical joke'.

Marie and Pierre enjoyed their work, despite the miserable shed. While they experimented to obtain a pure specimen of the new element radium they often wondered what it would look like and what colours its salts would be. The radium chloride they obtained surprised them by glowing in the dark. Madame Curie wrote, 'One of our joys was to go into our workroom at night; we then perceived on all sides the feebly luminous silhouettes of the bottles and capsules containing our products. It really was a lovely sight and always new to us. The glowing tubes looked like faint fairy lights.'

Why do the atoms of these elements, uranium, polonium and radium, give off the rays which Marie Curie named radioactivity? The explanation came from the British physicist, Lord Rutherford, in 1902. He suggested that radioactivity is caused by atoms splitting up. This was another revolutionary idea. For many years, scientists had accepted Dalton's view of the atom. The British chemist John Dalton had put forward his Atomic Theory in 1808. He had backed the theory that matter is composed of minute particles called atoms. Atoms, he said, cannot be created or destroyed or split. We know now that a large number of elements have atoms which are unstable and split up to form smaller atoms.

4.2 Protons, neutrons and electrons

Early in the twentieth century, the work of the Curies and of Rutherford and other scientists showed that the atom is made up of smaller particles. These are called **protons**, **neutrons** and **electrons**. These **subatomic particles** differ in mass and in electrical charge (see Table 4.1).

Table 4.1 Subatomic particles

Particle	Mass (in atomic mass units)	Charge (in elementary charge units)
Proton	1	+1
Neutron	1	0
Electron	1/1840	−1

Protons and neutrons both have the same mass as an atom of hydrogen. We call this mass 1 **atomic mass unit** ($1\ m_u$). ($1\ m_u = 1.67 \times 10^{-27}$ kg.) The mass of an atom depends on the number of protons and neutrons it contains. The number of protons and neutrons together is called the **mass number**. The electrons in an atom contribute very little to its mass.

Electrons carry a negative electric charge of 1 elementary charge unit ($-1\ e$). ($1\ e = 1.60 \times 10^{-19}$ coulombs.) Protons carry a positive electric charge of $+1\ e$. Neutrons are uncharged particles. The reason why whole atoms are uncharged is that **the number of electrons in an atom is the same as the number of protons**. The number of protons (which is also the number of electrons) is called either the **atomic number** or the **proton number**. This means that if you know the mass number and the atomic number you can find the number of neutrons:

$$\text{Number of neutrons} = \text{Mass number} - \frac{\text{Atomic (proton)}}{\text{number}}$$

For example, an atom of sodium has a mass of $23\ m_u$ and an atomic (proton) number of 11. The number of electrons is 11; the same as the number of protons. The number of neutrons in the atom is

$$23 - 11 = 12$$

Relative atomic mass

An atom of hydrogen consists of one proton and one electron. It is the lightest of atoms. Chemists compare the masses of other atoms with that of a hydrogen atom. They use **relative atomic masses**. The relative atomic mass A_r of a uranium atom is 238. This means that one uranium atom is 238 times as heavy as one atom of hydrogen.

$$\text{Relative atomic mass of an element} = \frac{\text{Mass of one atom of the element}}{\text{Mass of one atom of hydrogen}}$$

SUMMARY NOTE

The atoms of all elements are made up of three kinds of particles. These are
- protons, of mass $1\ m_u$ and electric charge $+1\ e$
- neutrons, of mass $1\ m_u$ (uncharged)
- electrons, of mass $1/1840\ m_u$ and electric charge $-1\ e$.

The number of protons (and the number of electrons) in an atom is the **atomic number**. The sum of the number of protons and the number of neutrons is the **mass number**.

JUST TESTING 14

Some relative atomic masses are
$A_r(H) = 1$, $A_r(He) = 4$, $A_r(C) = 12$, $A_r(O) = 16$, $A_r(S) = 32$.
Copy and complete the following sentences:
A sulphur atom is ____ times as heavy as an atom of hydrogen.
A sulphur atom is ____ times as heavy as an atom of helium.
One sulphur atom has the same mass as ____ oxygen atoms.
____ helium atoms have the same mass as two oxygen atoms.
Two carbon atoms have the same mass as ____ hydrogen atoms.

4.3 The arrangement of particles in the atom

How are protons, neutrons and electrons arranged in the atom? One idea was that they are all mixed up together, with the tiny electrons dotted about like currants in a bun, or plums in a plum pudding (see Fig. 4.1). This 'plum pudding' picture of the atom was disproved by the famous physicist, Lord Rutherford, in 1908.

Rutherford showed that most of the volume of an atom is space. The protons and neutrons occupy a tiny volume in the centre of an atom. Rutherford called this the **nucleus** (Fig. 4.2). The nucleus is minute in volume compared with the volume of the atom (see Fig. 4.3).

The electrons are present in the space around the nucleus. They guard the space by repelling the electrons of neighbouring

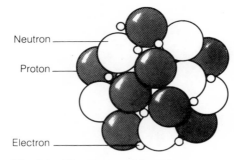

Fig. 4.1 The plum pudding atom

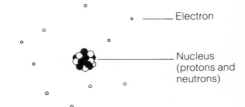

Fig. 4.2 The Rutherford atom

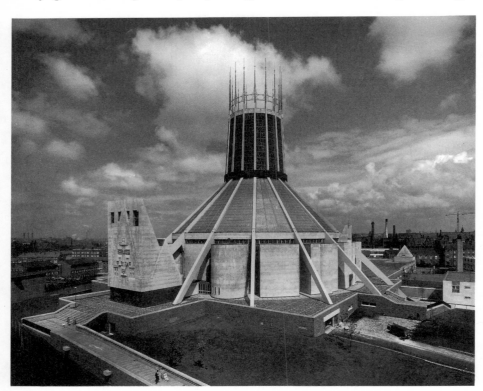

Fig. 4.3 If this building represents the space taken up by an atom, then a pea at the centre of the building represents the size of the nucleus

atoms. Remember that electrons are negatively charged. Repulsion will occur between the like charges on two electrons if they come close enough together. In Fig. 4.3 the electrons furthest away from the nucleus would be at the outside walls of the building.

JUST TESTING 15

1. Element A has atomic number 4 and mass number 9. How many protons, neutrons and electrons are present in one atom of A?

2. Copy this table into your book. Fill in the missing numbers.

Particle	Mass number	Atomic number	Number of protons	Number of neutrons	Number of electrons
Oxygen atom	16	8	—	—	—
Calcium atom	—	20	—	20	—
Bromine atom	80	—	35	—	—
Aluminium atom	27	13	—	—	—

3. State (i) the atomic number and (ii) the mass number of
   a. an atom with 15 protons and 16 neutrons
   b. an atom with 80 protons and 120 neutrons
   c. an atom with 27 protons and 33 neutrons.

4.4 Electrons in orbit

The electrons are in constant motion. They move round and round the nucleus in paths called orbits (see Fig. 4.4).

The scientist Niels Bohr added to our picture of the atom. He had the idea that the electrons in an atom do not all have the same energy. An electron needs more energy to circle the nucleus in an orbit of large radius than in an orbit close to the nucleus. The orbits are grouped together in **shells**. A **shell** is a group of orbits with similar energy. Each shell can hold up to a certain number of electrons (see Fig. 4.5). In any atom, the maximum number of electrons in the outermost group of orbits is 8.

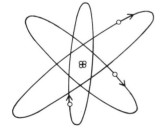

Fig. 4.4 Electrons orbiting the nucleus

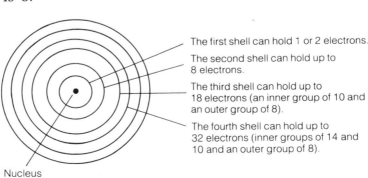

The first shell can hold 1 or 2 electrons.

The second shell can hold up to 8 electrons.

The third shell can hold up to 18 electrons (an inner group of 10 and an outer group of 8).

The fourth shell can hold up to 32 electrons (inner groups of 14 and 10 and an outer group of 8).

Fig. 4.5 Shells of electron orbits in an atom ▷   Nucleus

*4.5 How are the electrons arranged?

If you know the atomic (proton) number of an element, you know the number of electrons in an atom of the element. You can work out how the electrons are arranged by filling the lowest energy shells first. An atom of carbon has 6 electrons. The first shell will be filled by two of the electrons. The other four electrons will go into the second shell (see Fig. 4.6).

An atom of magnesium has atomic number 12. The first shell is filled by 2 electrons, the second shell is filled by 8 electrons, and 2 electrons occupy the third shell (Fig. 4.7). The arrangement of electrons can be written as (2.8.2). It is called the **electron configuration** of the element. Table 4.2 gives the electron configurations of the first 20 elements.

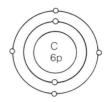

Fig. 4.6 Arrangement of electrons in the carbon atom

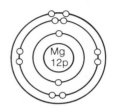

Fig. 4.7 The electron configuration of magnesium

Table 4.2 Electron configurations of the first 20 elements

Element	Symbol	Atomic (proton) number	1st shell	2nd shell	3rd shell	4th shell	Electron config-uration
Hydrogen	H	1	1				1
Helium	He	2	2				2
Lithium	Li	3	2	1			2.1
Beryllium	Be	4	2	2			2.2
Boron	B	5	2	3			2.3
Carbon	C	6	2	4			2.4
Nitrogen	N	7	2	5			2.5
Oxygen	O	8	2	6			2.6
Fluorine	F	9	2	7			2.7
Neon	Ne	10	2	8			2.8
Sodium	Na	11	2	8	1		2.8.1
Magnesium	Mg	12	2	8	2		2.8.2
Aluminium	Al	13	2	8	3		2.8.3
Silicon	Si	14	2	8	4		2.8.4
Phosphorus	P	15	2	8	5		2.8.5
Sulphur	S	16	2	8	6		2.8.6
Chlorine	Cl	17	2	8	7		2.8.7
Argon	Ar	18	2	8	8		2.8.8
Potassium	K	19	2	8	8	1	2.8.8.1
Calcium	Ca	20	2	8	8	2	2.8.8.2

SUMMARY NOTE

The protons and neutrons make up the nucleus at the centre of the atom. The electrons circle the nucleus in orbits. Groups of orbits with the same energy are called **shells**. The 1st shell can hold 2 electrons; the 2nd shell can hold 8 electrons; the 3rd shell can hold 18 electrons. The arrangement of electrons in an atom is called the **electron configuration**.

JUST TESTING 16

1. The electron configuration of phosphorus is (2.8.5). What does this tell you about the arrangement of electrons in the atom? Sketch the arrangement. (See Figs. 4.6 and 4.7 for help.)
2. Find the names of 4 boys and 2 girls in the names of the elements with the following atomic numbers: **a** 4, **b** 28, **c** 38, **d** 44, **e** 51, **f** 86. (See p. 316 for atomic numbers.)
3. Sketch the arrangement of electrons in the atoms of **a** B, **b** N, **c** F, **d** Al. (See p. 316 for atomic numbers.)

4.6 The repeating pattern of the elements

We have spent time looking at the structure of the atom and the electron configurations of the elements. How does this help with chemistry? If you look at the electron configurations of the atoms, some interesting points may strike you.

First, notice the elements with a full outer shell of electrons. These are helium (2), neon (2.8), argon (2.8.8) and, two elements which are not shown in Table 4.2, krypton and xenon. These elements are **the noble gases**, and are present in air (see Section 8.7). They are the least reactive of the elements. For many years, it seemed as though they took no part in any chemical reactions. Then, in 1960, krypton and xenon were made to combine with the very reactive element fluorine. The noble gases exist as single atoms. Their atoms do not combine in pairs to form molecules as do the atoms of most gaseous elements (e.g. O_2 and H_2). It seems logical to suppose that it is the full outer shell of electrons that makes the noble gases chemically unreactive.

Following each noble gas (that is, with atomic number 1 greater than the noble gas) is a metallic element with one electron in its outer shell. These elements are lithium (2.1), sodium (2.8.1), potassium (2.8.8.1) and (not shown in Table 4.2) rubidium and caesium. They are a set of very similar metallic elements. They all react with cold water to form hydrogen and a solution of the metal hydroxide (see Section 10.3). The strongly alkaline nature of the hydroxides gives these metals the name **the alkali metals**.

Preceding each noble gas (with atomic number 1 less than the noble gas) are the elements fluorine (2.7), chlorine (2.8.7) and (not shown in Table 4.2) bromine and iodine. These elements are a set of very reactive non-metallic elements. They are called **the halogens** because they react with metals to form salts. (Halogen comes from a Greek word and means salt-former.)

Some interesting patterns appear if the elements are first put into the order of their atomic numbers and then arranged in rows. A new row is started every time it is necessary to get a noble gas to fall into place under the preceding noble gas.

Table 4.3 A section of the Periodic Table

	Group 1	Group 2	Group 3	Group 4	Group 5	Group 6	Group 7	Group 0
Period 1	H (1)							He (2)
Period 2	Li (2.1)	Be (2.2)	B (2.3)	C (2.4)	N (2.5)	O (2.6)	F (2.7)	Ne (2.8)
Period 3	Na (2.8.1)	Mg (2.8.2)	Al (2.8.3)	Si (2.8.4)	P (2.8.5)	S (2.8.6)	Cl (2.8.7)	Ar (2.8.8)
Period 4	K (2.8.8.1)	Ca (2.8.8.2)						

This arrangement of the elements puts elements which have the same number of electrons in the outermost shell into vertical columns. It is called the **Periodic Table**. It was given this name in 1856 by a Russian chemist called Mendeleev. The Periodic Table simplifies the task of learning about the chemical elements. The eight vertical columns of elements are called **groups**. The noble gases are in Group 0. For the rest of the elements, the group number is the number of electrons in the outermost shell. The horizontal rows of elements are called **periods**. The first period contains only hydrogen and helium. The second period contains the elements lithium to argon. The complete Periodic Table is shown on p. 318. You will need to refer to it again in later chapters.

4.7 Isotopes

Sometimes an element has a relative atomic mass which is not a whole number, e.g. chlorine has a relative atomic mass of 35.5. Since

Mass number = Number of protons + Number of neutrons

the mass number *must* be a whole number. The element chlorine consists of two kinds of atoms with different mass numbers. One type has 17 protons and 18 neutrons and mass number 35. The other type of atom has 17 protons and 20 neutrons and mass number 37. All atoms of chlorine have 17 electrons. Since it is the number of electrons which decides the chemical behaviour, all chlorine atoms behave in the same way. The number of neutrons in the nucleus does not affect the chemical behaviour. These different forms of chlorine are called *isotopes*. Isotopes are forms of an element which differ in the number of neutrons in the atom. There are three chlorine atoms with mass 35 m_u for each chlorine atom with mass 37 m_u. The average atomic mass is therefore

$$\frac{(3 \times 35) + 37}{4} = 35.5 \; m_u.$$

This is why the relative atomic mass of chlorine is 35.5.

The isotopes of chlorine are written as $^{35}_{17}Cl$ and $^{37}_{17}Cl$. In general, isotopes are shown as

Mass number
                Symbol
Atomic number

The symbol $^{23}_{11}Na$ tells you that sodium has mass number 23 and atomic number 11. You then know that there are 11 protons, 11 electrons and 12 neutrons in the atom. The isotopes of hydrogen are $^{1}_{1}H$, $^{2}_{1}H$ and $^{3}_{1}H$. They are often referred to as hydrogen-1, hydrogen-2 and hydrogen-3. Hydrogen-2 is also called deuterium, and hydrogen-3 is also called tritium.

Carbon has three isotopes, carbon-12, carbon-13 and carbon-14. The isotope carbon-12 has replaced hydrogen as the reference point for the modern scale of relative atomic masses. Instead of

SUMMARY NOTE

The chemical properties of elements depend on the electron configurations of their atoms.
- The unreactive **noble gases** all have a full outer shell of electrons.
- The reactive **alkali metals** have a single electron in the outer shell.
- The reactive non-metallic elements called **halogens** have 7 electrons in the outer shell: they are 1 electron short of a full shell.

In the Periodic Table, elements are arranged in order of increasing atomic (proton) number in 8 vertical **groups**. The horizontal rows are called **periods**.

SUMMARY NOTE

Isotopes are atoms of the same element which differ in the number of neutrons. They contain the same number of protons (and therefore the same number of electrons).

$A_r(H) = 1.0000$, we use $A_r(C) = 12.0000$. There is very little difference numerically between the two scales. On the carbon-12 scale, $A_r(H) = 1.0080$.

JUST TESTING 17

1. Write the symbol with mass number and atomic number (as above) for each of the following elements:
   a aluminium with 13 protons and 14 neutrons
   b potassium with 19 protons and 20 neutrons
   c iron with 26 protons and 30 neutrons
   d uranium with 92 protons and 146 neutrons.

2. Three atoms are described below. Which two atoms have similar chemical properties?
   Atom A contains 5 protons and 6 neutrons.
   Atom B contains 9 protons and 10 neutrons.
   Atom C contains 13 protons and 14 neutrons.
   Explain your answer.

3. At one time, zinc beryllium silicate was used in the manufacture of fluorescent lights. When workers who came into contact with it developed a disease called 'berylliosis', the substance was banned. The body absorbs zinc beryllium silicate because beryllium follows the same chemical path through the body as a certain harmless element. Which element do you think this might be?

4.8 Radioactivity

Changes of the kind described by Madame Curie as radioactivity are **nuclear reactions**. In a nuclear reaction the atomic nucleus splits, and the protons and neutrons in it form two nuclei. The electrons divide themselves between the two new nuclei. Sometimes protons, neutrons and electrons fly out when the original nucleus divides (Fig. 4.8).

This type of change is quite different from a chemical reaction. In a chemical reaction, old bonds are broken and new bonds are made, but the atoms stay the same. In a radioactive change, new atoms of new elements are formed. Such changes are called **nuclear reactions**. Elements which undergo nuclear reactions are said to be **radioactive**. The breaking-up process is **radioactive decay**. In nuclear reactions, small particles and energy are given out or radiated. These particles and energy are called

SUMMARY NOTE

Marie Curie started the study of **radioactivity**. Lord Rutherford's explanation of radioactivity is that a **nuclear reaction** occurs. The nuclei of the radioactive element split, new atoms are formed and radiation is emitted. The radiation consists of α-particles, β-particles and γ-rays. The emission of radiation is called **radioactive decay**.

Nucleus of unstable radioactive atom

Nucleus starts to divide.

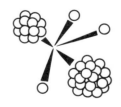
Two new nuclei are formed. Some subatomic particles fly out.

Fig. 4.8 A nuclear reaction ▷

radioactivity. Three types of radiation are given off: **α-particles and β-particles** and **γ-radiation**. Their characteristics are shown in Fig. 4.9. α-particles are the nuclei of helium atoms. They consist of 2 protons and 2 neutrons, and carry a positive charge. β-particles are electrons. γ-rays have extremely high penetrating power. They can pass through the skin, penetrate bone, and cause burns and cancers. People who work with sources of γ-rays protect themselves from the radiation by building a wall of lead bricks between themselves and the source (see Fig. 4.15).

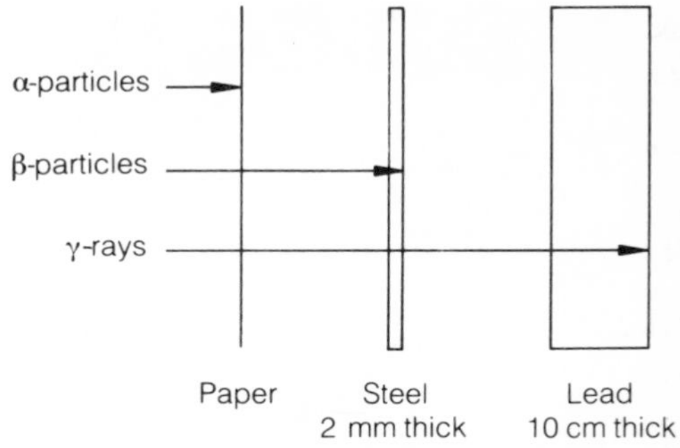

Fig. 4.9 The penetrating power of α, β and γ-rays

4.9 The oldest murder in history?

Stonehenge is the most famous ancient monument in the UK. The circle of massive stones has puzzled people for centuries. A ditch which used to surround the monument has become filled in over the years. Recently, an archaeologist, Dr John Evans, was excavating the ditch when, to his surprise, he spotted a skeletal foot sticking out. Carefully, he unearthed the rest of the skeleton. It was the remains of a young man. How long had he lain there? How had he met his end? What Evans saw made him suspect foul play.

The tip of a flint arrowhead was embedded in the breastbone, showing that it had entered the body from behind. Another flint arrowhead was lodged in one of the ribs. A second rib had been chipped by an arrow on its way through the body. Three arrowheads lay nearby. Two had missing tips. The angle of entry showed that the arrows had been fired at close range. It was murder!

Evans used a different kind of detective work to find out the age of the skeleton. He discovered that the skeleton had lain in the ditch since 2200 B.C. This must be the oldest murder ever investigated! How did Evans fix the date?

Fig. 4.10 Protection from radioactivity

The scientific method which Evans used is called **radiocarbon dating** or **carbon-14 dating**. This method has shown that parts of Stonehenge go back to 3500 B.C. Other parts are more recent, only 1200 B.C!

4.10 Making use of radioactivity

Radiocarbon dating

What is radiocarbon dating? Radioactivity is detected by the use of an instrument called a **Geiger–Müller counter** (named after its inventors). Placed near a source of radioactivity, a Geiger–Müller counter (Fig. 4.11) emits a series of clicks. The number of clicks per minute (or counts per minute, c.p.m.) measures the strength of the radioactivity. Alternatively, the radioactivity in c.p.m. can be registered on a digital electronic counter.

If radioactivity is measured at various times and then plotted against the time since measurements were started, the result is a curve like that shown in Fig. 4.12. From this curve you can see that

Fig. 4.11 A Geiger–Müller counter

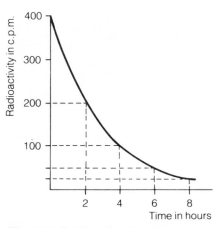

Fig. 4.12 Radioactive decay: a graph of count rate against time

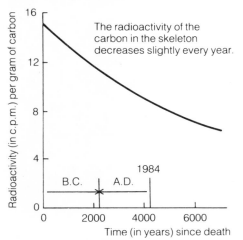

Fig. 4.13 Radiocarbon dating

- the time taken for the count to fall from 400 to 200 c.p.m. is 2 hours
- the time taken for the count to fall from 200 to 100 c.p.m. is 2 hours
- the time taken for the count to fall from 100 to 50 c.p.m. is also 2 hours.

The time for the rate of radioactive decay to fall to half its value is the same, no matter what the original rate of decay is. This time is called the **half-life**. The half-life of the element in Fig. 4.12 is 2 hours.

Carbon is made up of the isotopes carbon-12, carbon-13 and carbon-14. The isotope carbon-14 is radioactive. It has a half-life of 5700 years. A living tree takes in carbon-14 in the carbon dioxide which it uses for photosynthesis. When the tree dies, no more carbon-14 is taken in. The carbon-14 already present decays slowly; carbon-12 does not change. The ratio of the amount of carbon-14 left in the wood to the amount of carbon-12 present can be used to give the age of the wood.

Animals also take in carbon in their food while they are alive. After their death, the proportion of carbon-14 in their bones tells how long it is since they were alive. A carbon-14 decay curve is shown in Fig. 4.13. The radioactivity of the skeleton discovered by John Evans showed that the carbon-14 present had been decaying for 4200 years. This established the time of death as 2200 B.C. By 1984, the trail was too cold for the detective work to result in an arrest!

Other radioisotopes

Radioactive elements differ enormously in their half-lives. Some are listed in Table 4.4.

Table 4.4 Some radioactive isotopes

Isotope	Half-life	Isotope	Half-life
Uranium-238	5000 million years	Gold-198	3 days
Uranium-235	700 million years	Sodium-24	15 hours
Plutonium-239	24 000 years	Uranium-239	24 minutes
Carbon-14	5700 years	Fluorine-17	1 minute
Strontium-90	59 years	Fluorine-22	4 seconds
Cobalt-60	5 years	Polonium-214	1/5000 seconds
Iodine-131	8 days		

Scientists have devised many uses for radioactive isotopes. Some of these are illustrated in Fig. 4.14.

4.11 The dangers of radioactivity

Radioactivity can be of great use to us. If it is mismanaged, however, it can be a serious health hazard. People who work with radioactive sources have to protect themselves from the

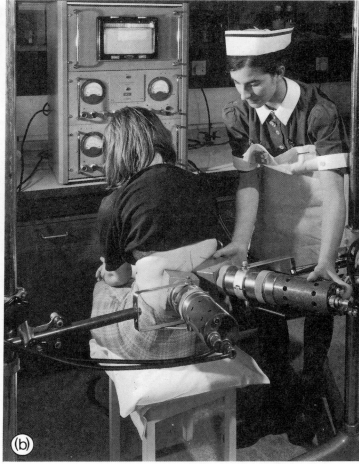

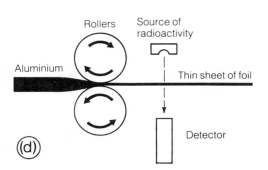

Fig. 4.14 Uses of radioactive isotopes: (a) this heart pacemaker contains plutonium-238; (b) measuring the uptake of iodine-131 in kidneys; (c) the technician is wrapping a belt of photographic film round a weld in the pipeline — if the weld is good, no radiation will penetrate it from the radioactive source inside the pipe to affect the film; (d) the diagram shows the production of metal foil (e.g. aluminium baking foil). The detector measures the amount of radiation passing through the foil. If the foil is too thick, the detector reading drops. The detector sends a message to the rollers, which move closer together to make the foil thinner.

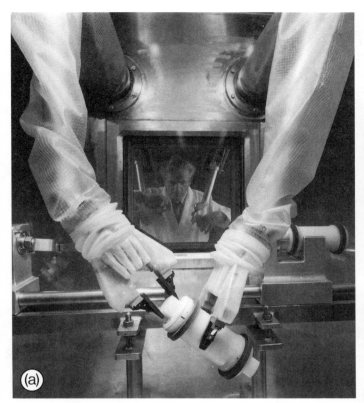

Fig. 4.15 Protection from radiation: (a) a long-handled tool is used to deal with a radioactive isotope; (b) a wall of lead bricks shields this scientist from γ-radiation; (c) pressurised suits protect workers from α-radiation

radiation they are using. They often use long-handled tools to separate themselves as far as possible from a radioactive source (see Fig. 4.15a). They build walls of lead bricks to protect themselves from γ-emitters (see Fig. 4.15b).

Workers use gloves and protective clothing when working with sources of radioactivity (Fig. 4.15c). This precaution prevents any radioactive material from getting on to their hands or clothes. If radioactive elements get inside the body, they are far more dangerous. They can irradiate the body organs from close quarters, and the risk of cancer is very great. Even α-rays, the rays with least penetrating power, can do damage if they are inside the body.

SUMMARY NOTE

The time taken for half the atoms present in a radioactive isotope to decay is called the **half-life** of the isotope. Half-lives vary from a fraction of a second to millions of years. Radioactive isotopes are used in medicine, in research and in industry. Workers handling radioactive materials take precautions to shield themselves from radiation.

JUST TESTING 18

1 Refer to the isotopes listed in Table 4.4. If you wanted to detect a leak in an underground water main, which radioisotope would you choose? The radioactivity must last long enough for you to drive along the route taken by the

5 mile long pipe to find out where the leak is. On the other hand, you do not want the radioactivity to contaminate the water for longer than is absolutely necessary.

2 A sample of radioactive isotope was put into a Geiger–Müller counter and the count rate was measured at various intervals.

Time (days)	Count rate (c.p.m.)
0	520
50	410
100	320
200	195
300	115
400	70

Plot a graph of the count rate against the time. Use the horizontal axis for time.
a Use the graph to find the count rate at 150 days and at 250 days.
b After what length of time was the count rate (i) 300 c.p.m., (ii) 150 c.p.m?
c What is the half-life of the radioactive element?

3 Explain the terms 'radioactive decay' and 'half-life'.
A sample of gold-198 has a count rate of 40 000 c.p.m. The half-life of this isotope is 3 days. How long will it take for the count rate to drop to 10 000 c.p.m?

4 Refer to the carbon-14 decay curve in Fig. 4.12.
What was the count rate of the skeleton just after death?
What is the half-life of carbon-14?
What was the count rate which dated the skeleton as 2200 B.C.?

5 Technetium-99 is a radioisotope used for medical diagnosis. It has a half-life of 6 hours. A scientist leaves 1 gram of technetium-99 in the laboratory at 6 p.m. on Friday. How much of the isotope will be left by 6 a.m. on Monday?

6 Strontium-90 is a radioactive isotope formed in nuclear reactors. One of the reasons it is dangerous is that it follows the same chemical pathway through the human body as another element X which is essential for health. By referring to the position of strontium in the Periodic Table, say which element you think is X.

4.12 The missing yellowcake

In November 1968 the cargo vessel Scheersberg sailed out of Antwerp in Belgium. She was loaded with a cargo of metal ore called **yellowcake**. For this voyage, the ship had taken on a new crew. She was bound for Genoa in Italy, but she never arrived. Two weeks later, she arrived at a Turkish port. The £3 million cargo was missing. Government secret agents tried to trace the cargo. The crew could not explain how 200 tonnes of yellowcake could have got lost. With the crew refusing to cooperate, the agents got nowhere. The mystery has never been solved.

Why is yellowcake so important? Why did the disappearance of the cargo cause so much worry? Why did the government take an interest, instead of leaving it to the owners? It was not only the £3 million the government was worried about. The government was afraid that the crew had sold the cargo to another country or to a group of terrorists. **Yellowcake** is an ore containing uranium oxide. There is something very special about uranium: it can be used to make atomic bombs.

4.13 The atomic bomb

In 1939 a German scientist called Otto Hahn did something sensational with uranium atoms. He split them. Hahn had been experimenting on firing neutrons at various elements. When he fired neutrons at uranium, some atoms of uranium-235 split into two new atoms and two neutrons. We call this atom-splitting process **fission** (see Fig. 4.16). When uranium-235 (U-235) atoms are split, an enormous amount of energy is released. Where does the energy come from? The sum of the masses of the atoms and neutrons produced is 0.2 m_u less than the mass of one atom of U-235. The missing mass has been converted into energy, **nuclear energy** (also called **atomic energy**).

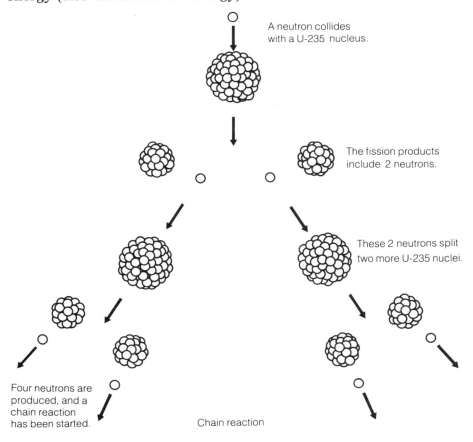

Fig. 4.16 The chain reaction in uranium fission

Shortly after Hahn's discovery the Second World War broke out. Scientists immediately started work on methods of making bombs which release this new form of energy.

Figure 4.16 shows what happens in a block of uranium-235. The fission of one U-235 atom produces two neutrons. These two neutrons split two more uranium atoms, to give four neutrons. A **chain reaction** is set off. In a large block of uranium-235, this results in an explosion. In a small block of uranium-235, neutrons escape from the surface of the block before producing fission, and an explosion does not occur. There is a **critical mass** below which a block of uranium will not make an atomic bomb. An atomic bomb consists of two blocks of uranium-235, each smaller than the critical mass. The bomb is detonated by firing one block into the other to make a single block which is larger than the critical mass. The detonation is followed by an atomic explosion.

Only two atomic bombs have been used in warfare. The two bombs destroyed two cities in 1945: Hiroshima and Nagasaki in Japan (see Fig. 4.17). The effects were devastating. Millions of

Fig. 4.17 Devastation

people were killed in the blast. Thousands were burned and thousands received a large dose of radiation from which they never recovered. There was a great outcry against atomic bombs. The campaign against them continues today. No country has used atomic weapons since 1945, although many nations have huge stockpiles of atomic weapons.

4.14 Nuclear reactors

In nuclear reactors, energy is obtained from the same reaction as in atomic bombs, the fission of uranium-235. This time, the reaction is carried out in a controlled way. The reactors use naturally occurring uranium, which is a mixture of uranium-235 and uranium-238. Only uranium-235 undergoes fission. Figure 4.18 shows a nuclear reactor. Graphite rods are used to slow down the neutrons. Rods of boron control the rate of fission by absorbing some of the neutrons. The nuclear fission generates heat. This heat is used to turn water into steam. The steam drives a turbine, which generates electricity.

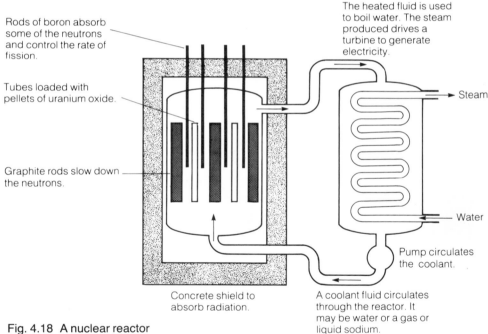

Fig. 4.18 A nuclear reactor

4.15 Nuclear fusion

When two atoms of hydrogen-2 (deuterium) collide at high speed, the nuclei fuse together:

$${}^{2}_{1}H + {}^{2}_{1}H \rightarrow {}^{3}_{2}He + {}^{1}_{0}n$$

An isotope of helium and a neutron are produced. The sum of the masses of the helium atom and the neutron is slightly less than the mass of the two hydrogen-2 atoms. The missing mass is converted into energy. The amount of energy released is enormous. This energy gives more hydrogen-2 atoms the energy they need to fuse with other atoms. Thus a chain reaction starts. As hydrogen-2 occurs in water as ^2H_2O, supplies are plentiful.

The fusion process may some day replace the fission of uranium-235 as a source of nuclear energy. It is very difficult, however, to give the hydrogen-2 atoms the energy they need to fuse in the first place. Scientists are still working on the problems of obtaining energy from fusion. The sun obtains its energy from the fusion of hydrogen-2 atoms. In the sun, the temperature is about 10 million °C, and the hydrogen-2 atoms have enough energy to fuse.

*4.16 The hydrogen bomb

The fusion of hydrogen-2 nuclei is the source of energy in the hydrogen bomb. The hydrogen-2 atoms are raised to the temperature at which they will fuse by the explosion of a uranium-235 bomb. The destruction caused by a hydrogen bomb would be so widespread that no nation has dared to use it in warfare.

SUMMARY NOTE

Energy is released in nuclear reactions. This **nuclear energy** can be released in an explosive manner in an **atomic bomb**. The **fission** (splitting) of uranium-235 was used in atomic bombs. The nuclear energy from the fission of uranium-235 is released in a safe, regulated manner in a nuclear power station. The **fusion** (joining together) of small nuclei also releases energy. The fusion of hydrogen-2 nuclei is the source of energy in the hydrogen bomb.

Exercise 4

1 An atom of the element X has atomic number 9 and mass number 19. Sketch the arrangement of electrons in an atom of X. How many protons and neutrons are present in the nucleus?

2 Atom A has atomic number 93 and mass number 239. Atom B has atomic number 94 and mass number 239. How many protons has atom A? How many neutrons has atom B? Are atoms A and B isotopes of the same element? Explain the meaning of the word 'isotope'.

3 Name the elements with the atomic numbers given in this sentence.
She had (29)-coloured hair, a laugh like a (47) bell and a heart of (79), but, as he gazed at her, his heart felt as heavy as (82) because he knew that she had a will of (26) and he would never persuade her to elope with him.

4 Copy and complete this table.

Element	Mass number	Atomic number	Protons	Neutrons	Electrons
X	11	-	-	12	-
Y	-	-	6	6	-
Z	-	8	-	-	16

5 What do a *group* of elements in the Periodic Table have in common?
What is a *period* of the Periodic Table?

*6 Explain why Mendeleev fitted the elements into vertical groups in his Periodic Table. Why did he fit lithium, sodium, potassium, rubidium and caesium into the same group? Why did he leave some gaps in his table?

7 **a** Explain the term 'radioactive element'. Give two examples.
  b Name the three types of radiation which are given off by radioactive isotopes. How do they differ in penetrating power? Which type of radiation is the most dangerous?

8 • Strontium-90 is radioactive and therefore dangerous.
  • Milk and bones contain more calcium than other tissues of the animal body.
  Why are babies especially at risk from damage by strontium-90 in the environment? (The clue is in the Periodic Table.)

9 Describe two examples of the way in which radioactive isotopes are used. What advice about safety would you give to someone who was about to use radioactive materials for the first time?

10 Lead-214 is radioactive, with a half-life of 27 minutes. A sample of lead-214 nitrate in a Geiger–Müller counter gives 6000 c.p.m. What will the count rate be after
   a 54 minutes, **b** 1 hour 48 minutes?

11 Complete this chart.

	Atomic number	Mass number	Number of protons	Number of neutrons	Number of electrons
a	3	7	-	-	-
b	32	41	-	-	-
c	20	-	-	16	-
d	-	-	-	45	35
e	-	-	8	8	-

12 A hospital is buying a radioactive source to irradiate cancer patients. Select a suitable isotope from the following list.

Isotope	Radiation	Half-life
Chromium-51	γ	26 days
Nickel-63	β	300 years
Caesium-135	β	2 million years
Caesium-136	β, γ	14 days
Caesium-137	β, γ	37 years
Cobalt-60	β, γ	5.3 years
Zinc-72	β, γ	49 hours
Zirconium-93	β	4 million years

Give reasons for your choice.

13 A doctor needs to put a radioactive isotope into the food of a patient in order to study his digestive system. The isotope chosen must have a suitable half-life. Which of the following list would be suitable?

Isotope	Half-life
Sulphur-35	87 days
Sulphur-37	5 minutes
Calcium-45	152 days
Sodium-24	15 hours
Sodium-25	58 seconds
Nitrogen-13	10 minutes

Give reasons for your choice.

14 Explain the difference between nuclear fusion and nuclear fission.

15 **a** The fission of uranium-235 is described as a 'chain reaction'. What is meant by a chain reaction?
   b Where does the enormous amount of energy liberated in the fission of uranium-235 come from?
   c Explain how nuclear energy is used to produce electricity.

16 Is there any real difference between the accident risk which society takes in using cars and the accident risk posed by atomic energy? Is one a worse killer than the other? Explain your answer.
   This is a question to which many people do not have an answer. Perhaps you would like to form a group to discuss the issue. The **nuclear debate** comes up again in Chapter 18.

CHAPTER 5 Electrochemical reactions

5.1 Electroplating

The tea service in Fig. 5.1 gleams like silver, but it is made of steel. The reason for the shine is that it has been given a coating of chromium. It will not rust as long as this surface layer is not scratched. Figure 5.2 shows how metal objects are chromium plated. The main reason for this treatment is to make the metal rustproof. Another benefit is that people find a bright shiny finish attractive.

During chromium plating, steel articles are immersed in a solution of a chromium salt. The steel is given a negative electric charge. For some reason, the result of this is that a layer of chromium becomes attached to the surface of the steel. To understand why this happens, you will have to study **electrochemical reactions**. There are two kinds. These are

- chemical reactions which take place when electricity flows through substances
- chemical reactions which produce electricity.

Fig. 5.1 Chromium plating gives this tea service a silvery shine

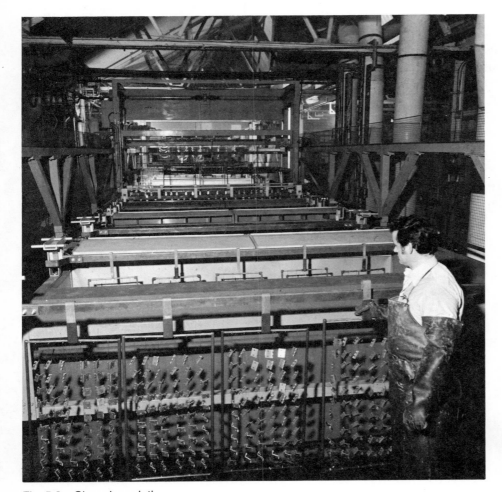

Fig. 5.2 Chromium plating

SUMMARY NOTE

Chromium plating is an electrochemical reaction.

5.2 Which substances conduct electricity?

You may like to do Experiment 5.1 to find out which substances conduct electricity. Afterwards, you will be able to divide the substances tested into four groups (see Tables 5.1 and 5.2).

Table 5.1 Electrical conductors and non-conductors: solids

Electrical conductors	Non-conductors, i.e. insulators
All metallic elements	Non-metallic elements, e.g. sulphur
All alloys	Many compounds, e.g. polyethene
One non-metallic element, the graphite allotrope of carbon	Crystalline salts, e.g. sodium chloride, copper(II) sulphate

Table 5.2 Electrical conductors and non-conductors: liquids

Electrolytes	Non-electrolytes
Solutions of acids and alkalis and salts. Such liquids are called **electrolytes**. Chemical changes occur at the electrodes. For example, copper(II) chloride solution changes into copper and chlorine and water.	The liquids which do not conduct electricity are organic compounds, such as ethanol, and other liquid compounds, such as water. They are called **non-electrolytes**.

> **SUMMARY NOTE**
>
> **Solids**: Metals, alloys and graphite conduct electricity.
> **Liquids**: Solutions of acids, alkalis and salts conduct electricity.

5.3 What happens when solutions conduct electricity?

Before you do Experiment 5.2, here are some new words:

A **cell** is a vessel in which *either* electricity produces a chemical change *or* a chemical change produces electricity.

The objects which conduct electricity into or out of the cell are called **electrodes**. The electrode connected to the positive terminal of the battery or other source of e.m.f. (electromotive force) is called the **anode**. The electrode connected to the negative terminal is called the **cathode**. The electrodes are usually made of elements such as platinum and graphite which do not react with electrolytes.

The results of Experiment 5.2 may make you wonder. Crystals of copper(II) chloride do not conduct electricity, and distilled water does not conduct electricity. Yet when the two substances are mixed to make a solution, the solution conducts. In the process the solute, copper(II) chloride, is split up. Copper appears as a layer on the negative electrode (the cathode). Chlorine is evolved at the positive electrode (the anode). Copper(II) chloride has been **electrolysed**, that is, split up by an electric current. The process is called **electrolysis**.

> **The electrodes...which is which?**
>
> How to remember:
> A̲no̲D̲e
> AD → ADD = +
> + is positive
> The anode is positive...and the cathode is negative.

> **SUMMARY NOTE**
>
> Check the meanings of:
> - cell
> - electrode
> - anode
> - cathode.

How can we explain electrolysis? Copper appears always at the negative electrode, never at the positive electrode. Is there a positive charge associated with copper in copper(II) chloride? We accept that copper, the element, is composed of copper atoms. Does the combined copper in copper(II) chloride consist of atoms with positive charges? Chlorine appears always at the positive electrode, never at the negative electrode. Is there a negative charge associated with the chlorine combined with copper in copper(II) chloride?

The scientist who did the first work on electrolysis was Michael Faraday. In 1834, he suggested a theory to explain his observations. He believed that the evidence of electrolysis pointed to the existence of charged particles of matter. He called them **ions**.

During electrolysis, positively charged copper ions travel to the negative electrode. When they reach the negative electrode, the ions lose their charge: they are **discharged** to form copper atoms. Negatively charged chloride ions travel to the positive electrode. When they reach the positive electrode, they lose their charge: they are discharged, and molecules of chlorine are formed. According to Faraday's theory, the compound copper(II) chloride consists of positively charged copper ions and negatively charged chloride ions.

Solid copper(II) chloride does not conduct electricity, yet when it is dissolved in water its aqueous solution does conduct. Is it the water in the solution that makes it an electrolyte? No, Experiment 5.1 shows that water is a very poor electrical conductor. To explain this difference, scientists suggested that, in a solid, the ions are fixed in position. They are held together by strong attractive forces between positive and negative ions. In a solution of a salt, the ions are free to move (Fig. 5.3). Another way to free the ions so that they can move to the electrodes might be to melt the solid (see Section 5.4).

The ions .. which is which?

How to remember:
think of the 5 Cs ...
'<u>C</u>urrent <u>c</u>arries <u>c</u>opper <u>c</u>ations to the <u>c</u>athode.'
<u>C</u>ations travel to the <u>c</u>athode ...
<u>A</u>nions travel to the <u>a</u>node.

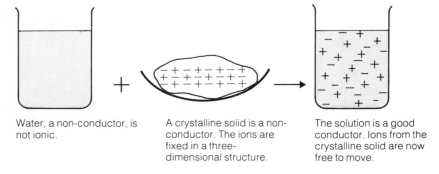

◁ Fig. 5.3 The ions must be free to move

Water, a non-conductor, is not ionic.

A crystalline solid is a non-conductor. The ions are fixed in a three-dimensional structure.

The solution is a good conductor. Ions from the crystalline solid are now free to move.

Non-electrolytes

Some of the liquids tested in Experiment 5.1 do not conduct electricity. It follows that these substances do not contain ions (see Section 6.3).

Weak electrolytes

Some substances conduct electricity to a very slight extent. In Experiment 5.1, you saw that ethanoic acid is a poor conductor,

SUMMARY NOTE

When a direct electric current is passed through solutions of some compounds, they are **electrolysed**, that is, split up by the current. Chemical changes occur in the **electrolyte**, resulting in the formation of new substances. Electrolysis is evidence that compounds which are electrolytes consist of positively and negatively charged particles called **ions**.

a weak electrolyte. A solution of ethanoic acid contains a small concentration of ions, which make it conduct (see Section 7.2).

5.4 Molten salts

The theory put forward to explain electrolysis is that some compounds consist of ions. In a solution of a salt, the ions must be free to move. Another way of enabling the ions to move might be to melt the salt. Figure 5.4 shows an apparatus which can be used to test this idea. Lead(II) bromide is a convenient salt to use because it has a fairly low melting point.

The experiment shows that molten lead(II) bromide is electrolysed. In the molten salt, ions are free to move.

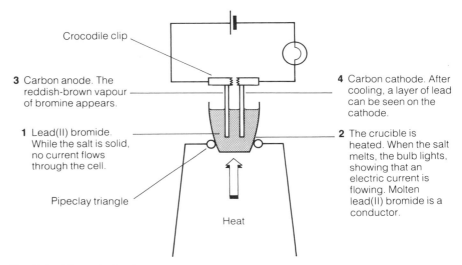

Fig. 5.4 Electrolysis of molten lead(II) bromide

> **SUMMARY NOTE**
>
> Some compounds conduct electricity when they are molten. As they do so, they are **electrolysed**. Their electrolysis is evidence that these compounds are composed of positive and negative **ions**.

5.5 What is the difference between an ion and an atom?

Atoms are uncharged. The number of protons in an atom is the same as the number of electrons (see Section 4.2). If an atom either gains or loses an electron, it will become electrically charged. Metal atoms and hydrogen atoms form positive ions (cations). For example, a sodium atom loses one electron to form a positively charged sodium ion:

sodium atom → electron + sodium ion
Na → e$^-$ + Na$^+$
(11 protons, (11 protons,
11 electrons: 10 electrons:
uncharged) charge = +1

The charge on a cation may be +1, +2 or +3.

- A copper atom loses 2 electrons to become a copper ion, Cu^{2+}.
- An aluminium atom loses 3 electrons to become an aluminium ion, Al^{3+}.

Non-metallic elements form negative ions (anions). They do this by gaining electrons. A chlorine atom gains one electron to become a chloride ion, Cl^-:

chlorine atom + electron → chloride ion
Cl + e⁻ → Cl^-
(17 protons, (17 protons,
17 electrons: 18 electrons:
uncharged) charge = −1)

An oxygen atom gains 2 electrons to become an oxide ion, O^{2-}. Some anions contain oxygen combined with another element. Examples are:

- hydroxide ion, OH^-
- nitrate ion, NO_3^-
- sulphate ion, SO_4^{2-}.

The ions you meet in this chapter are listed in Table 5.3. You will learn more about ions in Chapter 6.

Table 5.3 Symbols and formulas of some ions

Cations	Anions
Aluminium ion Al^{3+}	Bromide ion Br^-
Copper(II) ion Cu^{2+}	Chloride ion Cl^-
Hydrogen ion H^+	Iodide ion I^-
Lead(II) ion Pb^{2+}	Hydroxide ion OH^-
Sodium ion Na^+	Nitrate ion NO_3^-
Zinc ion Zn^{2+}	Sulphate ion SO_4^{2-}

JUST TESTING 19

1 a Divide the following list into (i) electrical conductors, (ii) non-conductors:
ethanol (alcohol), sodium chloride crystals, limewater, sugar solution, molten magnesium chloride, distilled water, solid sodium hydroxide, molten sodium hydroxide, sodium hydroxide solution.
 b What is the difference between the ways in which a copper wire and molten lead(II) bromide conduct electricity?
 c Some of the conductors in the list are electrolytes. Which are they?

2 Explain the words: cell, anode, cathode, anion, cation, electrolysis.

3 a Why do ions move towards electrodes?
 b Why is solid copper(II) chloride not an electrical conductor?

5.6 What happens at the electrodes?

Copper(II) chloride solution

Copper ions travel to the negative electrode, the cathode, where they form atoms of copper (see Fig. 5.5). When a copper ion reaches the cathode, this electrode, being negatively charged, can supply two electrons to the copper ion turning it into a copper atom. The equation for this electrode process is

copper(II) ion + 2 electrons → copper atom
Cu^{2+}(aq) + 2e⁻ → Cu (s)

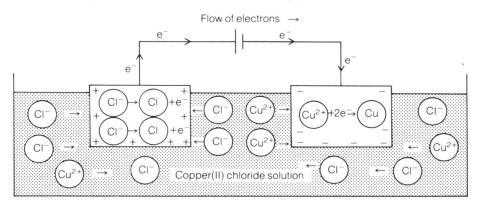

Fig. 5.5 Electrode processes in the electrolysis of copper(II) chloride at carbon electrodes

Chloride ions travel to the positive electrode, the anode, where chlorine gas is evolved. Being positively charged, the anode can take away an electron from a chloride ion. The chloride ion loses its electron — is **discharged** — to form a chlorine atom. Then two chlorine atoms join to form a chlorine molecule. The electrode process is

chloride ion → chlorine atom + electron
Cl^-(aq) → Cl (g) + e⁻

followed by

$2Cl(g) \rightarrow Cl_2(g)$

The electrons given up by the chloride ions at the positive electrode travel round the external circuit (the wires outside the cell) to the negative electrode. There they combine with copper ions to form copper atoms. A constant stream of electrons therefore flows through the external circuit (Fig. 5.5). This stream of electrons is an **electric current**.

SUMMARY NOTE

When electrolysis occurs, ions conduct the electric current through the cell. Positive ions (cations) travel to the negative electrode (the cathode). Negative ions (anions) travel to the positive electrode (the anode).

At the electrodes, ions are **discharged**. Positive ions take electrons from the negative electrode to become atoms. Negative ions give up electrons to the positive electrode to become atoms. Electrons travel from the positive electrode to the negative electrode through the circuit outside the cell.

JUST TESTING 20

1 Refer to Fig. 5.4.
 a Write the symbols for the lead(II) ion and the bromide ion.
 b At which electrode do you see (i) lead, (ii) bromine?
 c Write equations for the processes which occur at the two electrodes.
 d What is happening in the circuit which connects the electrodes outside the crucible?

> **2** Explain the way in which a solution of copper(II) chloride conducts electricity. What type of particle conducts electricity **a** through the cell and **b** through the circuit outside the cell?
>
> **3** Molten potassium iodide can be electrolysed. Name the products formed at **a** the positive electrode and **b** the negative electrode. Write equations for the electrode processes which result in the formation of these products.

*5.7 More complicated examples of electrolysis

Copper(II) sulphate solution

When copper(II) sulphate is electrolysed, copper is deposited on the cathode and oxygen is evolved at the anode. To explain how this happens, we have to think about the water present in the solution. Water is a covalent liquid, which is ionised to a very small extent into hydrogen ions and hydroxide ions:

water $\rightleftharpoons$ hydrogen ions + hydroxide ions
$H_2O(l) \rightleftharpoons H^+(aq) + OH^-(aq)$

At the positive electrode there are hydroxide ions, OH^-, as well as sulphate ions, SO_4^{2-}. Sulphate ions are more stable than hydroxide ions. It is easier for hydroxide ions to lose their electrons, and hydroxide ions are discharged. The OH groups which are formed exist for only a fraction of a second before rearranging to give oxygen and water:

$OH^-(aq) \rightarrow OH(aq) + e^-$
$4OH(aq) \rightarrow 2H_2O(l) + O_2(g)$

Although only a tiny fraction of water molecules are ionised, once hydroxide ions have been discharged, more water molecules ionise to replace them with fresh hydroxide ions.

Sodium chloride solution

When molten sodium chloride is electrolysed, the products are sodium and chlorine. When a solution of sodium chloride is electrolysed, chlorine forms at the positive electrode and hydrogen forms at the negative electrode. At the negative electrode, sodium ions and hydrogen ions are present. Sodium ions are more stable than hydrogen ions. It is easier for an electron to join a hydrogen ion than it is for it to join a sodium ion, so the discharge of hydrogen ions takes place. The hydrogen atoms formed join in pairs to form hydrogen molecules:

$H^+(aq) + e^- \rightarrow H(g)$
$2H(g) \rightarrow H_2(g)$

At the positive electrode, there is a small concentration of hydroxide ions from the ionisation of water molecules. Chloride ions are, however, discharged in preference to hydroxide ions.

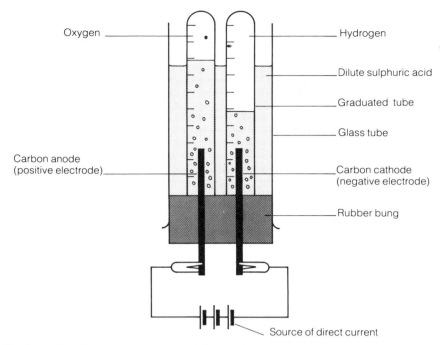

Fig. 5.6 The electrolysis of dilute sulphuric acid

Dilute sulphuric acid

The electrolysis of dilute sulphuric acid is shown in Fig. 5.6.

5.8 Which ions are discharged?

You will have realised by now that some ions are easier to discharge at an electrode than others.

Anions. Sulphate ions and nitrate ions are very difficult to discharge. In solutions of these ions, hydroxide ions are discharged instead, and oxygen is evolved.

Cations. The ions of very reactive metals, such as sodium, are difficult to discharge. Sodium atoms are very ready to react to form compounds containing sodium ions. It is difficult to reverse the process and force a sodium ion to accept an electron and become a sodium atom (see Fig. 5.7). The cations of less reactive metals, such as copper and lead, are easy to discharge.

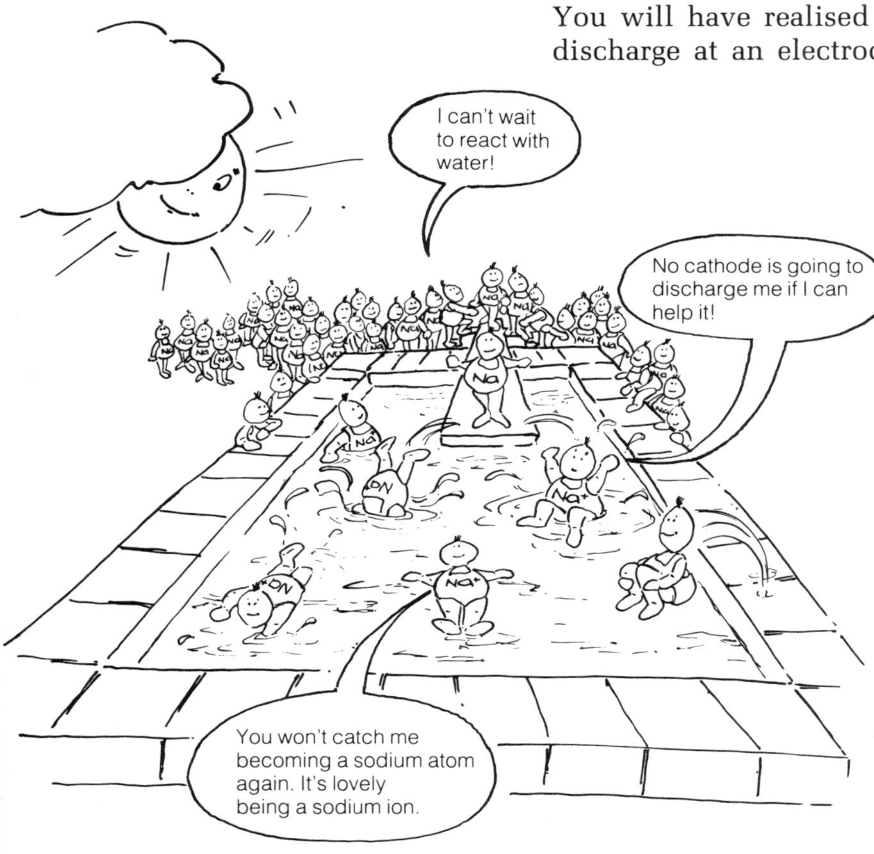

Fig. 5.7 Sodium atoms want to become sodium ions; sodium ions are happy as they are

*JUST TESTING 21

1. A solution of potassium bromide is electrolysed. A brown colour appears at one electrode.
   a At which electrode does the brown colour appear?
   b What is produced at the other electrode?
   c Write equations for the two electrode processes.

2. Sodium nitrate solution is electrolysed with carbon electrodes. Name the ions present. Which ions are discharged **a** at the positive electrode and **b** at the negative electrode? Write equations for the electrode processes.

3. An aqueous solution of sodium chloride is electrolysed.
   a What two kinds of ion are present at the positive electrode? Which is discharged? Which kind of ion accumulates?
   b What two kinds of ion are present at the negative electrode? Which is discharged? Which kind of ion accumulates?
   c As these changes take place, the solution of sodium chloride is turning into a solution of something else. What is it becoming?

4. In the electrolysis of acidified water (Fig. 5.6) the gases formed are in the ratio:

   2 volumes of hydrogen : 1 volume of oxygen

   When you have studied Section 11.4, return to this question. Use the information to work out the equation for the reaction:

   water $\rightarrow$ hydrogen + oxygen.

SUMMARY NOTE

Water is ionised to a very small extent to form $H^+(aq)$ ions and $OH^-(aq)$ ions. These ions are sometimes discharged when aqueous solutions are electrolysed.

The ions of very reactive metals, e.g. Na^+, are difficult to discharge. In aqueous solution, hydrogen ions are discharged and hydrogen is evolved at the negative electrode. The anions SO_4^{2-} and NO_3^- are difficult to discharge. In aqueous solution, OH^- ions are discharged instead, and oxygen is evolved at the positive electrode.

5.9 Electrolysis with reactive electrodes

In all the electrode processes that we have covered so far the electrodes have supplied electrons to the cell or removed electrons from the cell. They have been **inert**: they have not taken part in the chemical reactions. Sometimes, electrodes do take part in the cell reaction.

When copper(II) sulphate is electrolysed with carbon electrodes, as in Experiment 5.5, copper is discharged at the negative electrode and oxygen is evolved at the positive electrode. With copper electrodes, copper is discharged at the negative electrode as before.

Cathode (negative): $Cu^{2+}(aq) + 2e^- \rightarrow Cu(s)$

At the positive electrode, there are three possible processes which would release electrons. These are:

- the discharge of SO_4^{2-} ions
- the discharge of OH^- ions
- the ionisation of Cu atoms from the anode.

SUMMARY NOTE

Usually, **inert** electrodes of carbon or platinum are used. If the electrodes are not inert, that is, if they take part in the cell reaction, different products are obtained.

In fact, the third of these reactions happens:

Anode (positive): $Cu(s) \rightarrow Cu^{2+}(aq) + 2e^-$

The combined result of the reactions at the two electrodes is a transfer of copper from the positive electrode to the negative electrode.

JUST TESTING 22

1 Different products are obtained in the electrolysis of copper(II) sulphate solution with **a** carbon electrodes and **b** copper electrodes. What products are formed in each case? Write equations for the electrode processes which occur.

5.10 Applications of electrolysis

Electroplating

Often, people want to coat a relatively cheap metal with a more beautiful and more expensive metal. A thin layer of silver or gold can be used to beautify metal objects (Fig. 5.8). To limit the cost, a method of applying a thin layer only is needed. Electrolysis can be used to produce a thin even film of metal which adheres well to the surface being coated. The technique is called **electroplating** (Fig. 5.9).

Electroplating is used for protection as well as for decoration. You will remember from Section 5.1 that steel is chromium plated to prevent it from rusting. Chromium does not stick well to steel. Before steel is chromium plated, it is first electroplated with copper, which does adhere well to steel. It is then nickel plated. Nickel is not corroded by air or water. The final step is chromium plating. The result is a bright, non-corrodible surface.

The rusting of iron is a serious and expensive problem (see Section 10.11). In the manufacture of food cans, a layer of tin is electroplated on to iron. Tin is an unreactive metal, and the juices in foods do not react with it. A layer of zinc is applied to iron in the manufacture of **galvanised** iron. Electroplating is often employed. (See Experiments 5.7 and 5.8.)

Fig. 5.8 Silverplate

SUMMARY NOTE

One of the applications of electrolysis is **electroplating**. This method is used to coat a cheap metal with an expensive metal. It is used to coat iron and steel, which rust, with a metal which does not corrode.

1 The object to be plated is made the negative electrode (the cathode).

2 The electrolyte is a solution of one of the salts of the metal.

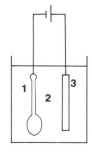

3 The positive electrode (the anode) is usually made of the plating metal. Metal atoms dissolve to form metal ions. This keeps the concentration of ions in solution constant.

Fig. 5.9 Electroplating

Extraction of metals from their ores

It is difficult to extract the most reactive metals from their ores. For many of them, e.g. sodium and aluminium, electrolysis is the only method which works (see Section 10.9).

An electrolytic method is also used to obtain pure copper from a slab of impure copper (see Section 10.9).

Manufacture of sodium hydroxide

If brine (sodium chloride solution) is electrolysed, hydrogen ions and chloride ions are discharged at the electrodes. Sodium ions and hydroxide ions accumulate in the solution. Gradually, the sodium chloride solution turns into a solution of sodium hydroxide. Figure 5.10 shows a diaphragm cell, an industrial method of carrying out this electrolysis continuously. The three important chemicals, sodium hydroxide (Section 7.3), chlorine (Section 15.7) and hydrogen (Section 9.17), are all obtained from the plentiful starting material, brine.

> **SUMMARY NOTE**
>
> Electrolysis is used for the extraction of some metals from their ores.

> **SUMMARY NOTE**
>
> The electrolysis of brine (sodium chloride solution) to give sodium hydroxide, chlorine and hydrogen is an important industrial process.

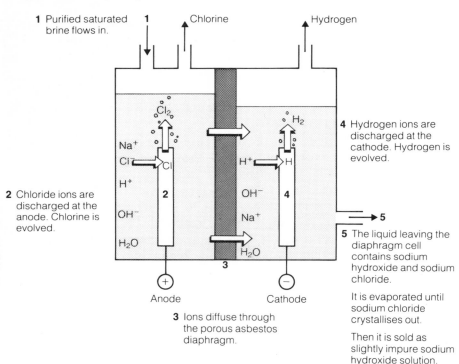

Fig. 5.10 The diaphragm cell for the electrolysis of brine (sodium chloride solution)

Career sketch: Chemical sales representative

Nigel's company has developed the ability to make a new product. Before they manufacture it, they want to know what the chances are of selling it. Nigel finds out where there are possibilities of sales. He does this by reading trade journals, market surveys etc. Then he visits the technical and purchasing staff of the companies he has spotted. He promotes the company's new product. He needs the ability to take in new information and also the ability to get on with people. Nigel has a degree in chemistry.

Fig. 5.11 Diaphragm cells used for the electrolysis of brine

JUST TESTING 23

1 **a** You need to plate an iron nail with nickel. Draw the apparatus and the circuit you would use. Say what the electrolyte is, what the electrodes are made of and what charge they carry.
  b Explain why it is worth while going to the trouble and expense of electroplating iron nails.

2 Some parts of a car body are protected from rusting by a coat of paint.
  a What advantage does chromium plating have over painting?
  b Why do you think the door handles are chromium plated, not painted?
  c What precautions are taken to ensure that the layer of chromium adheres well to the steel?

3 Why is the manufacture of gold-plated watches and jewellery big business? What two advantages do gold-plated articles have over articles made of cheaper metals? What is the advantage of gold plate over pure gold? Why is pure gold not used?

5.11 Chemical cells

There are two kinds of cells:

Electrolytic cells

Electric current flows into the cell. The result is that a chemical reaction takes place.

Chemical cells

A chemical reaction takes place in the cell. The result is that an electric current flows through the external circuit.

In the chemical cell shown in Fig. 5.12, the chemical reaction that takes place is

zinc + copper(II) ions → zinc ions + copper
$Zn(s) + Cu^{2+}(aq) \rightarrow Zn^{2+}(aq) + Cu(s)$

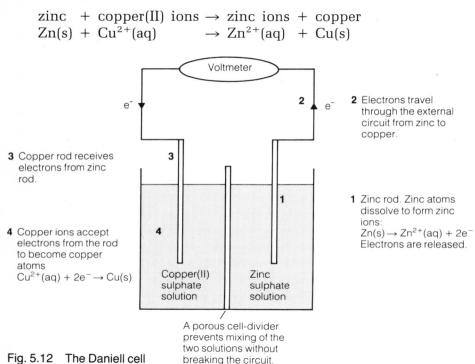

Fig. 5.12 The Daniell cell

As a result of this reaction, an electric current flows through the external circuit.

This cell is called a Daniell cell after its inventor. Many metals can be paired up to make a chemical cell. The more the metals differ in reactivity, the greater is the voltage supplied by the cell. (You will meet the reactivity series in Section 10.4).

Other chemical cells are

- the dry cells used in torch batteries and radios
- the lead–acid accumulators used in motor vehicles
- the fuel cells which power space vehicles

Fig. 5.13 Dry cells

SUMMARY NOTE

Electricity is obtained from chemical reactions in **chemical cells**.

Exercise 5

1. What are the electrical symbols for a bulb, a switch, a cell and a battery of three cells? Draw a circuit which you could use to test various materials to find out whether they are electrical conductors.
   Divide the following list into conductors and non-conductors:
   copper, steel, wood, PVC, brass, polythene, silver, candle wax, water, petrol, tetrachloromethane (CCl_4), sodium hydroxide solution, sugar solution.

2. Each of these clues has a one-word answer. (The number of letters in the word is given in brackets after the clue.) Write the answers down one beneath the other. What do the initial letters spell out?
   1. There are many of these in an electric current. (9)
   2. A metal which is easily discharged in electrolysis. (4)
   3. A substance which can be split up by electricity. (11)
   4. Every metal is one. (9)
   5. The number of charges on a sulphate ion. (3)
   6. This sort of metal is not discharged in electrolysis. (8)
   7. A gas which is often formed at a positive electrode. (6)
   8. The negative electrode. (7)
   9. A gas which may be formed at a negative electrode. (8)
   10. This may take electricity out of the cell. (9)
   11. This kind of salt conducts electricity. (6)
   12. This will not allow current to pass. (9)
   13. Make this if you want a salt to conduct. (8)
   14. Good electroplating needs plenty of this. (4)
   15. This tends to stop current flowing. (10)
   16. A ---- would be too long, even for clue 14. (4)

3. Name the substances A to H in this table. Write equations for the electrode processes which lead to their formation

Electrolyte	Positive electrode	Negative electrode
Dilute sulphuric acid	A	B
Copper(II) chloride solution	C	D
Sodium chloride solution	E	F
Calcium chloride solution	G	H

*4. In the electrolysis of potassium chloride solution, what element is formed **a** at the positive electrode and **b** at the negative electrode? Why does the solution around the negative electrode become alkaline?

5. Copper(II) sulphate solution can be electrolysed between platinum electrodes.
   a Describe what you see at both electrodes.
   b How could you test the product formed at the positive electrode?
   c Write equations for the electrode processes.

CHAPTER 6 *The chemical bond*

6.1 Getting in the swim with ions

How can we take the sting out of swimming? Chlorine does a good job of killing bacteria in pool water. Unfortunately, it also makes swimmers' eyes sting. It would be an advantage to replace chlorine with another **bacteriocide** which is also kind to the eyes.

A British firm has found an answer to the problem by inventing a new device. Two electrodes made of a silver–copper alloy dip into the water. A low voltage direct current is fed through the electrodes. The current is made to switch every half minute. The electrodes slowly release copper ions and silver ions into the water. Silver ions kill bacteria, and copper ions kill algae. The dead bacteria and algae are electrostatically charged. The charges make them clump together to form a solid which is easily removed by the pool filter.

The new device is priced at about £500 and costs the same to run as a 40 watt bulb. It can be used to purify water in hospitals. It could be used to provide safe drinking water in countries where sanitation is poor. The device makes use of **ions**. It is time to find out more about them.

6.2 The ionic bond

Sodium chloride

In Chapter 5, you studied electrolysis. You found that some compounds are electrolytes: they conduct electricity. Other compounds are non-electrolytes. To explain the results of electrolysis, we accepted the idea that electrolytes consist of small charged particles called **ions**. When sodium chloride is electrolysed, sodium is deposited at the negative electrode, and chlorine is given off at the positive electrode. This fits in with the theory that sodium chloride consists of positively charged sodium ions and negatively charged chloride ions, Na^+Cl^-.

Look at the arrangement of electrons in a sodium atom (Fig. 6.1). There is just one electron more than there is in an atom of the noble gas neon.

You have read in Section 4.6 that the noble gases are helium, neon, argon, krypton and xenon. These gases take part in hardly any chemical reactions. All the noble gases have full outer shells of electrons (Fig. 6.2). It seems logical to suppose that it is this electron arrangement that makes them stable, that is, chemically unreactive (Fig. 6.3).

When a sodium atom loses the lone electron from its outermost shell, the outer shell that remains contains 8 electrons. A sodium atom cannot just shed an electron. The attraction between the nucleus and the electrons prevents this. It can,

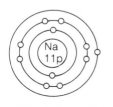

Fig. 6.1 Atoms of sodium and neon

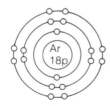

Fig. 6.2 The electron arrangements in helium, neon and argon

Fig. 6.3 The noble gases do not care to combine

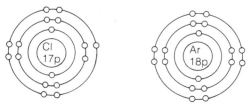

Fig. 6.4 The arrangement of electrons in chlorine and argon

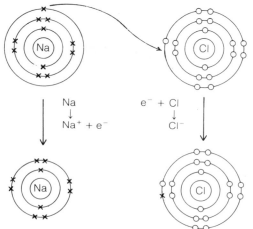

Fig. 6.5 The formation of sodium chloride

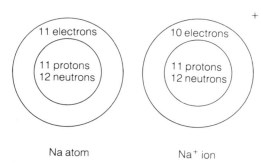

Fig. 6.6 The sodium atom and the sodium ion

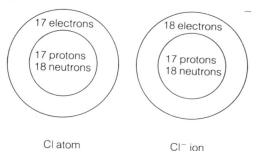

Fig. 6.7 A chlorine atom and a chloride ion

however, give an electron to a chlorine atom. In fact, sodium burns vigorously in chlorine to form sodium chloride.

Why should a chlorine atom want an electron? Look at the electron arrangement (Fig. 6.4). With one more electron, chlorine would have the same electron arrangement as argon. A full outer shell brings with it stability.

Figure 6.5 shows what happens when an atom of sodium donates an electron to an atom of chlorine. A full outer shell is left behind in sodium, and a full outer shell is created in chlorine. To make clear what happens, the sodium electrons have been shown as **x** and the chlorine electrons as **o**. This dot-and-cross diagram does not imply that there are two kinds of electrons!

A sodium atom has 11 electrons, 11 protons and 12 neutrons. After it loses one electron, it is left with 11 protons and 10 electrons. The total charge is $+11-10 = +1$ charge unit. It is not a sodium atom, Na, any longer: it is a sodium ion, Na^+ (see Fig. 6.6).

A chlorine atom has 17 protons and 17 electrons. After it gains an electron, making 17 protons and 18 electrons, the charge is $+17-18 = -1$ charge unit. The chlorine atom, Cl, has become a chloride ion, Cl^- (see Fig. 6.7).

Remember this from Chapter 5. An ion is an atom or group of atoms which carries an electric charge. A positively charged ion, e.g. Na^+, is called a **cation**. A negatively charged ion, e.g. Cl^-, is called an **anion**.

There is electrostatic attraction between oppositely charged ions (Fig. 6.8). This is what holds the ions, Na^+ and Cl^-, together. This electrostatic attraction is the **chemical bond** in the compound, sodium chloride. It is called an **ionic bond** or **electrovalent bond**. Sodium chloride is an ionic or electrovalent compound. The compounds which conduct electricity when they are melted or dissolved (Section 5.3) are electrovalent compounds.

A pair of ions, Na^+Cl^-, does not exist in isolation. Other ions are attracted to them. Sodium ions are attracted to chloride ions, and chloride ions are attracted to sodium ions. The result is a three-dimensional structure of alternate Na^+ and Cl^- ions (Fig. 6.9). This is a **crystal** of sodium chloride. The crystal is uncharged because the number of sodium ions is equal to the number of chloride ions. The forces of attraction between positive and negative ions are strong. This is why solid sodium chloride does not conduct electricity and is not electrolysed. Electricity is conducted by the ions of an electrolyte travelling to the electrodes. In solid sodium chloride, the ions cannot move out of their positions in the three-dimensional structure. When the salt is melted or dissolved, the ions are free to move and can travel towards the electrodes.

Fig. 6.8 There is an attraction between oppositely charged ions

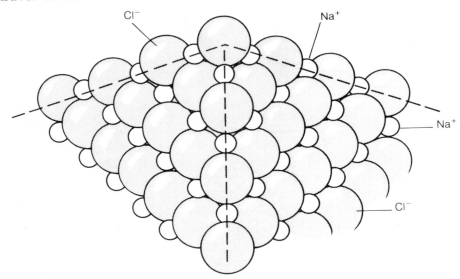

Fig. 6.9 The structure of sodium chloride

Magnesium fluoride

There are thousands of ionic compounds in addition to sodium chloride. Magnesium fluoride is one. Magnesium (2.8.2) has two electrons in the outermost shell. It has to lose these to attain a stable, full outer shell of electrons. Fluorine (2.7) needs to gain an electron. If a magnesium atom wants to lose two electrons and a fluorine atom needs to gain only one electron, a magnesium atom has to combine with two fluorine atoms (Fig. 6.10).

SUMMARY NOTE

The noble gases have full outer shells of 8 electrons (2 for helium). This is a very stable arrangement of electrons. A sodium atom has one electron more than a full shell. It can give its outermost electron to an atom of a non-metallic element, e.g. chlorine. In the process the sodium atom becomes a positively charged sodium ion. A chlorine atom is one electron short of a full outer shell. When a chlorine atom accepts one electron to fill its outer shell, it becomes a negatively charged chloride ion.

Once the sodium ion and chloride ion have been formed, they are held together by a strong electrostatic attraction. They have combined to form the compound sodium chloride. The electrostatic attraction between them is the type of chemical bond called the **ionic bond** or **electrovalent bond**. A three-dimensional crystal structure of ions is built up.

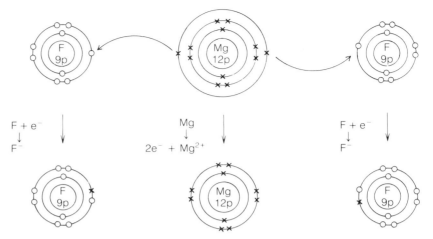

Fig. 6.10 The formation of magnesium fluoride

Magnesium oxide

Magnesium combines with oxygen to form magnesium oxide. One atom of magnesium, Mg (2.8.2), gives two electrons to one atom of oxygen, O (2.6). The ions Mg^{2+} (2.8) and O^{2-} (2.8) are formed. The electrostatic attraction between them is an ionic bond (Fig. 6.11).

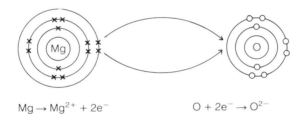

$Mg \rightarrow Mg^{2+} + 2e^-$ $O + 2e^- \rightarrow O^{2-}$

Fig. 6.11 The formation of an ionic bond between atoms of magnesium and oxygen

Ions and valency

Electrovalent compounds (ionic compounds) are formed when a metallic element combines with a non-metallic element. The charge on the ions formed by an element is the **valency** of that element. Magnesium forms Mg^{2+} ions: magnesium has a valency of 2. Sodium forms Na^+ ions: sodium has a valency of 1. Oxygen forms O^{2-} ions: oxygen has a valency of 2. Iodine forms I^- ions: iodine has a valency of 1. Some elements have a variable valency. Iron forms Fe^{2+} ions and Fe^{3+} ions: iron uses a valency of 2 in some compounds and 3 in others. In general,

Maximum valency employed by a metallic element = Number of electrons in the outermost shell

Maximum valency employed by a non-metallic element = 8 − (Number of electrons in the outermost shell)

Non-metallic elements often form oxo-ions (which contain oxygen), e.g. sulphate, SO_4^{2-}, which has a valency of 2, and nitrate, NO_3^-, which has a valency of 1.

> **SUMMARY NOTE**
>
> Metallic atoms can lose 1, 2 or 3 outer electrons, called **valency** electrons. The ions formed have a charge of +1, +2 or +3. Some non-metallic elements can gain 1, 2 or 3 electrons to form ions with charge −1, −2 or −3. The valency of an element is the charge on its ions. Ionic compounds are uncharged because the number of positive charges is equal to the number of negative charges.

JUST TESTING 24

1. Lithium, like sodium, is an alkali metal. It has the electron arrangement Li (2.1). Fluorine has the electron arrangement F (2.7). Draw the arrangement of electrons in the atoms of Li and F. Explain how an ionic bond is formed in lithium fluoride.

2. Sodium is a silvery-grey metal which reacts rapidly with water. Chlorine is a poisonous green gas. Sodium chloride is a white crystalline solid. Explain how the sodium in sodium chloride differs from sodium metal. Explain how the chlorine in sodium chloride differs from chlorine gas.

3. Which of the following elements give **a** positive ions, **b** negative ions: magnesium, potassium, barium, zinc, iron, sulphur, oxygen, iodine, fluorine? (The Periodic Table on p. 318 may help you.)

4. Use dot-and-cross diagrams to show how the following atoms combine:
   a Na (2.8.1) and O (2.6); **b** Al (2.8.3) and F (2.7); **c** Be (2.2) and F (2.7); **d** Ca (2.8.8.2) and O (2.6).

5. Name two differences between
   a a sodium ion, Na^+ (2.8), and a neon atom, Ne (2.8)
   b a chloride ion, Cl^- (2.8.8), and an argon atom, Ar (2.8.8).

6.3 The covalent bond

There are compounds which are not electrolytes. Since they do not conduct electricity, they cannot consist of ions. Non-electrolytes contain a type of chemical bond different from the ionic bond.

Two atoms of chlorine combine to form a molecule, Cl_2. Both chlorine atoms have the electron arrangement Cl (2.8.7). Neither chlorine atom wants to give an electron: both want to gain an electron. They achieve a full outer shell of electrons by sharing electrons. The pair of chlorine atoms shown in Fig. 6.12 is sharing the pair of electrons **ox**. One electron comes from each of the chlorine atoms. The atoms have come close enough together for their outer shells to overlap. Then the shared pair of electrons can orbit round both the chlorine atoms. Each chlorine atom therefore has eight electrons in the outermost orbit. **When two atoms share a pair of electrons, this is called a covalent bond**.

In hydrogen chloride, two electrons are shared between the hydrogen and chlorine atoms to form a covalent bond. The arrangement gives the hydrogen atom a full shell of 2 electrons, like the noble gas helium. The chlorine atom then has a full outer shell of 8 electrons, like the noble gas neon (Fig. 6.13).

The arrangement of electrons in water, H_2O, and in ammonia, NH_3, is shown in Fig. 6.14.

A molecule of water, H_2O, contains covalent bonds between one O atom and two H atoms. Each O atom has 8 electrons in

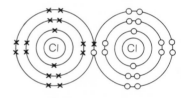

The pair of electrons **ox** is shared between the 2Cl atoms. One electron comes from each Cl.

Fig. 6.12 The chlorine molecule

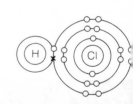

With 2 electrons, the outer shell of the H atom is full.

With 8 electrons, the outer shell of the Cl atom is full.

Fig. 6.13 A molecule of hydrogen chloride

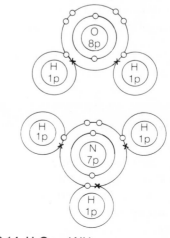

Fig. 6.14 H_2O and NH_3

its outer shell, and each H atom has two. Since the O atom shares two pairs of electrons, oxygen has a valency of 2. Since each H atom shares one pair of electrons, hydrogen has a valency of 1.

In ammonia, NH_3, each nitrogen atom shares three of its electrons, one with each of three hydrogen atoms. In this way, nitrogen obtains a full outer shell of 8 electrons, and hydrogen obtains a full shell of two. Nitrogen has a valency of 3, and hydrogen has a valency of 1.

*Double covalent bonds

In carbon dioxide, CO_2, the carbon atom shares 2 electrons with each of the 2 oxygen atoms. Each oxygen atom shares two of its electrons with the carbon atom (see Fig. 6.15). The two pairs of shared electrons are a **double bond**.

Fig. 6.15 CO_2

SUMMARY NOTE

Another type of chemical bond is the covalent bond. Atoms share electrons. By sharing, the bonded atoms both gain a full outer shell of electrons. The valency of the element equals the number of pairs of electrons which an atom of the element shares with other atoms in a covalent compound.

JUST TESTING 25

1 Sketch the arrangement of electrons in F (2.7) and H (1). Show by means of a sketch like those in the figures what happens when a covalent bond forms in HF.
2 Sketch the arrangement of electrons in the covalent compound CH_4 (C (2.4), H (1)).
*3 By means of a dot-and-cross diagram like those in the figures, sketch the arrangement of electrons in the outer shells of C and Cl in the molecule CCl_4 (C (2.4), Cl (2.8.7)).

6.4 Shapes of covalent molecules

The shapes of some covalent molecules are shown in Fig. 6.16. Covalent substances can be solids, liquids or gases. Gases consist of individual molecules.

In liquids, there are forces of attraction between the molecules. In some covalent substances, the attractive forces between molecules are strong enough to make the substances solids. Iodine is a shiny black crystalline solid. Forces of attraction hold molecules of iodine in a three-dimensional structure (Fig. 6.17). It is described as a **molecular structure**. There are two kinds of chemical bonds in it:
- the strong covalent I–I bonds
- the weak forces of attraction between I_2 molecules.

Diamond is an **allotrope** of carbon. Diamond has a structure (Fig. 3.4) in which every carbon atom is joined by covalent bonds to four other carbon atoms. Millions of carbon atoms form a giant molecule. The structure is **giant molecular** or **macromolecular**.

Graphite, the other allotrope of carbon, has a **layer structure** (Fig. 3.5). Strong covalent bonds join the carbon atoms within each layer. The layers are held together by weak forces of attraction.

SUMMARY NOTE

Some covalent substances are composed of individual molecules. Others are composed of larger units. These may be chains or layers or macromolecular structures.

Fig. 6.16 The shapes of some covalent molecules

Name	Formula	Structure	Molecular model
Chlorine	Cl_2	Cl – Cl	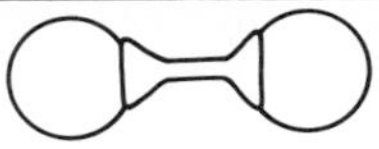
Hydrogen chloride	HCl	H – Cl	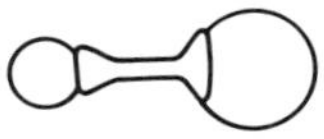
Oxygen	O_2	O = O	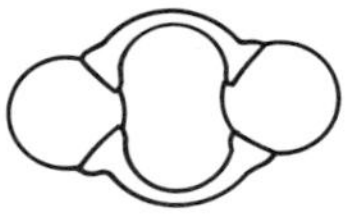
Water	H_2O	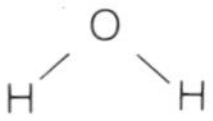 H–O–H	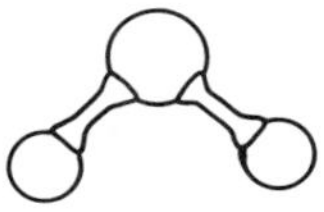
Ammonia	NH_3	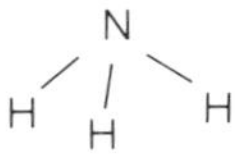 H–N(–H)–H	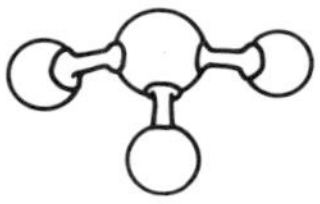
Methane	CH_4	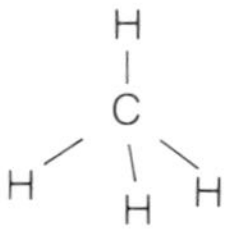 H–C(H)(H)–H	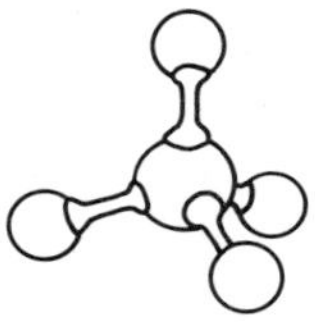
Ethene	$H_2C = CH_2$	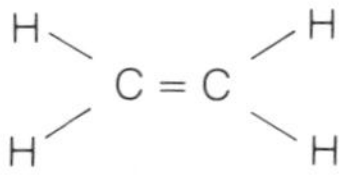 H₂C = CH₂	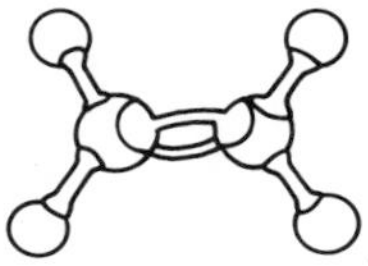
Carbon dioxide	CO_2	O = C = O	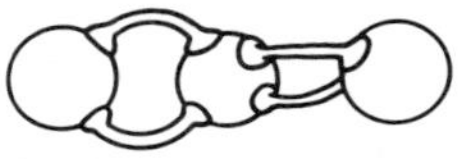

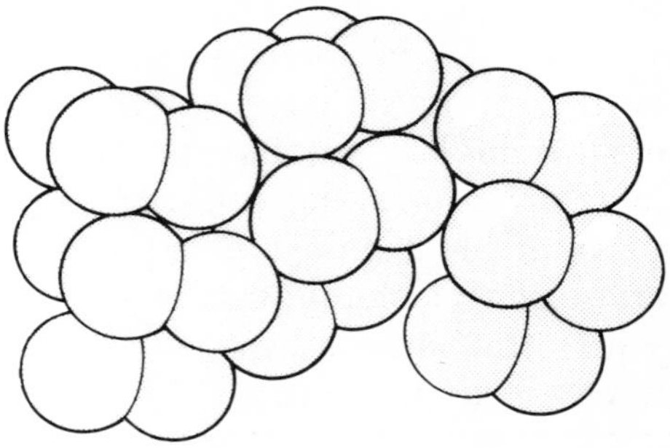

Fig. 6.17 The structure of an iodine crystal

JUST TESTING 26

1. Why is it difficult to rub away any carbon atoms from a diamond?
2. Why is graphite used as a lubricant?
3. Why does diamond have a high boiling point?
4. Solid iodine consists of shiny black crystals. Iodine vapour is purple. What is the difference in chemical bonding between solid and gaseous iodine?

6.5 Ionic and covalent compounds

The physical characteristics of substances, for example, whether they are solids, liquids or gases, depend on the type of chemical bonds in the substances. Chemical behaviour also depends on the types of bonds present.

Table 6.1 Differences between ionic and covalent compounds

Ionic bonding	Covalent bonding

1 Ionic compounds are formed when a metallic element (on the left-hand side of the Periodic Table) combines with a non-metallic element (on the right-hand side of the Periodic Table).
2 a Atoms of metallic elements form positive ions. Elements in Groups 1, 2 and 3 of the Periodic Table form ions with charges +1, +2 and +3, e.g. Na^+, Ca^{2+}, Al^{3+}.
  b Elements in Group 7 of the Periodic Table (non-metals) form anions, X^-, as each atom gains one electron from a metallic element.
  c Elements in Group 6 of the Periodic Table (non-metals) can form Y^{2-} ions as each atom gains 2 electrons from a metal atom.
3 There is a strong electrostatic attraction between ions of opposite charge. This is what gives ionic solids a close, regular structure. Ionic compounds are crystalline because of this regular structure. The strong force of attraction between ions of opposite charge makes it difficult to separate the ions. This is why ionic compounds have high melting points and high boiling points.
4 Ionic compounds are electrolytes. They conduct electricity when molten or in solution and are split up in the process (see Section 5.3).
5 Ionic compounds are insoluble in organic solvents, that is, solvents which have covalent bonds.
	1 Atoms of non-metallic elements combine with each other by sharing pairs of electrons in their outer shells. A shared pair of electrons is called a covalent bond.
	2 The maximum number of covalent bonds that an atom can form is equal to the number of electrons in the outer shell. Often, an atom does not use all its electrons in covalent bond formation.
	3 Many covalent compounds are composed of small molecules. The bonds which hold the atoms together are strong, but the attractions which exist between molecules are weak. Covalent substances are gases, liquids or solids with low melting points. The low melting point solids are molecular crystals, e.g. iodine (Fig. 6.17).
	   Atoms which form at least two covalent bonds can link in chains, in sheets (see graphite, Fig. 3.5) or in three-dimensional networks (see diamond, Fig. 3.4). Substances with a giant molecular structure have high melting and boiling points.
	4 Covalent compounds are non-electrolytes.
	5 Covalent compounds are soluble in organic solvents, e.g. ethanol, tetrachloromethane.

SUMMARY NOTE

The type of chemical bonds present, ionic or covalent, decides the properties of a compound, e.g. its physical state (solid, liquid or gas), boiling and melting points, solubility and electrolytic conductivity.

JUST TESTING 27

1. What kind of bonding would you expect between the following pairs of elements?
   a K and Cl; **b** Ca and Br; **c** Mg and S; **d** Cu and O; **e** S and O; **f** Na and H.
   (The Periodic Table on p. 318 may help you.)

2. From the information in the table, say what you can about the chemical bonds in A, B, C and D.

Solid	State	Melting point in °C	Does it conduct electricity when molten?
A	Solid	660	Conducts electricity when molten
B	Solid	720	Does not conduct electricity when molten
C	Solid	85	Does not conduct electricity when molten
D	Gas	−100	Does not conduct electricity

*6.6 Formulas of ionic compounds

You have seen how electrovalent compounds consist of oppositely charged ions. The compound is neutral because the charge on the positive ion (or ions) is equal to the charge on the negative ion (or ions). In calcium chloride, $CaCl_2$, one calcium ion, Ca^{2+}, is balanced in charge by two chloride ions, $2Cl^-$.

Fig. 6.18 The charges must balance

This is how you can work out the formulas of electrovalent compounds.

Compound:	Calcium chloride
Say which ions are present:	Ca^{2+}, Cl^-
Now balance the charges:	One Ca^{2+} ion needs two Cl^- ions
Ions needed are:	Ca^{2+} and $2Cl^-$ ions
The formula is:	$CaCl_2$

Table 6.2 The symbols of some common ions

Cation	Symbol
Aluminium	Al^{3+}
Ammonium	NH_4^+
Barium	Ba^{2+}
Calcium	Ca^{2+}
Copper(II)	Cu^{2+}
Hydrogen	H^+
Iron(II)	Fe^{2+}
Iron(III)	Fe^{3+}
Lead(II)	Pb^{2+}
Magnesium	Mg^{2+}
Mercury(II)	Hg^{2+}
Potassium	K^+
Silver	Ag^+
Sodium	Na^+
Zinc	Zn^{2+}

Anion	Symbol
Bromide	Br^-
Carbonate	CO_3^{2-}
Chloride	Cl^-
Hydrogencarbonate	HCO_3^-
Hydroxide	OH^-
Iodide	I^-
Nitrate	NO_3^-
Oxide	O^{2-}
Phosphate	PO_4^{3-}
Sulphate	SO_4^{2-}
Sulphide	S^{2-}
Sulphite	SO_3^{2-}

SUMMARY NOTE

The formula of an ionic compound is worked out by balancing the charges on the ions.

Compound:
Say which ions are present:
Now balance the charges:

Ions needed are:
The formula is:

Compound:
Ions present are:
Now balance the charges:

Ions needed are:
The formula is:

Compound:
Ions present are:
Now balance the charges:

Ions needed are:
The formula is:

Compound:
Ions present are:
Now balance the charges:

Ions needed are:
The formula is:

Sodium carbonate
Na^+, CO_3^{2-}
Two Na^+ are needed to balance one CO_3^{2-}
$2Na^+$ and CO_3^{2-}
Na_2CO_3

Copper(II) hydroxide
Cu^{2+}, OH^-
Two OH^- ions balance one Cu^{2+} ion
Cu^{2+} and $2OH^-$
$Cu(OH)_2$
(The brackets tell you that the 2 multiplies everything inside them. There are 2O atoms and 2H atoms, in addition to the 1Cu ion.)

Iron(II) sulphate
Fe^{2+}, SO_4^{2-}
One Fe^{2+} ion balances one SO_4^{2-} ion
Fe^{2+} and SO_4^{2-}
$FeSO_4$

Iron(III) sulphate
Fe^{3+}, SO_4^{2-}
Two Fe^{3+} would balance three SO_4^{2-} ions
$2Fe^{3+}$ and $3SO_4^{2-}$
$Fe_2(SO_4)_3$
(The brackets tell you that all the atoms inside them must be multiplied by the 3 that follows them. The formula contains 2Fe, 3S and 12O.)

You need to learn the symbols and charges of the ions in Table 6.2. Then you can work out the formula of any electrovalent compound containing these ions.

You will notice that the sulphates of iron are named iron(II) sulphate and iron(III) sulphate. The roman numerals II and III show which of its valencies iron is using. They are always used with the compounds of elements of variable valency.

Career sketch: Science librarian

Pat works in the library of a research centre. Scientists on the site ask him for books. If the centre does not stock a certain book, Pat asks the British Library to lend a copy. They send it to him, and he passes it on to the scientist who needs it. Pat records the loan on the computer. He left school after GCSE.

*JUST TESTING 28

1. Write the formulas of the following ionic compounds:
   a potassium chloride; **b** potassium nitrate; **c** silver nitrate; **d** ammonium bromide; **e** magnesium chloride; **f** copper(II) iodide; **g** zinc sulphate; **h** zinc nitrate; **i** calcium carbonate; **j** aluminium chloride; **k** sodium hydrogencarbonate.

2. Write the formulas of the following electrovalent compounds:
   a calcium hydroxide; **b** sodium sulphite; **c** iron(II) hydroxide; **d** iron(III) hydroxide; **e** aluminium oxide; **f** ammonium sulphate.

3. Name the following:
   a AlI_3; **b** $HgCl_2$; **c** $HgCl$; **d** $AgBr$; **e** $Cu(NO_3)_2$; **f** $CuCl$; **g** $FeBr_3$; **h** PbO_2; **i** PbO.

*6.7 Formulas of covalent compounds

In a covalent compound, pairs of electrons are shared between atoms. The number of electrons which an atom of an element shares in bond formation is the valency of the element. The formulas of covalent compounds are decided by the sort of dot-and-cross diagrams shown in Figs. 6.12 to 6.15.

> **SUMMARY NOTE**
>
> The formula of a covalent compound depends on the number of electrons shared by the bonded atoms.

*6.8 Equations: the balancing act

You began your work on equations in Section 3.11. Now you can tackle 'balancing' equations.

Example 1 Hydrogen and oxygen combine to form water. The word equation is

hydrogen + oxygen → water

The chemical equation could be

$$H_2(g) + O_2(g) \rightarrow H_2O(l)$$

This equation is not **balanced**. There are 2 oxygen atoms on the left-hand side (LHS) and only 1 oxygen atom on the right-hand side (RHS). To balance the O atoms, multiply H_2O on the RHS by 2:

$$H_2(g) + O_2(g) \rightarrow 2H_2O(l)$$

The O atoms are now balanced, but there are 4H atoms on the RHS and only 2H on the LHS. Multiplying H_2 on the LHS by 2,

$$2H_2(g) + O_2(g) \rightarrow 2H_2O(l)$$

The equation is now balanced (Fig. 6.19).

Number of atoms on LHS = 4H + 2O
Number of atoms on RHS = 4H + 2O

Number of atoms of each element on LHS = Number of atoms of each element on RHS

The total mass of the reactants = The total mass of the products

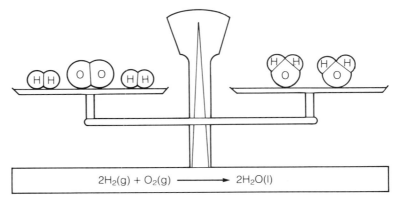

Fig. 6.19 A balanced equation; example 1

Example 2 Sulphur dioxide is oxidised by oxygen to sulphur trioxide:

sulphur dioxide + oxygen → sulphur trioxide
$SO_2(g)$ + $O_2(g)$ → $SO_3(g)$

You can see that the equation is not balanced. There are 4O on the LHS and 3O on the RHS. It is tempting to write O for oxygen on the LHS. You must not do this. Never change a formula. All you can do to balance an equation is to multiply formulas. Instead of changing O_2 to O, multiply SO_2 and SO_3 by 2:

$2SO_2(g) + O_2(g) \rightarrow 2SO_3(g)$

The equation is now balanced: 2S + 6O on the LHS; 2S + 6O on the RHS; see Fig. 6.20.

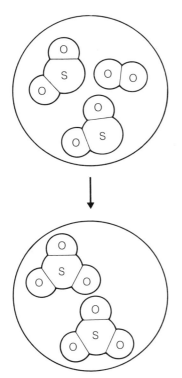

Fig. 6.20 A balanced equation; example 2

The steps in writing a chemical equation are
1 Write a word equation for the reaction.
2 Write the symbols and formulas for the reactants and products.
3 Add the state symbols.
4 Balance the equation. Multiply the formulas if necessary. Never change a formula.
5 Check again:
Number of atoms of each element on LHS = Number of atoms of each element on RHS

Take the reaction between sodium and water to form hydrogen and sodium hydroxide solution. Work through the five steps:

1 Sodium + water → hydrogen + sodium hydroxide solution

2 $Na + H_2O \rightarrow H_2 + NaOH$

3 $Na(s) + H_2O(l) \rightarrow H_2(g) + NaOH(aq)$

4 $2Na(s) + 2H_2O(l) \rightarrow H_2(g) + 2NaOH(aq)$

5 Number of atoms on LHS = 2Na + 4H + 2O
   Number of atoms on RHS = 2Na + 4H + 2O

JUST TESTING 29

*1 Copy these equations, and balance them.
 a $Fe_2O_3(s) + C(s) \rightarrow Fe(s) + CO(g)$
 b $Fe_2O_3(s) + CO(g) \rightarrow Fe(s) + CO_2(g)$
 c $NH_3(g) + O_2(g) \rightarrow NO(g) + H_2O(l)$
 d $Cr(s) + HCl(aq) \rightarrow CrCl_3(aq) + H_2(g)$
 e $Fe_3O_4(s) + H_2(g) \rightarrow Fe(s) + H_2O(l)$
 f $C_3H_8(g) + O_2(g) \rightarrow CO_2(g) + H_2O(l)$.

*2 Try writing equations for these reactions:
 a hydrogen + copper(II) oxide → copper + water
 b carbon + carbon dioxide → carbon monoxide
 c magnesium + sulphuric acid → hydrogen + magnesium sulphate
 d copper + chloride → copper(II) chloride
 e mercury + oxygen → mercury(II) oxide
 f sodium + oxygen → sodium oxide
 g iron + sulphur → iron(II) sulphide.

*3 Write balanced chemical equations for these reactions:
 a calcium + water → hydrogen + calcium hydroxide solution
 b iron + hydrochloric acid → iron(II) chloride solution + hydrogen
 c iron + chlorine → iron(III) chloride
 d aluminium + chlorine → aluminium chloride
 e zinc + steam → zinc oxide + hydrogen.

Exercise 6

Note You will need the electron arrangements shown on p. 46.

1 What happens to their electrons when atoms combine by means of an ionic bond? Illustrate your answer with a diagram showing the valency electrons (the outermost electrons) in **a** sodium fluoride, **b** magnesium chloride, **c** lithium oxide.

2 What happens to their electrons when atoms combine by means of a covalent bond? Illustrate your answer with a dot-and-cross diagram of the valency electrons in **a** HF; **b** CH_4; **c** PH_3; **d** CH_2Cl_2; **e** H_2S.

3 **a** Explain why the ionic compound $PbBr_2$ melts at a higher temperature than the covalent compound CH_2Br_2.
 b You can smell the compound $CHCl_3$, chloroform. You cannot smell the compound KCl. Explain the difference in terms of the chemical bonds in the two compounds.

4 Write the formulas of **a** mercury(II) oxide, **b** copper(II) nitrate, **c** aluminium nitrate, **d** aluminium sulphate, **e** barium nitrate, **f** calcium carbonate, **g** calcium hydrogencarbonate.

Note Question 5 is a revision question. You should tackle it *after* you have met the reactions.

*5 Copy out these equations into your book, and then balance them.
 a $H_2O_2(aq) \rightarrow H_2O(l) + O_2(g)$
 b $Fe(s) + O_2(g) \rightarrow Fe_3O_4(s)$
 c $Mg(s) + N_2(g) \rightarrow Mg_3N_2(s)$
 d $P(s) + Cl_2(g) \rightarrow PCl_3(s)$
 e $P(s) + Cl_2(g) \rightarrow PCl_5(s)$
 f $SO_2(g) + O_2(g) \rightarrow SO_3(g)$
 g $Na_2O(s) + H_2O(l) \rightarrow NaOH(aq)$
 h $KClO_3(s) \rightarrow KCl(s) + O_2(g)$
 i $NH_3(g) + O_2(g) \rightarrow N_2(g) + H_2O(g)$
 j $Fe(s) + H_2O(g) \rightarrow Fe_3O_4(s) + H_2(g)$
 k $H_2S(g) + O_2(g) \rightarrow H_2O(l) + SO_2(g)$
 l $H_2S(g) + SO_2(g) \rightarrow H_2O(l) + S(s)$.

CHAPTER 7 Acids, bases and salts

7.1 The acid taste

The Californian wine industry owes much of its success to a device called a pH meter. 'pH' is a term chemists use to describe how acidic or alkaline a solution is, and a pH meter measures acidity. In California, pH meters are used to measure how acidic the grape juice is. Picking the grapes at the right pH is one of the most important factors in producing a good wine. The right degree of acidity gives wine a fresh tart taste. Too much acidity results in an undrinkable sour wine, tasting like vinegar. Too little acidity gives wine a flat uninteresting taste.

As grapes ripen, their sugar content increases, and their acidity decreases. In the hot sun of California, grapes ripen early in the year, and their sugar content is high. Sometimes the pH meter tells growers that their grapes lack acidity. The acids in grapes include tartaric acid and citric acid. Producers use the reading on the pH meter to tell them how much tartaric acid they should add to bring the natural concentration of tartaric acid in grape juice up to the level that produces a good wine.

Fig. 7.1 A pH meter

7.2 Acids

Thousands of years ago, people found out how to ferment grapes to make wine. They also found out that if the fermentation process went on for too long they got vinegar (sour wine). Centuries ago, chemists distilled vinegar and obtained a sour-tasting substance which we now call ethanoic acid. They obtained other acids also from plant and animal sources: methanoic acid from ants, citric acid from fruits and lactic acid from sour milk. Acids such as these, which were obtained from plant and animal sources, were named **organic acids**.

Later, chemists found ways of making sulphuric acid, hydrochloric acid and nitric acid from minerals. They called these acids **mineral acids**. They are much more reactive than organic acids. We describe mineral acids as **strong** acids and organic acids as **weak** acids. Mineral acids played a big part in the advance of chemistry. They made it possible for chemists to carry out new reactions and make new compounds.

What chemical reactions are typical of acids? Experiment 7.1 will enable you to answer this question.

1 Acids have a sour taste. You can test the truth of this statement with lemon juice (which contains citric acid) or vinegar (which contains ethanoic acid). **Do not try any of the strong acids**.

2 Acids can change the colour of certain substances, called **indicators**. For example, acids turn blue litmus red.

3 Acids react with many metals to produce the gas hydrogen.

> **SUMMARY NOTE**
>
> The names and formulas of some acids are
> - hydrochloric acid, $HCl(aq)$, a strong acid
> - sulphuric acid, $H_2SO_4(aq)$, a strong acid
> - nitric acid, $HNO_3(aq)$, a strong acid
> - ethanoic acid, $CH_3CO_2H(aq)$, a weak acid.

Some metals, e.g. copper, are slow to react with acids. Some other metals, e.g. sodium, react dangerously fast. Metals which react at a safe speed are magnesium, aluminium, zinc, iron and others. Chapter 10 will tell you more about metals.

4 When aqueous solutions of acids are electrolysed, hydrogen is evolved at the negative electrode (see Section 5.9).

5 Acids neutralise **bases** (see Section 7.3).

6 Acids react with carbonates and hydrogencarbonates to give the gas carbon dioxide. Carbonates are compounds which contain carbonate ions, CO_3^{2-}, e.g. calcium carbonate, $CaCO_3$, and sodium carbonate, Na_2CO_3. Hydrogencarbonates contain the ion HCO_3^-, e.g. sodium hydrogencarbonate, $NaHCO_3$.

What is an acid?

A Swedish chemist called Svante Arrhenius made a deduction from these observations. He theorised that aqueous solutions of acids contain hydrogen ions, $H^+(aq)$. This theory explains many reactions of acids, including the two listed below.

1 On electrolysis, hydrogen gas is evolved at the negative electrode. This can be explained by the discharge of hydrogen ions:

hydrogen ions + electrons → hydrogen molecules
$$2H^+(aq) + 2e^- \rightarrow H_2(g)$$

2 In a reaction with a reactive metal M, hydrogen gas is given off and a salt of the metal is formed:

metal atoms + hydrogen ions → metal ions + hydrogen molecules
$$M(s) + 2H^+(aq) \rightarrow M^{2+}(aq) + H_2(g)$$

Arrhenius gave this definition of an acid:
An acid is a substance that releases hydrogen ions when dissolved in water.

7.3 Bases

Another class of substances has been known for centuries, the substances we call **bases**. Sodium hydroxide, NaOH, is an example. Bases have a number of properties in common (see Experiment 7.2).

1 Soluble bases can change the colour of indicators; for example, soluble bases turn red litmus blue.

2 Soluble bases have a soapy feel to the skin. This is because bases actually convert some of the oil in your skin into soap (see Experiment 7.2). Soluble bases, e.g. ammonia, are used in the home as degreasing agents because they have this ability to react with oil and grease to form a soluble soap.

3 Bases react with acids. They counteract the acid — **neutralise** the acid. The product of the reaction is neither an acid nor a base; it is a neutral substance called a **salt**. The reaction is called **neutralisation**. It is the basis of the definition of a base. **A base**

> **SUMMARY NOTE**
>
> Acids
> - have a sour taste
> - change the colour of indicators
> - react with metals to give hydrogen
> - react with carbonates and hydrogencarbonates to give carbon dioxide.

> **SUMMARY NOTE**
>
> All acids have a number of reactions in common. The reason is that solutions of all acids contain a high concentration of hydrogen ions. It is the hydrogen ions that are responsible for the reactions that typify acids.

SUMMARY NOTE

Examples of bases are:
- **metal oxides**, including
  copper(II) oxide, CuO
  zinc oxide, ZnO
  aluminium oxide, Al_2O_3;
- **metal hydroxides**, including
  sodium hydroxide, NaOH
  and magnesium hydroxide,
  $Mg(OH)_2$.
  Most metal oxides and
  hydroxides are insoluble.
  The soluble ones are called
- **alkalis**. They are:
  sodium hydroxide, NaOH
  potassium hydroxide, KOH
  and calcium hydroxide,
  $Ca(OH)_2$.
  Ammonia solution, $NH_3(aq)$,
  is an alkali.

Table 7.1 Coloured insoluble metal hydroxides

copper(II) hydroxide, $Cu(OH)_2(s)$	blue
iron(II) hydroxide, $Fe(OH)_2(s)$	green
iron(III) hydroxide, $Fe(OH)_3(s)$	rust
magnesium hydroxide, $Mg(OH)_2(s)$	white

SUMMARY NOTE

Bases neutralise acids to form salts. Soluble bases change the colours of indicators.

Metal oxides and hydroxides are bases. Soluble bases are called alkalis. Solutions of alkalis contain high concentrations of hydroxide ions, $OH^-(aq)$. They are degreasing agents, and have a soapy 'feel'.

is a substance that reacts with an acid to form a salt and water only.

4 When an alkali dissolves in water, it produces hydroxide ions, $OH^-(aq)$. This is why an alkali will react with metal salts. Most metal hydroxides are insoluble. When hydroxide ions from a solution of an alkali meet metal ions from a solution of a metal salt, an insoluble metal hydroxide is **precipitated** from solution. A solid which forms when two liquids are mixed is called a **precipitate**.

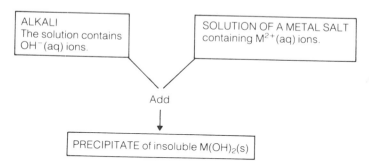

For example, in solution

copper(II) sulphate + sodium hydroxide → copper(II) hydroxide + sodium sulphate

$CuSO_4(aq) + 2NaOH(aq) \rightarrow Cu(OH)_2(s) + Na_2SO_4(aq)$

*Ionic equations

There is another way of writing this equation. The reaction is the combination of copper ions and hydroxide ions. The sodium ions and sulphate ions take no part in the reaction. They are called **spectator ions**. The equation can be written without the spectator ions:

$Cu^{2+}(aq) + 2OH^-(aq) \rightarrow Cu(OH)_2(s)$

This is called an **ionic equation**.

The colours of insoluble metal hydroxides can be of help in finding out what metal cations are present in a solution.

JUST TESTING 30

1 Alkalis are a subset of bases. Copy the Venn diagram (Fig. 7.2). Fill in the names of the four common alkalis. Write the names of four insoluble bases.

***2** Write formulas for these bases:
iron(II) hydroxide,
iron(III) hydroxide,
barium oxide, magnesium hydroxide.

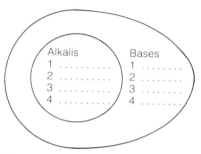

Fig. 7.2

7.4 Neutralisation

What takes place in neutralisation, the reaction between an acid and a base? Consider the reaction between hydrochloric acid and the alkali (soluble base), sodium hydroxide solution:

hydrochloric acid	+	sodium hydroxide	→	sodium chloride	+ water
HCl(aq)	+	NaOH(aq)	→	NaCl(aq)	+ H_2O(l)
acid	+	**alkali**	→	**salt**	**+ water**

When the solutions of acid and alkali are mixed, hydrogen ions, H^+(aq), and hydroxide ions, OH^-(aq), combine to form water molecules:

$$H^+(aq) + OH^-(aq) \rightarrow H_2O(l)$$

Sodium ions, Na^+(aq), and chloride ions, Cl^-(aq), remain in the solution, which becomes a solution of sodium chloride. If you evaporate the solution, you obtain solid sodium chloride. **Neutralisation is the combination of hydrogen ions from an acid and hydroxide ions from a base to form water. In the process, a salt is formed.**

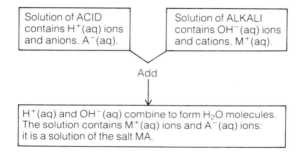

What happens when an acid neutralises an insoluble base (Experiment 7.3)? Consider sulphuric acid and magnesium oxide:

sulphuric acid	+	magnesium oxide	→	magnesium sulphate	+ water
H_2SO_4(aq)	+	MgO(s)	→	$MgSO_4$(aq)	+ H_2O(l)
acid	+	**base**	→	**salt**	**+ water**

Hydrogen ions and oxide ions, O^{2-}, combine to form water:

$$2H^+(aq) + O^{2-}(s) \rightarrow H_2O(l)$$

The resulting solution contains magnesium ions and sulphate ions: it is a solution of magnesium sulphate. If you evaporate it, you will obtain magnesium sulphate crystals.

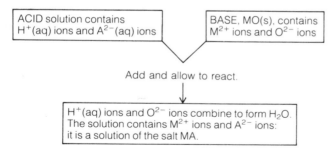

7.5 Strengths of acids and bases

Some acids are stronger than others. A strong acid reacts faster than a weak acid with metals, bases and carbonates (see Experiment 7.1). Some bases are stronger than others. Ammonia is a weak base. A strong base will drive ammonia out of an ammonium salt:

ammonium salt + strong base → ammonia + salt of strong base

For example,

ammonium chloride + calcium oxide → ammonia + calcium chloride

This reaction is used as a laboratory preparation of ammonia and as a test for ammonium salts (see Section 14.4).

Universal Indicator can distinguish between strong and weak acids. It turns different colours in strongly acidic and weakly acidic solutions. It can also distinguish between strongly basic and weakly basic solutions (see Table 7.2). Each shade of Universal Indicator colour corresponds to a **pH number**. The pH number is a measure of the acidity or alkalinity of the solution. For a neutral solution, pH = 7; for an acidic solution, pH < 7; for an alkaline solution, pH > 7 (see Experiment 7.4). Table 7.3 lists the colours of some other indicators.

> **SUMMARY NOTE**
>
> Some acids are stronger than others. The mineral acids are stronger than organic acids, e.g. ethanoic acid. Some bases are stronger than others. Ammonia is a weak base. Universal Indicator turns different colours in strongly acidic, weakly acidic, neutral, weakly alkaline and strongly alkaline solutions. For each shade of colour, there is a pH number. The pH of a solution is a measure of its acidity or alkalinity.

Table 7.2 The colour of Universal Indicator in different solutions

pH	Strongly acidic	Weakly acidic	Neutral	Weakly alkaline	Strongly alkaline
	0 1 2 3	4 5 6	7 8	9 10 11	12 13 14
Colour	← Red →	Orange Yellow	Green Blue	← Violet →	

Table 7.3 The colours of some common indicators

Indicator	Acidic colour	Neutral colour	Alkaline colour
Litmus	Red	Purple	Blue
Phenolphthalein	Colourless	Colourless	Pink
Methyl orange	Red	Orange	Yellow

> **Career sketch: Laboratory assistant 2**
>
> Paul is a laboratory assistant. He works for a manufacturer of agricultural chemicals. Chemists in the company's laboratory are constantly making new compounds. It is Paul's job to try them out to see whether they will do what the chemists hope they will do. He may spray a selection of plants in the laboratory with a new compound which the chemists hope will make plants grow faster or more slowly. He measures the plants' growth and keeps a careful record. If the new chemical does what the chemists hope, then Paul sends it on for large-scale trials. Paul left school after GCSE.

Table 7.4 Some common acids and bases

	Where you find it and why you use it
Acid	
Carbonic acid	In fizzy drinks
Citric acid	In fruit juices
Ethanoic acid	In vinegar
Hydrochloric acid	Used for 'pickling' metals before they are coated
Nitric acid	Used for making fertilisers and explosives
Phosphoric acid	In rust inhibitor; used for making fertilisers
Sulphuric acid	In car batteries; used for making fertilisers (Chapter 13)
Base	
Ammonia	In cleaning fluids as a degreasing agent; used in the manufacture of fertilisers
Calcium hydroxide	Spread on soil which is too acidic
Calcium oxide	Used in the manufacture of cement, mortar and concrete
Magnesium hydroxide	In 'antacid' indigestion tablets and 'milk of magnesia'
Sodium hydroxide	In powerful oven cleaners as a degreasing agent; used in the manufacture of soap

Fig. 7.3 Some common acids

Table 7.5 pH values of some solutions

pH	Solution
0	Bench hydrochloric acid
1	Digestive fluids in the stomach
2	Lemon juice
3	Vinegar Orange juice
4	Soda water
5	
6	Rain water Cows' milk
7	Pure water Blood
8	
9	Sodium hydrogencarbonate solution and sea water
10	
11	Bench ammonia solution
12	Sodium carbonate (washing soda) solution
13	Calcium hydroxide solution (limewater)
14	Bench sodium hydroxide solution

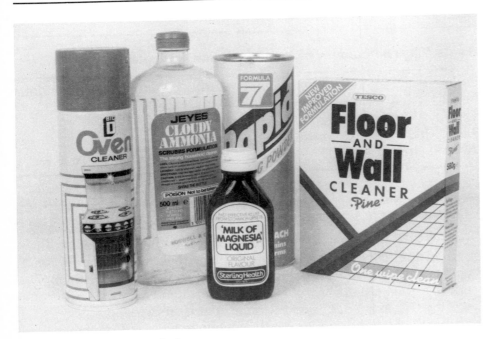

Fig. 7.4 Some common alkalis

Table 7.5 shows the pH values of solutions of some common acids and alkalis. Some salts do not have a pH value of 7. Carbonates and hydrogencarbonates are alkaline in solution.

SUMMARY NOTE

We use many acids and bases in everyday life. Some common acids and bases are listed. The pH values of their solutions are given in Table 7.5.

JUST TESTING 31

1. Give an example of **a** a weak acid, **b** a strong acid, **c** a weak base, **d** a strong base. Where might you find each of these substances outside the laboratory?
2. Say whether these substances are strongly acidic, weakly acidic, neutral, weakly basic or strongly basic:
   a grapefruit juice, pH = 3.0; **b** blood, pH = 7.4; **c** beer, pH = 5.0; **d** rain, pH = 6.2; **e** sea water, pH = 8.5; **f** saliva, pH = 7.0.
3. Why is sodium hydrogencarbonate used to treat indigestion while sodium carbonate is not?
4. Why does stewing apples in an aluminium saucepan leave it clean?
5. Sherbet contains sodium hydrogencarbonate and citric acid (a solid). When you put it in your mouth, it generates carbon dioxide and fizzes. Why does it do this only when wet?
6. Indigestion tablets are of many kinds. They may contain one of the following compounds:
   $NaHCO_3$, $Al(OH)_3$, Al_2O_3, $MgCO_3$, $Mg(OH)_2$, MgO, $CaCO_3$.
   Which would react fastest with the excess of acid in the stomach?
7. A greasy oven can be cleaned by spraying the inside with a concentrated solution of sodium hydroxide. How does it work?

*7.6 Acids and bases redefined by Brönsted and Lowry

A Danish chemist called Brönsted and a British chemist called Lowry were dissatisfied with the Arrhenius definition of acids and bases. Working separately, they reached the same conclusion. They realised that no substance could dissociate to give hydrogen ions unless there was another substance present to combine with hydrogen ions as they were formed. They both formulated the same new definition of an acid. **An acid is a substance that can donate a proton to another substance.** When hydrogen chloride dissolves in water, a chemical reaction takes place (Fig. 7.5). Oxonium ions, $H_3O^+(aq)$, and chloride ions, $Cl^-(aq)$, are formed:

hydrogen chloride + water → oxonium ions + chloride ions
$HCl(g)$ + $H_2O(l)$ → $H_3O^+(aq)$ + $Cl^-(aq)$

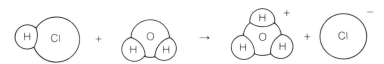

Fig. 7.5 Formation of an oxonium ion

The acid does not simply dissociate into ions: it reacts with water. Water is the proton acceptor. Hydrogen chloride cannot give a proton unless there is a substance present to accept a proton. **A substance which can accept a proton from another substance is a base.**

The oxonium ion, $H_3O^+(aq)$, is a hydrogen ion combined with a water molecule, $H^+ \cdot H_2O(aq)$. For convenience, it will be written as $H^+(aq)$ and called a hydrogen ion.

Bases are proton acceptors. Figure 7.6 shows ammonia accepting a proton from a water molecule:

ammonia + water → ammonium ion + hydroxide ion
$NH_3(aq) + H_2O(l) \rightarrow NH_4^+(aq) + OH^-(aq)$

Fig. 7.6 Formation of a hydroxide ion

> ### *SUMMARY NOTE
> According to the Brönsted–Lowry definition
> - an acid is a substance which can donate a proton to another substance,
> - a base is a substance which can accept a proton from another substance.

*JUST TESTING 32

1. Complete the following equation:
   $NH_3(aq) + H_2O(l) \rightarrow \ldots + \ldots$
   Explain why ammonia, NH_3, is said to be acting as a Brönsted–Lowry base.
2. What is the difference between hydrogen chloride gas and hydrochloric acid?
3. Write the equation for the formation of oxonium ions from hydrobromic acid, $HBr(aq)$.

*7.7 Weak acids and bases

Why are some acids stronger than others? Why are some bases stronger than others? A strong acid is completely ionised in solution. A solution of hydrochloric acid contains oxonium ions (hydrogen ions) and chloride ions. There are no hydrogen chloride molecules present. A weak acid is only partially ionised in solution. A solution of ethanoic acid contains chiefly molecules of ethanoic acid. Very few of the molecules (only 1 per cent) ionise:

ethanoic acid + water ⇌ oxonium ions + ethanoate ions
$CH_3CO_2H(aq) + H_2O(l) \rightleftharpoons H_3O^+(aq) + CH_3CO_2^-(aq)$

The reverse arrows, ⇌, mean that the ionisation is incomplete. The solution contains oxonium ions (hydrogen ions), ethanoate ions and ethanoic acid molecules.

Ammonia is a weak base. In water, it ionises to a small extent:

ammonia + water ⇌ ammonium ions + hydroxide ions
$NH_3(aq) + H_2O(l) \rightleftharpoons NH_4^+(aq) + OH^-(aq)$

The concentration of ammonium ions and hydroxide ions is small, but it is sufficient to account for many of the reactions of

> ### *SUMMARY NOTE
> Strong acids are completely ionised; weak acids are partially ionised.
>
> Strong bases are completely ionised; weak bases are partially ionised.

ammonia solution. The hydroxide ions present will precipitate insoluble metal hydroxides from solution, as other alkalis do (see Section 7.3).

> ### *JUST TESTING 33
>
> 1 Hydrogen cyanide, HCN, is a poisonous gas which dissolves in water to form a solution of the weak acid HCN(aq), hydrocyanic acid. Write an equation to show the dissociation of hydrocyanic acid into oxonium ions and cyanide ions.

*7.8 Amphoteric oxides and hydroxides

All metal oxides and hydroxides are basic. They all react with acids to form salts. The oxides and hydroxides of some metals are also acidic: they react with bases to form salts. For example:

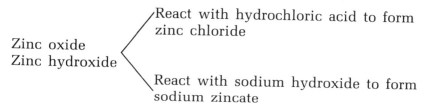

Oxides and hydroxides of this kind are **amphoteric**, that is, both basic and acidic. In sodium zincate, Na_2ZnO_2, the metal zinc is part of the anion, ZnO_2^{2-}. The formation of the zincate salt is

zinc hydroxide + sodium hydroxide → sodium zincate + water
$Zn(OH)_2(s) + 2NaOH(aq) \rightarrow Na_2ZnO_2(aq) + 2H_2O(l)$

Other amphoteric oxides and hydroxides are aluminium oxide and hydroxide, Al_2O_3 and $Al(OH)_3$, and lead(II) oxide and hydroxide, PbO and $Pb(OH)_2$.

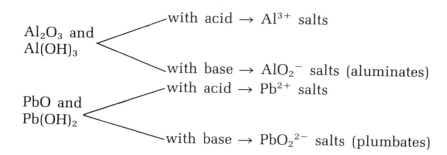

> ### JUST TESTING 34
>
> 1 Describe what you see when sodium hydroxide solution is added gradually to a solution of aluminium sulphate.
>
> 2 A solution of sodium hydroxide is added to a solution of a metal salt. A white precipitate is formed. The white precipitate dissolves when an excess of sodium hydroxide is added. Could the metal salt be **a** magnesium hydroxide or **b** zinc hydroxide? Explain your answer.

7.9 Salts

Silver bromide, AgBr, is a salt which can give you a lot of fun. The reason why you can take photographs is that light affects silver bromide. Photographic film consists of a paper and cellulose base covered with a film of emulsion. In black and white film, the emulsion consists of grains of silver bromide suspended in gelatin. When the film is exposed, light falls on to some areas of the emulsion. It changes the grains of silver bromide which it meets. In each exposed grain, some silver ions, Ag^+, are converted into silver atoms, which are black. These black atoms form a picture of the object called a **latent image** (Fig. 7.7). When the film is **developed**, the process started by light continues. More silver ions in the areas affected by light are converted into silver atoms. The result is a **negative**. A **fixer** is used to remove unchanged silver bromide. The negative has to be placed on a piece of photographic paper and exposed to light to give a **print** (see Fig. 7.7).

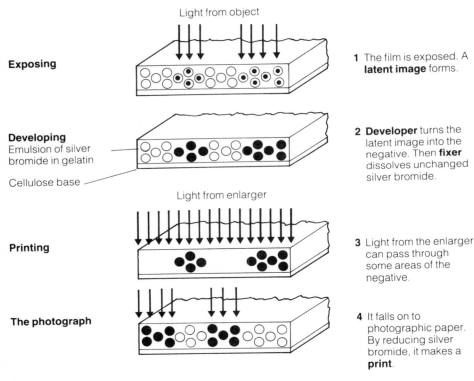

Fig. 7.7 Photography

Occurrence of salts

This useful salt, silver bromide, does not occur naturally. Many other salts do occur naturally in the Earth's crust. One of the most widespread and useful is sodium chloride, NaCl, which is known as **common salt** (see Chapter 15). Other examples are calcium fluoride, CaF_2 (called **fluorite** or **Blue John**), magnesium sulphate, $MgSO_4$ (called **Epsom salts**), lead(II) sulphide, PbS (called **galena**).

Other salts, including silver bromide, must be made in the laboratory. Many can be made by neutralising acids. In a salt, the hydrogen ions in the acid have been replaced by metal ions or by the ammonium ion. Salts of hydrochloric acid are called **chlorides**. Salts of sulphuric acid are called **sulphates**. Salts of nitric acid are called **nitrates**. (See Table 7.6.)

Table 7.6 Salts of some common acids

Acids	Salts
Hydrochloric acid, HCl(aq)	Sodium chloride, NaCl
	Zinc chloride, $ZnCl_2$
	Iron(II) chloride, $FeCl_2$
	Ammonium chloride, NH_4Cl
Sulphuric acid, H_2SO_4(aq)	Sodium sulphate, Na_2SO_4
	Magnesium sulphate, $MgSO_4$
	Iron(III) sulphate, $Fe_2(SO_4)_3$
Nitric acid, HNO_3	Sodium nitrate, $NaNO_3$
	Zinc nitrate, $Zn(NO_3)_2$

Salts may be soluble or insoluble. When soluble salts are made, the final step is often to allow them to crystallise from solution. The crystals of some salts contain water. Examples are

- copper(II) sulphate-5-water, $CuSO_4 \cdot 5H_2O$
- cobalt(II) chloride-6-water, $CoCl_2 \cdot 6H_2O$
- iron(II) sulphate-7-water, $FeSO_4 \cdot 7H_2O$

In Experiment 7.7 you can see how the water content of crystals gives crystals their shape and their colour. It is called **water of crystallisation**. When blue crystals of copper(II) sulphate-5-water are heated at 100 °C, the water of crystallisation is driven off. The white powdery solid that remains is **anhydrous** (without water) copper(II) sulphate:

copper(II) sulphate-5-water → water + copper(II) sulphate
blue crystals　　　　　　　　as steam　　white powder, **anyhdrous**

$CuSO_4 \cdot 5H_2O(s)$　　→　$5H_2O(g)$ + $CuSO_4(s)$

SUMMARY NOTE

There are thousands of salts. Many of them are extremely useful to us. Some of them occur naturally, but most of them must be made by the chemical industry.

Some soluble salts crystallise with water of crystallisation. This gives the crystals their shape and, in some cases, their colour.

JUST TESTING 35

1 Name the salts with the following formulas: KI, KNO_3, NaBr, $BaCO_3$, Na_2SO_4, $NiCl_2$, $CrCl_3$, $BaSO_4$, MgI_2, $ZnCl_2$.
*2 Write formulas for these salts: potassium sulphate, potassium chloride, copper(II) chloride, lead(II) sulphate, sodium bromide, ammonium sulphate.

7.10 Uses of salts

We use many different salts for a multitude of different purposes. Photography is one activity which depends on a salt, the salt silver bromide. There are many other applications for salts. A few examples will be given.

Agricultural uses. Ammonium nitrate and ammonium sulphate are used as fertilisers (Section 14.7). Copper(II) sulphate solution is used on vines and potatoes to kill pests (see Sections 13.9 and 14.7).

Domestic uses. Sodium carbonate-10-water is 'washing soda', used as a water softener (Section 9.13).

Medical uses. Calcium sulphate is mined as calcium sulphate-2-water. When heated it forms calcium sulphate-½-water. This is plaster of Paris. It combines with water to form a strong material which is used to support broken bones (Fig. 7.8). It is also used for plastering walls.

Iron(II) sulphate-7-water is given in the form of 'iron tablets' to people who suffer from anaemia.

Barium sulphate is given to patients who are suspected of having a stomach ulcer. After a 'barium meal', an X-ray shows the path taken by the salt. Being large, barium ions give good X-ray photographs.

Industrial uses. Anhydrous calcium chloride is used as a drying agent. Sodium chloride has many uses (see Section 15.1).

Some of these salts are mined, e.g. sodium chloride and calcium sulphate. Others must be made by the chemical industry, e.g. silver bromide and copper(II) sulphate.

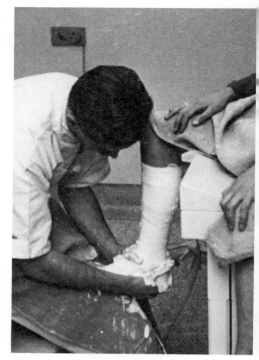

Fig. 7.8 Plaster of Paris

SUMMARY NOTE

Salts are used in agriculture, in the home, in medicine and in industry. Methods of making salts are therefore important.

7.11 Methods of making salts

The methods of making salts fall into three classes:
- Direct combination of the elements, called **synthesis**
- Methods for soluble salts
- Methods for insoluble salts.

Table 7.7 summarises the facts about the solubility of salts.

Table 7.7 Soluble and insoluble salts

Salts	Soluble	Insoluble
Chlorides	Most are soluble	Silver chloride, AgCl Lead(II) chloride, $PbCl_2$
Sulphates	Most are soluble	Barium, calcium and lead(II) sulphates, $BaSO_4$, $CaSO_4$ and $PbSO_4$
Nitrates	All are soluble	None
Carbonates	Sodium and potassium carbonates, Na_2CO_3 and K_2CO_3	Most are insoluble
Ethanoates	All are soluble	None
Sodium, potassium and ammonium salts	All are soluble	None

SUMMARY NOTE

You will need to refer to this information about soluble and insoluble salts.

SUMMARY NOTE

Some salts can be made by direct combination of the elements.

Direct combination of elements

A reactive metal and a reactive non-metallic element may combine directly (Experiments 3.1–3.4). For example,

aluminium + iodine → aluminium iodide

iron + sulphur → iron(II) sulphide

These are examples of **synthesis**. This method can be used for soluble and insoluble salts. You will meet the synthesis of oxides in Chapter 10 and the synthesis of chlorides in Chapter 15.

Methods for soluble salts

1 *Reaction between a metal and a dilute acid*. A reactive metal will displace hydrogen from a dilute acid to give a salt (see Experiment 7.5):

Metal + Acid → Salt + Hydrogen

For example,

zinc + sulphuric acid → zinc sulphate + hydrogen

Figure 7.9a shows how the reaction can be used to prepare a salt.

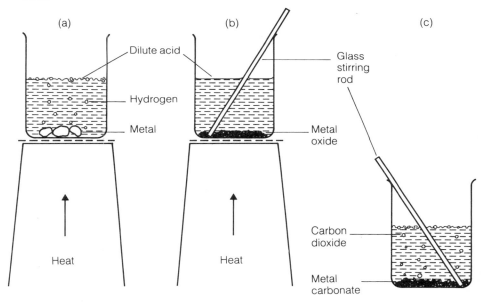

1 Use an excess of metal. When no more hydrogen gas is evolved, the reaction is over: all the acid has reacted.

Use an excess of metal oxide. When the solution no longer turns blue litmus red, the reaction is over.

Use an excess of metal carbonate. When the evolution of carbon dioxide stops, the reaction is over.

2 Remove the excess of metal in (a) or metal oxide in (b) or metal carbonate in (c) by filtration.

3 Evaporate the filtrate until crystallisation begins. Crystals of the metal salt form.

Fig. 7.9 Preparation of soluble salts

2 *Reaction between an insoluble base and a dilute acid*. A base reacts with an acid to give a salt and water (see Experiment 7.6):

Acid + Base → Salt + Water

For example,

copper(II) oxide + sulphuric acid → copper(II) sulphate + water

Figure 7.9b shows how the reaction can be used for the preparation of a salt.

3 *Reaction between a metal carbonate and a dilute acid.*
A metal carbonate reacts with a dilute acid to give a salt and carbon dioxide and water:

Acid + Carbonate → Salt + Carbon dioxide + Water

For example,

barium carbonate + hydrochloric acid → barium chloride + carbon dioxide + water

Figure 7.9c shows how the reaction can be used for the preparation of a salt.

SUMMARY NOTE

Soluble salts are made by the reactions:

Acid + Reactive metal → Salt + Hydrogen

Acid + Insoluble base → Salt + Water

Acid + Metal carbonate → Salt + Water + Carbon dioxide

In these three methods, an excess of the solid reactant is used, and unreacted solid is removed by filtration.
The Acid + Alkali method follows later.

JUST TESTING 36

Refer to Fig. 7.9.
1 How do you know that all the acid has been used up?
2 Why is it necessary to ensure that all the acid is used up?
3 If some acid were left unneutralised in step **1**, what would happen to it in step **3**? How would it affect the result?
4 Why is it easier to remove an excess of base than an excess of acid?

4 *Reaction between an acid and an alkali.* When an alkali reacts with an acid

Acid + Alkali → Salt + Water

For example,

potassium hydroxide + hydrochloric acid → potassium chloride + water

An alkali is a soluble base. You cannot remove an excess by filtration, as in methods **1**, **2** and **3**. You must add exactly the right amount of alkali to neutralise the acid. **Titration** can be used. You can use this method of adding one reactant to another in a measured way in Experiment 7.8. Figure 7.10 summarises the method.

SUMMARY NOTE

Soluble salts can be made by the reaction:

Acid + Alkali → Salt + Water

The method of titration is used to add exactly the right amounts of acid and alkali.

Career sketch: Analytical chemist

Gwynneth is an analytical chemist who works for an agricultural advisory service. One of her jobs is to measure the protein content of different foods. She works in a laboratory which is equipped with many kinds of apparatus. She has an automatic instrument to do titrations for her. She must understand how her instruments work and be able to interpret the results they give. Gwynneth has A-level chemistry.

2 *Read* the burette.

3 *Titrate.* Add acid slowly from the burette. Swirl the conical flask. Stop when the indicator changes to its neutral colour.

4 Read the burette again. What volume of acid has been used?

1 The conical flask holds a measured volume of alkali and some indicator. It sits on a white tile.

5 Now take the same volume of alkali *without indicator*. Add the *same volume* of acid as you used in the titration.

6 Evaporate the salt solution until it begins to crystallise.

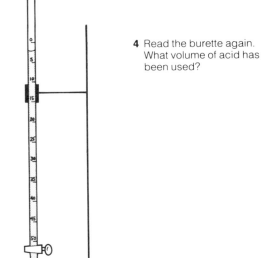

Fig. 7.10 Titration

JUST TESTING 37

1 Complete the following word equations:
   a potassium hydroxide + sulphuric acid → potassium .... + ...
   b nickel oxide + hydrochloric acid → .... chloride + ...
   c calcium hydroxide + hydrochloric acid → .... + ...
   d chromium oxide + hydrochloric acid → .... + ...
   e aluminium oxide + sulphuric acid → .... + ...
   f magnesium oxide + nitric acid → .... nitrate + ...
   g sodium hydroxide + nitric acid → .... + ...

*2 Write chemical equations for the reactions listed in **1a**, **1b** and **1c**.

Method for insoluble salts: precipitation

An insoluble salt can be made by **precipitation** (see Experiment 7.9). When two chosen solutions are added, the insoluble salt **precipitates**. For example, to make the insoluble salt lead(II) sulphate you add two solutions. The reaction that has occurred is

lead(II) nitrate + sodium sulphate → lead(II) sulphate + sodium nitrate

$Pb(NO_3)_2(aq) + Na_2SO_4(aq) \rightarrow PbSO_4(s) + 2NaNO_3(aq)$

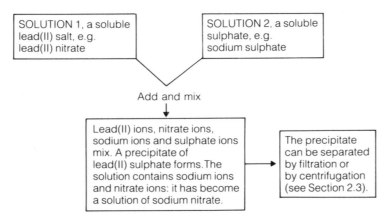

*When lead(II) ions and sulphate ions meet, they cannot remain in solution because lead(II) sulphate is insoluble. The sodium ions and nitrate ions play no part in the reaction: they remain in solution. They are called **spectator ions**. The equation for the reaction can be written without the spectator ions:

lead(II) ions + sulphate ions → lead(II) sulphate
$Pb^{2+}(aq) + SO_4^{2-}(aq) \rightarrow PbSO_4(s)$

This is called an **ionic equation**.

In general, to make an insoluble salt MA you have to choose a soluble salt of M and a soluble salt of A. Table 7.7 will help you. Remember that dilute sulphuric acid can be used to provide sulphate ions, $SO_4^{2-}(aq)$, and dilute hydrochloric acid can be used to provide chloride ions, $Cl^-(aq)$.

Remember:
- All nitrates are soluble.
- Dilute sulphuric acid contains sulphate ions; dilute hydrochloric acid contains chloride ions.
- All sodium, potassium and ammonium salts are soluble.

SUMMARY NOTE

Insoluble salts are made by precipitation.

JUST TESTING 38

(See Table 7.7 for solubility.)

1 To make lead(II) sulphate, which lead(II) salt other than lead(II) nitrate could you use? Which sulphate other than sodium sulphate could you use?

2 Silver chloride is insoluble. You can make it by adding solutions of a soluble choride and a soluble silver salt.
   a Name a soluble chloride.
   b Name a soluble silver salt.
   c Write a word equation for the reaction that happens when you add the two solutions.
   d Describe how you would obtain a dry specimen of silver chloride.
   e Write an ionic equation for the reaction.

3 Barium sulphate is insoluble.
   a Name two solutions which you could mix to give a precipitate of barium sulphate.
   b Describe what you would do to obtain barium sulphate from the mixture.
   c Write a word equation for the reaction.
   d Write a chemical equation.

4 Nickel carbonate is insoluble.
   a What two solutions would you use to prepare it?
   b Write a word equation for the reaction.

Exercise 7

1. What type of container is used for strong acids?

2. List three salts and their uses.

3. You are given five solutions in bottles labelled A, B, C, D and E. One bottle contains water and the others contain solutions of citric acid (from lemons), sulphuric acid, sodium hydroxide and sodium hydrogencarbonate (which is used as a remedy for indigestion). Given a bottle of Universal Indicator, how could you find out which solution is which?

4. • Teeth become coated with a layer of food and bacteria called **plaque**.
   • Toothpaste contains solid aluminium hydroxide.
   How does solid aluminium hydroxide work to remove plaque from teeth?
   How does it remove the acids which form in plaque?
   Why do we need to remove mouth acids?
   Why is a detergent added to toothpaste to make it foam?
   Why is flavouring, such as mint, added to toothpaste?

5. Scouring powders are strong cleaners for jobs like oven cleaning. They contain an alkali, often sodium hydroxide, and powdered stone.
   How does the powder help in cleaning?
   How does the alkali help in cleaning?

6. a Name the acid used in car batteries.
   b If you spilt some of this acid, what household substance could you use to neutralise it?
   c Name the acid present in vinegar.
   d Name the compound present in bath salts.
   e What gas is formed when vinegar reacts with bath salts?

7. Magnesium sulphate crystals, $MgSO_4 \cdot 7H_2O$, can be made from magnesium oxide and dilute sulphuric acid by the following method.
   Step 1 Add an excess of magnesium oxide to dilute sulphuric acid. Warm.
   Step 2 Filter.
   Step 3 Partly evaporate the solution from Step 2. Leave it to stand.
   Step 4 Filter the solution from Step 3.

   a Explain why an excess of magnesium oxide is used in Step 1.
   b Name (i) the residue and (ii) the filtrate in Step 2.
   c Explain why the solution is partly evaporated in Step 3.
   d What is removed by filtering in Step 4?
   e Would you dry the product by strong heating or by gentle heating? Explain your answer.
   f Write a word equation for the reaction. Write a symbol equation.

8. How would you make potassium nitrate? What starting materials would you use, and what steps would you take to obtain crystals of the product?
   Explain why you have chosen this method. (For solubility information, see Table 7.7.)

9. What method would you use to make copper(II) chloride? Explain why you have chosen this method. Say what starting materials you would need, and what you would do to obtain solid copper(II) chloride. (For solubility information, see Table 7.7.)

10. How would you make barium carbonate? Say what starting materials you would need, and describe what you would do with them. Explain why you chose this method. (For solubility information, see Table 7.7.)

11. When you add lemon juice to tea, the tea changes colour. Suggest a reason for the change.

12. Arrange the following pH values in order of increasing acid strength. Which values show basic, neutral and acidic solutions?
    a 5.0; b 1.5; c 9.5; d 7.0; e 12.0; f 6.5; g 7.4.

13. Silver bromide, the light-sensitive salt in photographic film, is made by mixing solutions of silver nitrate and potassium bromide. Write a word equation and a chemical equation for the reaction. What precaution would you have to take to obtain a useful product?

14. • Sauerkraut is made by allowing cabbage to ferment.
    • Cabbage juice has a pH of 5. Sauerkraut has a pH of 3.
    What kind of substance must be produced when cabbage is fermented?

15 Barium ions are highly poisonous. A person who swallows barium chloride solution should be given sodium sulphate solution to drink. Explain why this is an antidote. Write a word equation and a chemical equation for the reaction that occurs.

16 The label on a bottle of baking soda says that one of its uses is to relieve acid indigestion. The acid in the stomach is hydrochloric acid. Explain how baking soda, sodium hydrogencarbonate, acts as an antacid. Give a word equation for its reaction with stomach acid.

CHAPTER 8 Air

8.1 Saving lives with heroism and oxygen

The clock on the wall of the Credito Mexicano bank in the centre of Mexico City stopped at 7.20 a.m. on Thursday, 19th September, 1985. The windows broke, the floors heaved, the walls swayed, but the bank stood. Next door, the Hotel Regis collapsed into a mountain of rubble, crushing or trapping 500 tourists. Schools, houses and apartment blocks collapsed. By 8 a.m., everyone who was left in Mexico City knew that they were lucky to be alive. They had survived the worst earthquake in their city's history. In the centre of the city, 250 buildings had collapsed and 10 000 people had died.

People who had escaped injury started to search for survivors. Thousands of people lay trapped under masses of twisted steel and fallen concrete. There was a second danger for those who had survived the earthquake: lack of air. The air was heavy with dust and smoke. Broken gas mains caught fire. If there were any survivors in the ruins of the Hotel Regis, the fire smouldering through the ruins took away their oxygen and their chance of being rescued. When survivors were found under the rubble, rescuers fed air to them through plastic pipes, to prevent them suffocating in the dust and smoke-filled ruins.

There were some near-miraculous rescues. When the Juarez Hospital collapsed, its 12 stories folded in, trapping a thousand patients and staff between the floors. Rescuers tunnelled between the collapsed floors finding survivors and removing bodies. After 7 days, rescuers brought out alive some tiny newborn babies who had survived for a week without food. They were rushed to a hospital. Many of the babies survived. In incubators, they were warmed slowly back to a normal temperature, and given food and oxygen (see Fig. 8.1). Pure oxygen from a cylinder can save the life of someone who has been deprived of air.

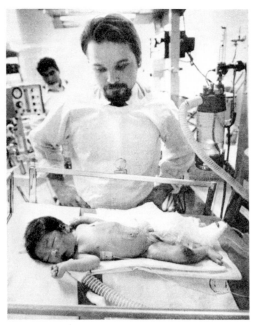

Fig. 8.1 One of Mexico City's infant miracles

8.2 Air

For people in normal health, there is enough oxygen in the air they breathe. The percentage by volume of oxygen in clean, dry air is 21%. Figure 8.2 shows the percentages of nitrogen and other gases in pure, dry air. Water vapour and pollutants may be present also.

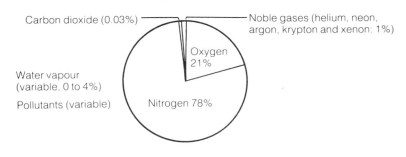

Fig. 8.2 Composition of clean, dry air in percentage by volume

8.3 How to obtain oxygen and nitrogen from air

The method used to separate oxygen and nitrogen from air is fractional distillation (see Section 2.5) of liquid air (Fig. 8.3). Drastic cooling is needed to liquefy air at −200 °C. Two physicists called Joule and Thompson noticed that when a compressed gas is allowed to expand suddenly its temperature drops. This cooling effect is used to liquefy air (Fig. 8.3). Liquid air is distilled in an insulated fractionating column. The temperature is −190 °C at the top of the column and −200 °C at the bottom. Nitrogen, with the lower boiling temperature of −196 °C, can exist as a gas at the top of the column. Oxygen, with the higher boiling temperature of −183 °C, remains as a liquid and is run off from the bottom of the column. Argon is drawn off from the middle of the column. The fractions, oxygen, nitrogen and argon, can be redistilled if very pure gases are required. They are stored under pressure in strong steel cylinders. Many important uses are found for them.

SUMMARY NOTE

Air is a mixture of oxygen, nitrogen, noble gases, carbon dioxide, water vapour and pollutants. Oxygen and nitrogen are obtained by the fractional distillation of liquid air.

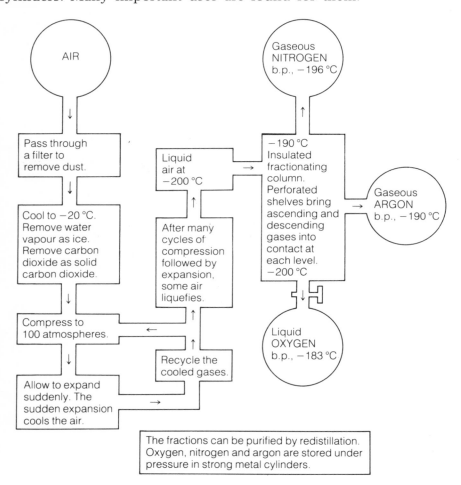

Fig. 8.3 Fractional distillation of liquid air

8.4 Uses of oxygen

Hospitals need oxygen for patients who have difficulty in breathing. Tiny premature babies often need oxygen, and people with lung diseases are kept alive by daily sessions with their

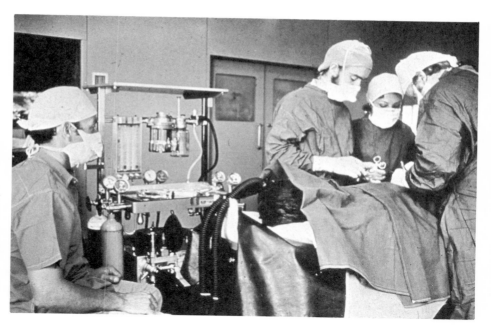
Fig. 8.4 Oxygen in hospitals

Fig. 8.5 Oxygen in industry

SUMMARY NOTE

There are many uses for pure oxygen. Hospitals use it to treat patients who have breathing difficulties. Industries, especially the steel industry, use oxygen. Mountaineers, deep-sea divers, high altitude pilots and astronauts all carry oxygen. Oxygen is used to combat pollution.

Fig. 8.6 A diver needs oxygen

breathing machines. Patients who are having operations are given an anaesthetic mixed with oxygen (Fig. 8.4).

Industry has many uses for oxygen. The 'oxy-acetylene' torch burns the gas ethyne (formerly called acetylene) in oxygen to produce a temperature of about 4000 °C. This torch is used for cutting and welding metal plates (Fig. 8.5).

The steel industry uses oxygen to burn off impurities, such as carbon, sulphur and phosphorus. The removal of these impurities turns brittle cast iron into strong steel (see Section 10.10). The consumption of oxygen is high: one tonne of oxygen for every tonne of cast iron turned into steel. Many steel plants make their own oxygen on site.

Aeroplanes which fly at high altitude must carry oxygen. At a height of 5 km, the air pressure is only half that at sea level. Climbers who are tackling high mountains take oxygen with them.

Divers carry oxygen cylinders when they are working under water (Fig. 8.6). They breathe a mixture of oxygen and helium (see Section 8.7). At one time, divers used to carry cylinders of air. Under water, at high pressure, nitrogen dissolves in the blood. When a diver surfaced, nitrogen came out of solution, and if the diver surfaced quickly the bubbles collected in the bloodstream, causing a painful condition called the **bends**. The use of helium instead of nitrogen to dilute oxygen avoids the bends because very little helium dissolves in the blood, even under pressure.

Astronauts must carry oxygen, and even unmanned space flights need oxygen. To lift spacecraft into orbit, rockets burn kerosene in oxygen. After lift-off, the spacecraft burns hydrogen in oxygen (Fig. 8.7).

Oxygen is used to combat pollution. Air or oxygen is used in the treatment of sewage (see Chapter 9). If it were not for this treatment, many rivers and lakes would be spoiled by sewage (Fig. 8.8). Oxygen is used to restore life to polluted lakes and rivers (see Chapter 9).

Fig. 8.7 Space rockets carry oxygen

Fig. 8.8 Oxygen fights pollution

Fig. 8.9 These foods are packaged under nitrogen

SUMMARY NOTE

Nitrogen is used when an inert atmosphere is needed.

8.5 Nitrogen

Nitrogen is used by the food industry. When foods such as nuts and bacon are processed, they need to be protected from oxygen. Otherwise, oxygen in the air will oxidise oils in the foods to rancid products. To exclude oxygen, a stream of nitrogen is passed over the processed foods while they are being packed.

Many uses of nitrogen depend on its unreactive nature. Liquid nitrogen is used when an unreactive refrigerant is needed. The food industry uses it for the fast freezing of foods. Hospitals use it to preserve kidneys from a donor until they are transplanted into a patient. Vets use liquid nitrogen to preserve semen, ready for use in artificial insemination. They can transport the semen of a prize bull or a famous racehorse around the country and use it to improve a dairy herd or racing stable. There is a firm in the USA which obtains semen from human donors, stores it in liquid nitrogen and sells it to single women who are anxious to have babies.

8.6 Carbon dioxide and water vapour

Although carbon dioxide is only 0.03% by volume of air, we could not do without it. It is essential for photosynthesis in plants (see Section 12.6). Figure 8.10 shows how you can test for the presence of carbon dioxide in air. The test depends on the white precipitate formed when carbon dioxide reacts with a solution of calcium hydroxide, **limewater** (see Section 12.6).

The content of water vapour in the air varies from 0% in a desert to 4% in a rain forest. Figure 8.10 shows how to test air for water vapour. Any substance which contains water turns anhydrous copper(II) sulphate from white to blue (see Section 7.9).

SUMMARY NOTE

Air contains
- carbon dioxide
- water vapour.

This is why air
- turns limewater cloudy
- turns anhydrous copper(II) sulphate from white to blue.

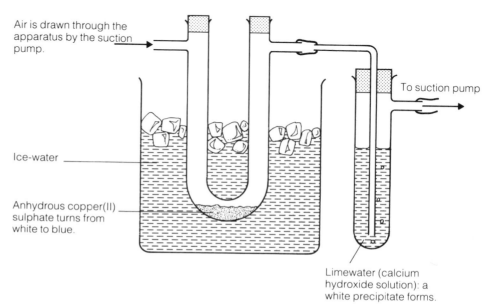

Fig. 8.10 Testing for carbon dioxide and water vapour in air

8.7 The noble gases

In the 1930s, airships were a popular means of transport. These 'lighter than air' ships contained big gas bags filled with hydrogen. Air is 14 times denser than hydrogen, and the gas bags were able to lift into the air a 200 tonne airship with 50 passengers. Diesel engines attached to the outside of the airships could propel them at up to 140 kilometres per hour (90 m.p.h). Passengers enjoyed the spacious shipboard-type accommodation, together with the comfort of the smooth ride and the excitement of fine views. Transatlantic crossings were particularly popular until disaster struck in 1937. The pride of Germany's fleet of airships, the Hindenburg, caught fire when landing in the USA. Of the 50 passengers, only a few were able to leap to safety. Investigation showed that the fire had been started by an act of sabotage, not by any design fault. Nevertheless, Germany decided to stop using airships, and other nations followed suit.

Now, nearly 50 years later, the UK-based firm, Airships Industries, has pioneered the return of the airship. In June 1985, an airship flew over Lords during the Test Match, over Wimbledon during the tournament and over Henley Regatta. It carried an advertisement (see Fig. 8.11). The manufacturers, besides selling airships for advertising, also make airships which carry passengers. Both civil and military uses have been found for the new airships.

Fig. 8.11 Skyship 500 carries 2 tonnes at 90 km per hour. It is only 1/100 of the size of a 1930s airship

The reason why the new airships are safe is that they use the gas helium. Although helium is denser than hydrogen, it is still much less dense than air. Helium does not burn; in fact, helium takes part in no chemical reactions whatsoever. It is one of the **noble gases**.

The noble gases are a set of elements which make up 1% by

Fig. 8.12 Neon lights

volume of air (Fig. 8.2). They are helium, neon, argon, krypton and xenon. So unreactive are these elements that for a long time people believed that they formed no compounds, and called them the **inert gases**. Argon is the most abundant of the noble gases (0.09% of air). It can be obtained by the fractional distillation of liquid air (Fig. 8.3). You met the noble gases in Chapter 4.

Neon lights are well known for their brightness and long life (Fig. 8.12). When an electric discharge is passed through neon at low pressure, it glows brightly. Argon, krypton and xenon are also used in artificial lights. The tungsten filament in an electric light bulb reaches 2500 °C in operation. If oxygen were present, it would oxidise the filament. During manufacture, air is flushed out and the bulb is filled with argon. So many uses have been found for helium (e.g. for divers as well as for airships and balloons) that the supply from air is insufficient. It is obtained from natural gas (Chaper 19).

> **SUMMARY NOTE**
>
> The noble gases take part in few chemical reactions. Helium is used to fill airships. Neon and other noble gases are used in illuminated signs.

> ### JUST TESTING 39
>
> 1. Why does an Olympic cyclist fill his tyres with helium instead of air?
> 2. The Coastguard Service has bought one of the new airships. What advantages does an airship have over a boat for coastal patrol?
> 3. What would form if a light bulb containing a tungsten filament were filled with air?

8.8 How can oxygen be prepared in the laboratory?

Hydrogen peroxide is often used as a source of oxygen. It decomposes to give oxygen and water:

hydrogen peroxide → oxygen + water
$$2H_2O_2(aq) \rightarrow O_2(g) + 2H_2O(l)$$

Solutions of hydrogen peroxide are kept in stoppered brown bottles to slow down the rate of decomposition. To speed up the

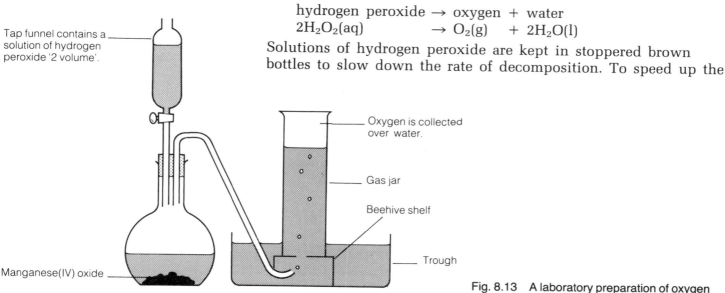

Fig. 8.13 A laboratory preparation of oxygen

reaction, you can either warm the solution or add a catalyst (see Chapter 19). The catalyst manganese(IV) oxide, MnO_2, is often used (see Experiments 8.1 and 8.5).

Test for oxygen

Oxygen is a colourless, odourless gas which is only slightly soluble in water. It is neutral. Since only one-fifth of air is oxygen, substances burn much faster in pure oxygen than in air. One test for oxygen is to lower a glowing wooden splint into the gas. If the splint starts to burn brightly, the gas is oxygen. **Oxygen relights a glowing splint** (see Experiment 8.1).

8.9 The reactions of oxygen with some elements

Nearly all elements combine with oxygen. Many elements burn in oxygen. The products of the reactions are **oxides**. An oxide is a compound of oxygen with one other element. Table 8.1 shows the results of the study you made in Experiment 8.1.

Table 8.1 The reactions of some elements with oxygen

Element	Observation	Product	Action of product on water
Copper (metal)	Does not burn; turns black	Copper(II) oxide, CuO (black solid)	Insoluble
Magnesium (metal)	Burns with a bright white flame	Magnesium oxide, MgO (white solid)	Dissolves slightly to give an alkaline solution, $pH = 9$
Iron (metal)	Burns with yellow sparks	Iron oxide, Fe_3O_4 (blue–black solid)	Insoluble
Sodium (metal)	Burns with a yellow flame	Sodium oxide, Na_2O (yellow–white solid)	Dissolves readily to form a strongly alkaline solution, $pH = 10$
Sulphur (non-metal)	Burns with a blue flame	Sulphur dioxide, SO_2, a fuming gas with a choking smell	Dissolves readily to form a strongly acidic solution, $pH = 2$
Carbon (non-metal)	Glows red	Carbon dioxide, CO_2, an invisible gas	Dissolves slightly to give a weakly acidic solution, $pH = 4$
Phosphorus (non-metal)	Burns with a yellow flame	Phosphorus(V) oxide, P_2O_5, a white solid	Dissolves to give a strongly acidic solution, $pH = 2$

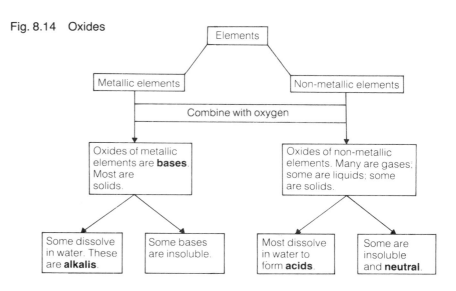

Fig. 8.14 Oxides

A pattern can be seen in the nature of oxides (see Fig. 8.14). The oxides of metallic elements are **bases**. If the bases dissolve in water, they are called **alkalis**. The oxides of non-metallic elements are mostly **acids**, but some are **neutral**. Acids react with bases to form **salts**. (See Chapter 7.)

When an element or compound combines with oxygen, we say it has been **oxidised**. Oxygen **oxidises** carbon to carbon dioxide. This reaction is an **oxidation**

$$\overset{\longleftarrow \text{Carbon is oxidised} \longrightarrow}{\underset{\longleftarrow \text{This reaction is oxidation} \longrightarrow}{\begin{array}{l} \text{carbon} + \text{oxygen} \rightarrow \text{carbon dioxide} \\ \text{C(s)} \quad + \text{O}_2\text{(g)} \quad \rightarrow \quad \text{CO}_2\text{(g)} \end{array}}}$$

The opposite of oxidation is **reduction**. When a substance loses oxygen, it is **reduced**. See Chapter 9 for more about reduction.

SUMMARY NOTE

Oxygen can be made by the catalytic decomposition of hydrogen peroxide. A test for oxygen is to see whether it relights a glowing splint. Many elements burn in oxygen. Most elements react with oxygen to form oxides. The oxides of metals are bases. The oxides of non-metallic elements are acidic or neutral.

JUST TESTING 40

1 Describe two differences in physical properties between metallic and non-metallic elements. (See Chapter 3 if you need to revise.)
  State two ways in which metallic and non-metallic elements differ in their chemical reactions. (See Chapter 6 as well as this chapter.)

2 **a** Why is magnesium used in distress flares?
  b The fireworks called 'sparklers' are coated with iron filings. Write a word equation for the reaction that takes place when you light one.

*3 Write word equations and balanced chemical equations to show what products are formed when the following elements are burned in oxygen:
  a sodium,
  b barium,
  c silicon,
  d aluminium.

> *4 The oxides of barium, lithium and radium are soluble in water.
>   a Give the formulas of the bases formed when these oxides react with water.
>   b Write balanced chemical equations for the reactions.
>   (Refer to the Periodic Table to find out the valencies of the elements.)
> *5 Find tellurium and iodine in the Periodic Table. Would you expect the aqueous solutions of their oxides to be acidic or basic or neutral?

8.10 Combustion

Compounds, as well as elements, can be oxidised. An oxidation reaction in which energy is given out is called a **combustion** reaction. If there is a flame, the combustion is described as **burning**. Substances which undergo combustion are called **fuels**.

The combustion of fuels is one of the most important reactions. We need fuels to heat our homes, to cook our food, to run our cars and to generate electricity. Many of our most important fuels, e.g. petrol, kerosene (paraffin), diesel fuel and natural gas, are petroleum oil derivatives (Chapter 16). These fuels are mixtures of **hydrocarbons**. Hydrocarbons are compounds of carbon and hydrogen only. It is important to know what products are formed when these fuels burn. Figure 8.15 shows how you can test the products of combustion of kerosene (paraffin), which is sold for use in domestic heaters. **Do not use petrol in this apparatus.** You can burn a candle instead of kerosene. Candle wax is another hydrocarbon fuel obtained from crude oil.

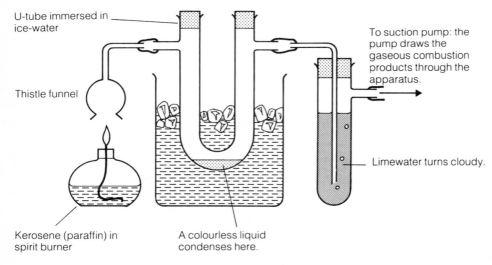

Fig. 8.15 Testing the combustion products of a hydrocarbon fuel, e.g. kerosene or candle wax (NOT petrol)

As you see from Figure 8.15, the combustion products are carbon dioxide and water. The same products are obtained with other hydrocarbon fuels:

hydrocarbon + oxygen → carbon dioxide + water vapour

Some fuels, e.g. candle wax, also leave a deposit of carbon (soot) on the inside of the thistle funnel. This proves that the air supply was insufficient to oxidise all the carbon in the hydrocarbon fuel to carbon dioxide, CO_2. There is another product of incomplete combustion which you cannot see or smell. This is the poisonous gas carbon monoxide, CO. When petrol is burned in the cylinders of a vehicle engine, the exhaust gases contain some carbon monoxide, some unburnt hydrocarbons and some soot in addition to the harmless products, carbon dioxide and water. There have been many instances of people being poisoned by carbon monoxide while running a car engine in a closed garage.

For the fire triangle and fire extinguishers, see Chapter 12.

JUST TESTING 41

1 Kerosene (called paraffin by retailers) makes a convenient fuel for use in room heaters. Why is it important that a room should be well ventilated if you are using a kerosene (paraffin) heater in it?
2 People complain of headaches when they are forced to spend time in slow-moving traffic jams. Which of the exhaust gases do you think is responsible?

8.11 Breathing and respiration

Another process which involves air is **breathing**. We **inhale** (breathe in) and **exhale** (breathe out) air to keep alive. We inhale between 15 000 and 20 000 litres of air each day. The differences between inhaled air and exhaled air are shown in Table 8.2.

Table 8.2 The composition of air in percentage by volume

Air	Oxygen	Carbon dioxide	Water vapour	Nitrogen	Noble gases
Inhaled	21%	0.03%	0%	78%	1%
Exhaled	16%	4%	1%	78%	1%

The reason for the differences is that inhaled air takes part in the biochemical reactions called **respiration**. Inhaled air dissolves in the blood supplied to the lungs. In the bloodstream, some of the oxygen in the dissolved air oxidises food materials. The oxidation products are carbon dioxide and water. They are carried by the bloodstream to the lungs. During the oxidation of foods, energy is released. Carbohydrates (sugars and starches) and fats are the main energy-providing foods. The combustion of foods in living tissues is **respiration**.

carbohydrate + oxygen → carbon dioxide + water + energy
(e.g. glucose, a sugar, $C_6H_{12}O_6$)

$C_6H_{12}O_6(aq) + 6O_2(g) \rightarrow 6CO_2(g) + 6H_2O(l) +$ energy

SUMMARY NOTE

Combustion:
Substance + oxygen react to form an oxide or oxides + energy
Burning:
Combustion, accompanied by a flame
Respiration:
Combustion which takes place in living tissues.

The energy released is used as heat energy, to keep us warm, and as kinetic energy (energy of motion) to allow us to be active.

JUST TESTING 42

1. What happens when you blow on to a cold window? Explain.
2. What do you see when you breathe out on a very cold, frosty day? Explain.
3. Describe an experiment you could do to show that the air you breathe out contains more carbon dioxide than the air you breathe in.

8.12 Rusting

Another process which involves air is **rusting** (see Experiments 8.3 and 8.4). Rust is the reddish brown solid which appears on the surface of iron and steel. Other metals also can be corroded (Section 10.3), but only iron and steel rust. Rusting is a nuisance. Millions of pounds a year are spent on replacing rusty iron and steel structures. Methods of preventing rust save money. To find methods of preventing rust, you must first find out what causes it.

Your experiments showed that air and water are needed for iron to rust. There must also be a trace of acid present; the carbon dioxide content of the air is sufficient. Of the air, 21% is used up. The remaining gas will not allow a splint to burn in it, so the part of the air that reacts with iron must be oxygen. Like other reactions studied in this chapter, rusting is an oxidation reaction:

iron + oxygen → iron(III) oxide (rust)

Methods of preventing iron from rusting are covered in Section 10.11.

JUST TESTING 43

1. a Describe experiments which you could do to find what percentage by volume of the air is used in (i) the rusting of iron, (ii) the oxidation of magnesium and (iii) respiration.
   b What results would you expect to find in each of the three experiments?

SUMMARY NOTE

- The combustion of fuels, respiration and the corrosion of metals, e.g. the rusting of iron and steel, are all oxidation reactions.
- The combustion of hydrocarbon fuels is a vital source of energy in our economy. These fuels burn to form carbon dioxide and water. If the supply of air is insufficient for complete combustion, the products carbon monoxide (a poisonous gas) and carbon (soot) are formed.
- Respiration is the source of our energy. Food materials are oxidised in the bloodstream with the release of energy.
- Rusting is the oxidation of iron and steel to iron(III) oxide.

8.13 Pollution

The problem

We are constantly exposed to air. Each day we breathe in 15 000 to 20 000 litres of air, to pass over the sensitive tissues of our lungs. The lung diseases of cancer, bronchitis and emphysema are common illnesses in regions where air is highly polluted. Through the lungs, pollutants can reach every part of our bodies. The main air pollutants are shown in Table 8.3.

Table 8.3 The main pollutants in air

Pollutant	Emission*	Source
Carbon monoxide, CO	100	Vehicle engines and industrial processes
Sulphur dioxide, SO_2	33	Combustion of fuels in power stations and factories
Hydrocarbons	32	Combustion of fuels in factories and vehicles
Dust	28	Combustion of fuels; mining; factories
Oxides of nitrogen, NO and NO_2	21	Vehicle engines and fuel combustion

*Emissions are given in millions of tonnes per year in the UK.

This is an alarming list of pollutants. We are now going to look into the following questions:
- Where do the pollutants come from?
- What harm do they do?
- What can be done to reduce them?

First, we shall look at how we manage to survive at all with all these pollutants being released into the air.

Air currents and temperature inversions

The sun beating down on the Earth keeps the surface of the Earth warm, and the Earth warms the lower atmosphere. The air in the upper atmosphere is cooler. Since warm air is less dense than cold air, warm air has a tendency to rise and cold air to sink (Fig. 8.16a). These currents of air mix the lower atmosphere. The dirty air we produce is carried upwards and dispersed quickly into the vast upper atmosphere.

Sometimes these air currents disappear for a time. Then the air

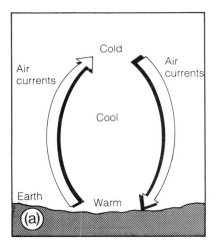
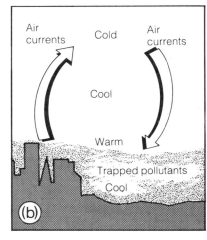

Fig. 8.16 (a) Air currents; (b) a temperature inversion

in the higher regions of the atmosphere, warmed by the sun, becomes warmer than air near the ground. There is a **temperature inversion** (Fig. 8.16b). The cooler air near the ground is the denser, and stays put. Pollutants poured into a stagnant layer do not become dispersed. They accumulate near ground level, mainly over cities. Temperature inversions force us to breathe our own pollutants, and sometimes these make us ill. Temperature inversions are most likely to occur in places which have a hot climate and still air.

> **SUMMARY NOTE**
>
> Pollutants are sent into the air mainly by motor vehicles, by industry and by electric power stations. Fortunately, they are carried upwards by rising currents of warm air. A **temperature inversion** stops the dispersal of pollutants.

Carbon monoxide

Worldwide, the emission of carbon monoxide is 350 million tonnes a year. The chief source is the internal combustion engine. As you read in Section 8.10, if the combustion of petrol is incomplete, the exhaust gases contain carbon monoxide and carbon as well as the harmless combustion products, carbon dioxide and water. Incomplete combustion occurs when there is insufficient oxygen for full oxidation of the hydrocarbons in petrol. Car designers find that maximum performance is obtained when fuel-rich mixtures are burned in the cylinders. This design leads to the discharge of carbon monoxide, carbon and unburnt hydrocarbons.

What harm does carbon monoxide do? Carbon monoxide combines with haemoglobin. Haemoglobin is the red susbtance in blood which transports oxygen round the body. It combines with oxygen in the lungs, carries oxygen as oxyhaemoglobin round the body and releases oxygen to the tissues.

In the lungs:

haemoglobin + oxygen → oxyhaemoglobin
$Hb(aq)$ + $O_2(g)$ → $O_2Hb(aq)$

In the tissues:

oxyhaemoglobin → oxygen + haemoglobin
$O_2Hb(aq)$ → $O_2(aq)$ + $Hb(aq)$

The freed haemoglobin then circulates back to the lungs to pick up more oxygen. If carbon monoxide is present in the lungs, it combines with haemoglobin:

carbon monoxide + haemoglobin → carboxyhaemoglobin
$CO(g)$ + $Hb(aq)$ → $COHb(aq)$

Carbon monoxide is 200 times better at combining with haemoglobin than oxygen is. It therefore ties up haemoglobin so that haemoglobin can no longer combine with oxygen. At a level of 1%, carbon monoxide will kill quickly; at lower levels, it causes headaches and dizziness and affects reaction times.

Being colourless and odourless, carbon monoxide gives no warning of its presence. Measurements have shown fairly high levels of carbon monoxide in crowded streets during rush hours. Since carbon monoxide is concentrated at street level, it is most likely to affect people when they are driving and need sound judgment and quick reflexes.

SUMMARY NOTE

Carbon monoxide is produced when hydrocarbon fuels burn incompletely in an inadequate supply of air. It is a poison which combines with haemoglobin in the blood, thus starving the body of oxygen. Catalytic converters can be fitted in the exhaust pipes of cars to bring about the oxidation of carbon monoxide to the harmless product, carbon dioxide.

What can be done? Nature can deal with carbon monoxide. Soil organisms can convert it into carbon dioxide or methane. In city streets, where the concentration of carbon monoxide is high, there is little soil to remove it.

People are working on a number of solutions to the problem.

- The tuning of vehicle engines can be altered to provide more air and produce only carbon dioxide and water. A snag is that this increases the formation of oxides of nitrogen (see Section 13.6).
- Many cars are now fitted with **catalytic converters**. The hot exhaust gases pass over a catalyst, which helps to oxidise carbon monoxide to carbon dioxide (see Acid Rain). Catalytic converters only work when the petrol contains no lead compounds.
- In the future, vehicles may be powered by fuels which burn more cleanly than hydrocarbons (see Section 13.9).

Sulphur dioxide

Where does sulphur dioxide come from? Almost all the sulphur dioxide in the air comes from industrial sources. Industrial smelters, which obtain metals from sulphide ores, produce tonnes of sulphur dioxide daily. Half of the output of sulphur dioxide comes from the burning of coal. All coal contains between 0.5 and 5 per cent sulphur.

sulphur in coal + oxygen in air → sulphur dioxide
$$S(s) + O_2(g) \rightarrow SO_2(g)$$

The sulphur dioxide escapes up chimneys into the atmosphere. Worldwide, 150 million tonnes a year are emitted. The emission is growing as countries become more industrialised. Most of the coal is burned in power stations. These are always situated in densely populated regions where the demand for electricity is high and where there are plenty of people to breathe in the sulphur dioxide (see Fig. 8.17).

Fig. 8.17 Sulphur dioxide emission

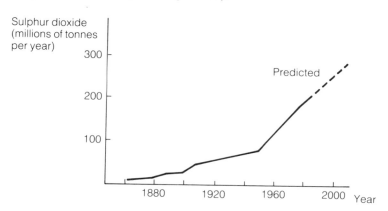

What harm does sulphur dioxide do? Sulphur dioxide is a colourless gas with a very penetrating and irritating smell. It is poisonous; at a level of 0.5%, it will kill. The immediate effects of inhaling sulphur dioxide are coughing, chest pains and shortness of breath. Sulphur dioxide is thought to contribute to bronchitis and lung diseases.

What can be done about it?
- Tall chimneys carry sulphur dioxide away from factories (Fig. 8.18). Unfortunately it comes down to earth again as **acid rain**.
- A number of processes for removing sulphur dioxide from the emission of power plants are being tried (see below).

Fig. 8.18 The chimneys of a power station

SUMMARY NOTE

Factories and power stations burn coal, which contains sulphur. They send sulphur dioxide into the air. Metal smelters oxidise sulphides to sulphur dioxide. Sulphur dioxide causes bronchitis and lung diseases. In the upper atmosphere, it reacts with water to form **acid rain**.

Acid rain

Tall chimneys carry away the sulphur dioxide and other pollutant gases. They are not allowed to pollute the air breathed by the factory workforce or the local residents.

Unfortunately, the acid gases do not disappear. In the air, sulphur dioxide meets water vapour, and reacts to form sulphurous acid, H_2SO_3. This is slowly oxidised by oxygen in the air to sulphuric acid, H_2SO_4. When it rains or snows, down comes the sulphuric acid as acid rain or acid snow. Oxides of nitrogen are converted into nitric acid, HNO_3 (see Fig. 8.19).

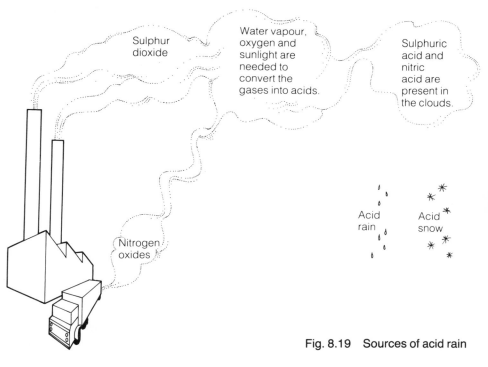

Fig. 8.19 Sources of acid rain

SUMMARY NOTE

Sulphur dioxide and oxides of nitrogen are present in the exhaust gases of coal- and oil-fired power stations. In the atmosphere, these gases react with water vapour to form acid rain.

Rain is naturally weakly acidic. Carbon dioxide from the air dissolves in it to form the weak acid, carbonic acid, H_2CO_3. Sulphuric acid is a thousand times stronger than carbonic acid (see Fig. 8.20).

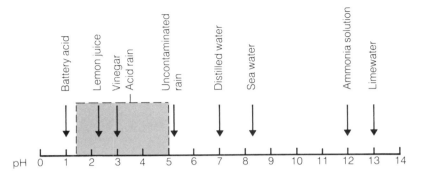

Fig. 8.20 The pH values of rain, acid rain and other solutions (for pH, see Chapter 7)

What harm does acid rain do? The main damage so far has been to lakes and fish, to trees and to building materials such as stone, concrete, metal and cement.

Fish cannot survive in acidic lake water. Thousands of lakes in Canada, Norway and Sweden are now empty of fish. These countries produce very little of the acid rain and snow that falls on them. The pollution is carried to Scandinavia by the prevailing winds from the UK and from Germany, France and other highly industrialised European countries. Canada receives much of its pollution from the USA.

Some countries suffer more than others from attack by acid rain. If the soil contains basic substances, these will partially neutralise the rain as it trickles through the soil. Cold countries like Sweden receive acid snow for many months. During the spring thaw, a huge volume of acidic water is released into the lakes. The soil has no chance to neutralise it. The acidic lake water destroys water plants. As a result, light can penetrate further through the water. Acid lakes look beautifully clear and blue.

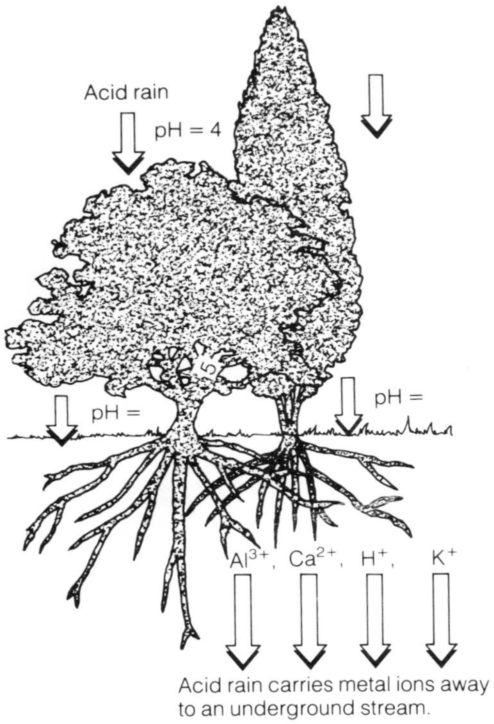

The effect of acid rain on trees is shown in Fig. 8.21. It dissolves minerals from the soil, removing the ions Ca^{2+}, K^+, Al^{3+} and Pb^{2+}. Some of these ions are nutrients, which should be absorbed by the roots of the trees. Acid rain deprives the trees of these nutrients.

Building materials suffer from attack by acid rain. Concrete and stone react with acids. Many stone buildings are made of limestone, calcium carbonate. You saw how this reacts with acids in Section 7.2. Cement and concrete contain calcium hydroxide, which reacts with carbon dioxide in the air to form calcium carbonate. So when cement and concrete 'weather', calcium carbonate is formed (see Experiment 12.2). Acid rain attacks all these substances.

Metals are attacked by acids. Iron railings, steel window-frames, motor vehicles and bridges are all corroded faster when rain water is acidic.

What can be done about it? All this damage costs money. Replacing rusty motor vehicles costs money. Restoring stonework

Fig. 8.21 Acid rain falls on trees

> SUMMARY NOTE

Acid rain makes lakes too acidic for fish to live in. It removes minerals from the soil. Trees are deprived of essential minerals and die. Acid rain attacks construction materials such as limestone, cement, concrete and metals.

costs money. The Canadian timber industry is suffering as the growth of trees is affected. Tourism is suffering in Canada and Sweden: tourists do not want to fish in 'dead' lakes. Sweden has adopted the practice of spraying calcium oxide (quicklime) into acid lakes. Being a base, calcium oxide neutralises acids:

calcium oxide + sulphuric acid → calcium sulphate + water
$CaO(s)$ + $H_2SO_4(aq)$ → $CaSO_4(aq)$ + $H_2O(l)$

American companies have started neutralising sulphur dioxide at its source. They use jets of wet lime (calcium oxide) to bombard the combustion gases before they leave the chimney. This process removes sulphur dioxide, nitrogen dioxide and other acid gases from the exhaust gases. Cleaning the exhaust gases in this way costs money. The product is calcium sulphate. Some of this by-product can be used as hardcore in road-making. The rest must be removed and dumped. Another method of removing sulphur dioxide is to 'scrub' the exhaust gases with ammonia. The end product of this process is ammonium sulphate. It can be sold in large quantities as a fertiliser.

In 1979, 31 European countries signed the Convention on Long Range Transboundary Air Pollution. In this agreement, each nation agreed to reduce the amount of pollution that it allows to sweep across its frontiers. When the nations put this agreement into practice, Norway and Sweden will no longer be damaged by pollutants from central European countries and the UK. Canada and the USA have signed a similar agreement.

Fig. 8.22 Stonework attacked by acid rain

JUST TESTING 44

1. Why do lakes in a limestone region become less acidic as a result of acid rain than lakes in a granite region?
2. Tall chimneys carry sulphur dioxide away from factories. Explain why this is not a final solution to the problem of pollution.
3. What is formed when marble statues are attacked by air containing sulphur dioxide as well as the gases normally present in air?
4. A thousand Swedish schoolchildren wrote to the Government of West Germany to complain about the pollution of lakes in Sweden. Why did the children think West Germany could help?
5.
   - A smelter emits 500 tonnes of sulphur dioxide daily. This can produce 800 tonnes of sulphuric acid a day.
   - To make rain acidic (pH = 4) needs 0.01 g of sulphuric acid per kg of water.

   a How many tonnes of rain can be made acidic by this plant in a year? (Round off your answer to give an approximate figure. 1 tonne = 1000 kg.)

   b A yearly rainfall brings 2 million tonnes of rain water to each square mile of the Earth's surface. How many square miles of the Earth's surface would the acid rain in **a** cover?

Note See Exercise 11 for calculations on pollution.

Fig. 8.23 Cashing in on the acid rain scare

SUMMARY NOTE

The damage caused by acid rain is costly. European countries have agreed to stop exporting their pollutants. Industries and power stations are trying methods of reducing their emission of acidic gases.

Particles: smoke, dust and grit

The atmosphere contains millions of tonnes of tiny solid and liquid particles. **Particles** are droplets of liquid and small bits of solids. Dust storms, forest fires and volcanic eruptions send particles into the air. Man-made sources of dust and smoke are mining, land-clearing, coal-burning power stations, industries and vehicles. These sources are increasing.

What harm do particles do? Particles make our environment dirty. They scatter light and give air a 'smoggy' appearance. Grime falls on to clothing, buildings and plants.

Particles make our air unhealthy. They increase the danger of **smog**. Smoke consists of particles. Fog consists of water droplets. When smoke and fog mix, the fog prevents the smoke from blowing away. **Smog** is formed (Fig. 8.24). The sulphur dioxide in the smoke reacts with water droplets in the fog to form sulphurous acid and sulphuric acid. When we breathe smog, the acid irritates our lungs and makes them produce mucus. This makes it more difficult to breathe, and can cause the death of people who already suffer from a respiratory disease. In the smog which hit London in 1952 and lasted for 5 days, 5000 people died. In 1956, there was another fatal smog, and in that year Parliament passed the **Clean Air Act**. This created **smokeless zones** in which only low-smoke and low-sulphur fuels can be burned.

There is another effect of dust in the atmosphere: it can affect the climate (see Chapter 12).

Fig. 8.24 Smog

How can particles be removed? Some industries use sprays of water to wash out particles from their waste gases. Others circulate air through filters. Electrostatic precipitators are used to remove small particles. They depend on the fact that most particles carry a small amount of electrostatic charge (Fig. 8.25).

The exhaust gases of vehicles are not treated by any of these methods.

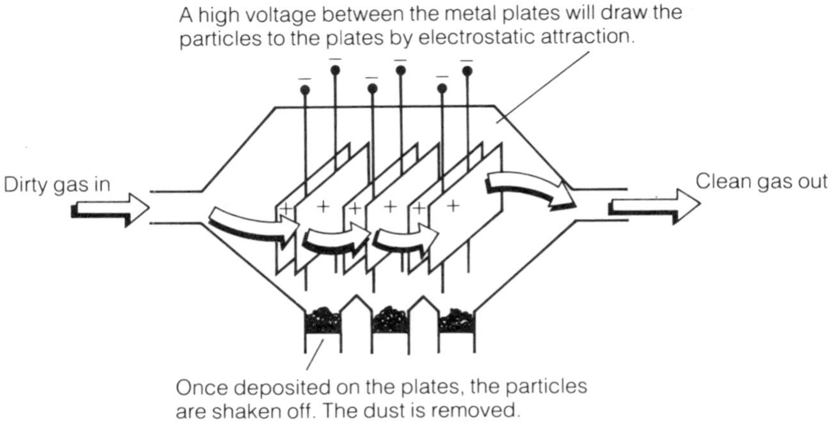

A high voltage between the metal plates will draw the particles to the plates by electrostatic attraction.

Dirty gas in

Clean gas out

Once deposited on the plates, the particles are shaken off. The dust is removed.

Fig. 8.25 An electrostatic precipitator

SUMMARY NOTE

Factories and power stations send smoke, dust and grit into the air. This dirt damages buildings and plants. It pollutes the air we breathe. When it mixes with fog, **smog** is formed.

Oxides of nitrogen

When fuels are burned in the oxygen of the air, nitrogen is also present. At the high temperatures that result, some nitrogen combines with oxygen:

nitrogen + oxygen → nitrogen monoxide
$N_2(g) + O_2(g) \rightarrow 2NO(g)$

nitrogen monoxide + oxygen → nitrogen dioxide
$2NO(g) + O_2(g) \rightarrow 2NO_2(g)$

In consequence, the gases nitrogen monoxide, NO, and nitrogen dioxide, NO_2, are released from the chimneys of power stations and factories and in the exhaust gases from motor vehicles. This mixture of gases is sometimes shown as NO_x.

What harm do they do? Nitrogen monoxide, NO, is only moderately toxic. It is soon converted into nitrogen dioxide in air. Nitrogen dioxide, NO_2, is highly corrosive and toxic. It irritates the breathing passages. It reacts with oxygen and water to form nitric acid, an ingredient of **acid rain**.

What can be done? One hope of reducing the quantity of nitrogen monoxide in exhaust gases is to bring about the reaction

nitrogen monoxide + carbon monoxide → nitrogen + carbon dioxide
$2NO(g) + 2CO(g) \rightarrow N_2(g) + 2CO_2(g)$

The reaction takes place in the presence of a catalyst. **Catalytic converters** are now being fitted in the exhausts of some American cars to reduce the emission of oxides of nitrogen. If lead compounds are present in the exhaust gases, they stop the catalyst working.

> ### SUMMARY NOTE
>
> Oxides of nitrogen, NO and NO_2 (NO_x), are formed in the engines of motor vehicles. They react with air and water to form the strong acid, nitric acid. This strongly irritates the respiratory passages. Catalytic converters can be fitted in the exhaust pipes of cars to convert nitrogen oxides into nitrogen.

JUST TESTING 45

1. In petrol engines, hydrocarbons are burned to form a number of products. What are these products? How does the ratio of carbon dioxide to carbon monoxide in the exhaust gases change as the engine temperature increases?

    Nitrogen combines with oxygen in the combustion chamber. What is formed? How does the proportion of these gases in the exhaust gas alter as the engine temperature increases?

Hydrocarbons

Where do they come from? There are natural sources of hydrocarbons. Methane, CH_4, is produced when dead plant material decays in the absence of air. Although only 15 per cent of the hydrocarbons in the air come from our activities, they affect human health because they are concentrated in city air.

What harm do they do? Hydrocarbons by themselves cause little damage. In sunlight, they react with the oxygen and oxides of nitrogen in the air. As a result of these **photochemical** reactions

(reactions which are brought about by light), irritating and toxic compounds are formed. The irritating smog of Los Angeles in the USA is mostly of photochemical origin. It is known as **photochemical smog**.

What can be done? The hydrocarbons emitted by petrol engines can be reduced by increasing the oxygen supply so as to burn the petrol completely. There is a snag. The problem is this:

increase oxygen supply → decrease in hydrocarbons and carbon monoxide in exhaust gas
→ increase in oxides of nitrogen in exhaust gas

Chemists are doing research on the problem. The solution may be to run engines at a lower temperature (in order to reduce the combination of oxygen and nitrogen) and employ a catalyst (to ensure the complete combustion of hydrocarbons at the lower temperature).

Other pollutants

Heavy metals and their compounds are serious air pollutants. **Heavy** metals are metals with a density greater than 5 g/cm^3. We need small quantities of some heavy metals, e.g. copper and zinc, in our food. Others, e.g. lead and mercury, are toxic.

Mercury. Air contains a small concentration of mercury vapour from natural sources. We add greatly to this by earth-moving activities, such as mining and road-making. By disturbing soil and rock, we allow the mercury which they contain to escape into the air. Many metal ores contain mercury compounds, which are released during the winning of the metals from their ores. The combustion of coal and oil also sends mercury vapour into the air. Both mercury and its compounds cause kidney damage, nerve damage and death.

Lead. Lead is not naturally present in air. Lead and its compounds are discharged by the combustion of coal, by the roasting of metal ores and by vehicle engines (Fig. 8.26). Unlike the other pollutants in exhaust gases, lead compounds have been purposely added to the fuel. Tetraethyl lead, TEL, is added to improve the performance of the engine. City dwellers take in lead from many sources. Many people have blood levels of lead which are nearly high enough to produce the symptoms of lead poisoning. Mild lead poisoning symptoms are depression, tiredness, irritability and headache. Higher levels of lead cause damage to the brain, liver and kidneys.

There is a cure for this type of pollution: the elimination of lead additives to petrol. The UK is committed to eliminating lead from petrol by 1990. The USA is already using lead-free petrol. The catalytic converters now fitted to American cars are 'poisoned' by lead compounds.

> **SUMMARY NOTE**
>
> Hydrocarbons from vehicle exhausts are an ingredient of **photochemical smog**. A solution to the problem of hydrocarbons may be found by running petrol engines at a lower temperature with a catalyst.

Fig. 8.26 Pedestrians breathe exhaust fumes

> **SUMMARY NOTE**
>
> Heavy metals are serious air pollutants. Levels of mercury and lead and their compounds in the air are increasing.

The price of clean air

Of the pollutants mentioned in this chapter, carbon monoxide, oxides of nitrogen, hydrocarbons and lead compounds all come from motor vehicles. The UK, like many other countries, has decided to stop adding TEL (tetraethyl lead) to petrol. The addition of TEL turns a low octane fuel into a high octane (four-star) fuel on which engines run smoothly. When TEL is banned in 1990, we shall need to use the more expensive high octane fuels. This will increase the cost of motoring. Research chemists in the petrochemical industry are trying to find other ways of managing without TEL. They are working on methods of turning low octane fuels into high octane fuels.

When British cars run on lead-free petrol, they will be able to use catalytic converters. These will turn oxides of nitrogen, carbon monoxide and hydrocarbons in the exhaust gas into harmless products. The cost of fitting converters will increase the price of cars. The use of catalytic converters increases the petrol consumption of cars. Pollution control will therefore increase both the purchase price and the running cost of a car. Many people believe that the extra cost will be worthwhile if it improves the quality of the air we breathe. Other people resent the expense. They prefer not to take atmospheric pollution seriously.

Sulphur dioxide and oxides of nitrogen are sent into the air by power stations, metal smelters and factories. Chemists have the means to remove them from the mixtures of exhaust gases produced, but this costs money. Building cleaning systems into the chimneys costs money. The chemicals consumed every day in running a cleaning system cost money. The result is an increase in the price of the electricity or metal or article which is produced. British manufacturers do not want to put up their prices. They are afraid that their competitors in other countries will have an advantage over them.

JUST TESTING 46

1 Why is carbon monoxide such a dangerous pollutant? How does nature remove it from the air?

2 Why are the oxides of sulphur so obnoxious as air pollutants? How does nature remove them?

3 Why are particulate pollutants dangerous, even though they may be chemically unreactive solids?

4 How do lead and its compounds get into the air? What damage do they cause? What is the solution to this problem?

5 Imagine you are a British manufacturer. You run a factory which sends tonnes of sulphur dioxide into the air every day. After reading some books on pollution, you decide to do your bit towards clean air. You look into the cost of installing and running a clean-up system. You calculate how much you will have to raise the price of your products to cover the cost.

> Then the factory foreman points out that your chief competitor is doing nothing about pollution control. The foreman is afraid that you will price your product out of the market. You could go bankrupt. Then the workforce will be out of their jobs.
>   **a** Write a letter to your competitor, explaining to him why sulphur dioxide is a pollutant. Try to persuade him to cut down the release of sulphur dioxide from his factory if you do the same.
>   **b** Write a letter to your Member of Parliament. Tell him or her that you think Parliament should pass a law to make all manufacturers cut pollution by the same percentage. Explain why you think this would be a good idea.
>   **c** It is possible that you do not agree with **a** and **b**. In this case, explain what your views are.
> **6** A question for discussion:
> As pollution control devices are added to cars, the number of miles we can get from a litre of petrol decreases. This means that we have to burn more petrol to travel the same number of miles. In this way, we produce more pollutants which have to be removed. Does this make pollution control self-defeating?

Exercise 8

1 The boiling points of the main components of air are nitrogen, −196 °C, and oxygen, −183 °C.
  a Which of these two elements is the easier to liquefy?
  b Describe briefly how air is liquefied.
  c Describe how oxygen and nitrogen are obtained from liquid air.
  d Give one large-scale industrial use for each of these gases.
  e Name another gas which is obtained from liquid air, and say what it is used for.

2 A sample of air is passed slowly through sodium hydroxide solution and then over heated copper.
  a Which gas is removed by sodium hydroxide? What product is formed?
  b Which gas is removed by copper? What product is formed?
  c If 200 cm^3 of air are treated in this way, what volume of gas will remain?

3 Why is carbon monoxide such a dangerous pollutant? How does it get into the air? How does nature remove it from the air?

4 When air is bubbled slowly through pure water, the pH of the water changes gradually from 7.0 to 5.7.
  a Which of the gases present in air is responsible for the change?
  b Why is the presence of this gas in air so important?

5 Why are the oxides of sulphur so unpleasant as air pollutants? How does nature remove them?

6 (i) Say whether the oxides of the following elements are acidic or basic or neutral:
     a iron, **b** sulphur, **c** carbon, **d** calcium, **e** hydrogen.
   (ii) Describe the appearance of the flame when the following burn in oxygen:
     a magnesium, **b** iron, **c** sulphur, **d** kerosene, **e** carbon.

7 A woman drives a round trip of 40 miles to work and home again on 250 days a year. Her car does 40 miles to the gallon, and the petrol contains 2 g of lead per gallon.
  a How much lead does this driver use in a year?
  b What happens to the lead?
  c What do you think should be done to reduce this source of lead in the environment?

8 The average mercury content of coal is 0.2 parts per million. Assuming all the mercury escapes into the air through the chimney of the power plant, calculate the mass of mercury that will be released each year by a large power plant burning 10 000 tonnes of coal per day.

9 Rush-hour traffic into and out of cities is much greater than the traffic during the rest of the day. To cope with rush-hour traffic, extra lanes must be built in the roads. Extra road maintenance is needed. Extra police are needed. The slow-moving rush-hour traffic produces massive pollution.

Bus passengers produce much less pollution per head.

Imagine that it is your job to economise on road maintenance and to cut down air pollution. What steps would you take to discourage motorists from driving and encourage them to use public transport?

10 a A power plant consumes 10 000 tonnes of coal per day. The coal contains 3% sulphur. (This is high-sulphur bituminous coal.) What mass of sulphur would be sent into the air each day in the absence of controls?
   b If there were a tax on pollution, power plants would be encouraged to discharge less sulphur into the air. At a tax rate of 10p/tonne of sulphur, what would this power station have to pay?
   c What tax rate do you think would be high enough to make a power station remove the sulphur from its waste gases?
   d How could the power station recoup some of the expense?

11 Lead compounds are accumulating in the Arctic, far away from the cities which produce them. Figure 8.27 shows how the level of lead in Arctic ice has increased this century.

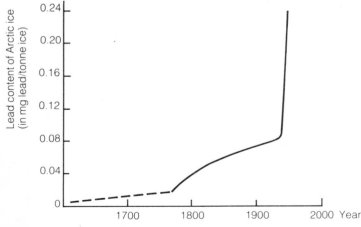

Fig. 8.27 Lead content of Arctic ice

a What social change happened around the year 1750 and had the effect of increasing atmospheric lead?
b Deduce from the graph when people started adding tetraethyl lead to petrol. Why did this petrol additive cause a steep increase in the level of lead in the atmosphere? How did it cause an increase in the level of lead in Arctic ice? What advantage does the treated petrol have? Why have steps been taken to stop adding TEL to petrol?

12 A mixture of gases is produced by the combustion of petrol in a car engine. The composition of the mixture depends on the ratio of petrol vapour to air in the cylinders. Figure 8.28 shows how the relative amounts of nitrogen monoxide, hydrocarbons and carbon monoxide alter as the fuel-to-air ratio changes from a rich mixture to a lean mixture.
   a Explain why the emission of nitrogen monoxide increases from left to right on the graph.
   b Explain why the emission of hydrocarbons decreases from left to right on the graph.
   c Explain why the emission of carbon monoxide decreases from left to right.
   d What other substances are present in the exhaust gas?

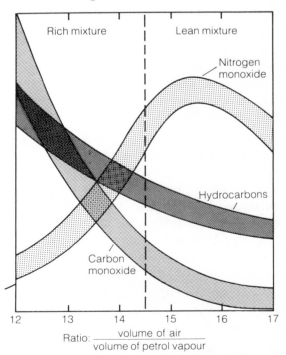

Fig. 8.28 Relative emissions of exhaust gases in rich and lean mixtures

CHAPTER 9 Water

9.1 We need it

Three out of five people in Third World countries have no easy access to clean water. Women and children often spend much of their day collecting water from rivers and wells (Fig. 9.1).

Often, rivers are used for sewage disposal as well as for drinking. Many diseases are water-borne (Fig. 9.2). Four-fifths of the diseases in the Third World are linked to dirty water and lack of sanitation.

The United Nations have called the 1980s the Water Decade. The UN target is clean water and sanitation for all by 1990. International aid agencies (UN, Oxfam, the World Bank and others) are giving funds for pumps and irrigation channels (to provide water for agriculture) and for disinfection kits (to provide water for drinking).

Fig. 9.1 She fills her bucket from the pump

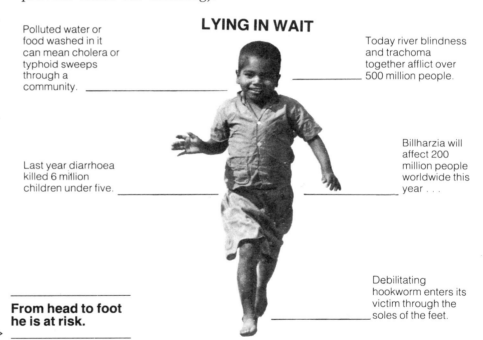

Fig. 9.2 This Oxfam poster shows where water-borne diseases attack

9.2 Why is water so important?

Life began in water. The first plants and animals were aquatic (water-living). The chemical reactions that gave them life must have been reactions that could take place in water. As more complex terrestrial (land-living) creatures evolved, water kept its importance. Aqueous (in water) solutions fill the cells of the human body. Food materials and oxygen are transported round the body in an aqueous solution — blood. Digestion takes place in an aqueous solution. The body of an adult contains about 40 litres of water. In plants too, nutrients are transported round the structure in aqueous solution.

SUMMARY NOTE

Many people in the world do not have safe drinking water. Water is essential for all life processes.

Water carries away our wastes: our sewage, our industrial grime, the waste chemicals from our factories and the waste heat from our power stations. River water decomposes some of these wastes, dilutes others until they are harmless and carries the rest out of sight, away to the sea. (See Section 9.4.)

Fig. 9.3 Water for relaxation

Fig. 9.4 Water: an eyesore and a health hazard

9.3 The water cycle

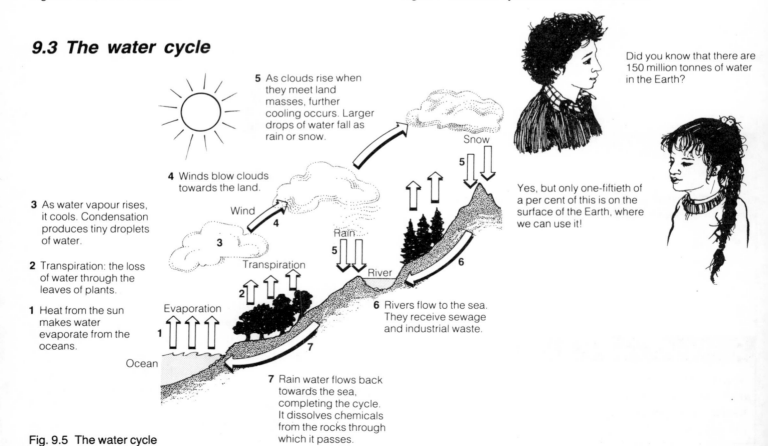

1 Heat from the sun makes water evaporate from the oceans.
2 Transpiration: the loss of water through the leaves of plants.
3 As water vapour rises, it cools. Condensation produces tiny droplets of water.
4 Winds blow clouds towards the land.
5 As clouds rise when they meet land masses, further cooling occurs. Larger drops of water fall as rain or snow.
6 Rivers flow to the sea. They receive sewage and industrial waste.
7 Rain water flows back towards the sea, completing the cycle. It dissolves chemicals from the rocks through which it passes.

Fig. 9.5 The water cycle

Did you know that there are 150 million tonnes of water in the Earth?

Yes, but only one-fiftieth of a per cent of this is on the surface of the Earth, where we can use it!

One litre of sea water contains:
- 26 g of sodium chloride (common salt)
- 7 g of magnesium sulphate (Epsom salts)
- 2 g of calcium sulphate (related to plaster of Paris)
- 1 g of potassium chloride
- 30×10^{-9} g of gold chloride.

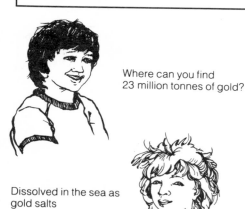

Where can you find 23 million tonnes of gold?

Dissolved in the sea as gold salts

Four-fifths of the world's surface is covered by water. The heat of the sun makes water evaporate from oceans, rivers and lakes. As it rises, water vapour cools and condenses to form clouds of tiny droplets. Further cooling of the clouds produces larger drops of water which fall to the ground as rain or, if it is cold enough, as snow. Rain water trickles through soil and porous rocks until eventually it finds its way back to the sea. Then the cycle begins all over again (Fig. 9.5).

As rain water falls through the air, it dissolves oxygen, nitrogen and carbon dioxide. When carbon dioxide dissolves, it forms a solution of the weak acid, carbonic acid. As rain water trickles through porous rocks, it dissolves chemicals from the rocks. The salts which rain water dissolves find their way to the sea. When water evaporates from the sea, the dissolved salts remain (Fig. 2.5).

In industrial regions, the air contains pollutant gases. These dissolve in rain water to make **acid rain** (see Section 8.13).

9.4 Dissolved oxygen

The oxygen that dissolves in water has a vital role to play. Oxygen is not very soluble: water is saturated with oxygen when the concentration of dissolved oxygen is only 10 p.p.m. (parts per million; 10^{-3}%). Water-living creatures and plants and all fish depend on this dissolved oxygen. When the level of dissolved oxygen falls below 5 p.p.m., aquatic plants and animals start to suffer.

Water has a natural ability to purify itself when it becomes contaminated. This ability depends on the dissolved oxygen and the bacteria present in water. When organic matter (plant and animal debris) gets into the water, **aerobic** bacteria (bacteria which need oxygen) use this matter as food. They use dissolved oxygen to oxidise debris to harmless products, such as carbon dioxide and water, with the release of energy:

$$\text{oxidation reactions inside bacteria}$$
organic pollutants + oxygen → carbon dioxide + water + energy
(plant and animal debris) (dissolved) (harmless products) (needed by bacteria)

If water is to remain life-supporting, the oxygen which the bacteria use must be replaced by the dissolving of more oxygen. Sometimes rivers are overburdened with organic debris, for example, when untreated sewage is discharged into them. When this happens the dissolved oxygen is used up more rapidly than it is replaced, and the aerobic bacteria die. Then **anaerobic** bacteria (bacteria which do not need oxygen) take over. They attack the organic matter to produce unpleasant decay products, such as ammonia, hydrogen sulphide and methane.

For some synthetic materials, e.g. plastics, there are no bacteria able to oxidise them. Such materials are **non-biodegradable**, and they persist for a very long time in water.

SUMMARY NOTE

There is a natural **water cycle**. Water vapour enters the atmosphere by **evaporation**. Water vapour leaves the atmosphere by **condensation**, followed by precipitation as rain or snow. Air dissolves in rain water. The oxygen that dissolves in water is vital to fish. If too much organic matter, e.g. sewage, is discharged into a river, the dissolved oxygen is used up in oxidising the organic matter, and the water becomes unable to support life.

9.5 Water treatment

The water that we use is taken mainly from lakes and rivers. It is purified to make it safe to drink. Water treatment plants use filtration and chlorination (see Fig. 9.6).

This safe drinking water is very different from the disease-carrying water which killed thousands of people in the early nineteenth century. Chlorination of drinking water is the biggest single factor that has made the difference. It is time that people in underdeveloped countries had safe water too. (For chlorine, see Section 15.7; for fluoridation of drinking water, see Section 15.11.)

In some areas, water is obtained from underground sources. As water trickles down from the surface through porous rocks, suspended matter is left behind. This **ground water** does not need the full treatment shown in Fig. 9.6. It is pumped out of the ground and chlorinated before use.

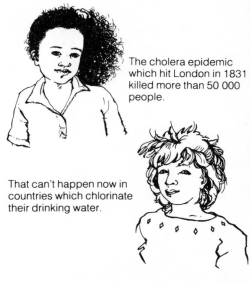

The cholera epidemic which hit London in 1831 killed more than 50 000 people.

That can't happen now in countries which chlorinate their drinking water.

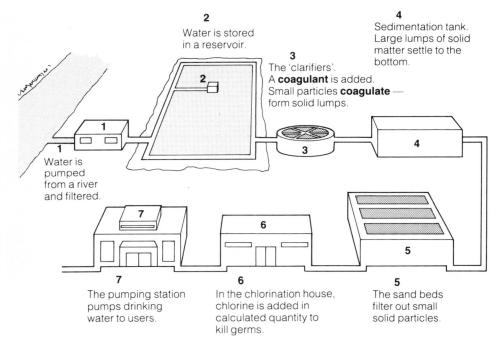

2 Water is stored in a reservoir.

3 The 'clarifiers'. A **coagulant** is added. Small particles **coagulate** — form solid lumps.

4 Sedimentation tank. Large lumps of solid matter settle to the bottom.

1 Water is pumped from a river and filtered.

7 The pumping station pumps drinking water to users.

6 In the chlorination house, chlorine is added in calculated quantity to kill germs.

5 The sand beds filter out small solid particles.

Fig. 9.6 A water treatment works

9.6 Sewage works

After use, water flows into sewers which take it to a sewage works. There, it is purified sufficiently to be discharged into a river (see Fig. 9.7).

Not all sewage is treated in this way. A great deal of raw sewage is still discharged into rivers and into the sea (see Section 9.16). The reason is the cost of building new sewage works. As the population increases, some Water Authorities do not have enough equipment to treat all their area's sewage. They treat as much as they can and hope that rivers and the sea will break down the rest. With the volume of sewage increasing, this solution to the problem is becoming less and less satisfactory.

Fig. 9.7 A sewage works ▷

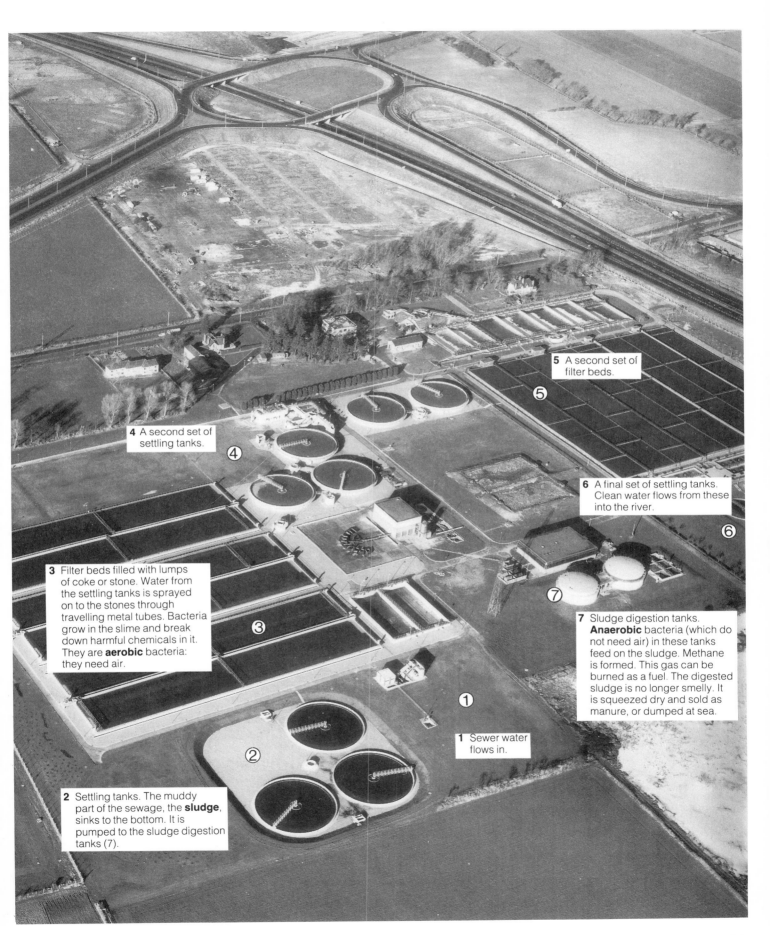

9.7 What do we use it for?

The uses shown in Fig. 9.8 represent only 10% of our total use of water. The other 90% is used:

- in agriculture to grow food
- by manufacturing industries (used as a solvent, for cleaning and for cooling)
- by power stations to generate electricity (used for cleaning and for cooling).

In industrialised countries, the total water consumption amounts to around 80 000 litres (80 tonnes) a year per person. The manufacture of

- 1 tonne of steel uses 75 tonnes of water
- 1 tonne of paper uses 2500 tonnes of water
- 1 motor car uses 45 tonnes of water.

This water is used, purified and recycled.

Fig. 9.8 Water: 160 litres a day

9.8 Chemically speaking

What is water, chemically speaking? You can split up water by passing a direct electric current through it; that is, by **electrolysis** (see Section 5.9). The gases hydrogen and oxygen are the only products you obtain. The volume of hydrogen is twice that of oxygen. This shows (see Section 11.3) that the formula is H_2O:

$$\text{water} \xrightarrow{\text{electrolyse}} \text{hydrogen} + \text{oxygen}$$
$$2H_2O(l) \longrightarrow 2H_2(g) + O_2(g)$$

Figure 9.9 shows what happens when you burn hydrogen in air and collect the products of combustion. The only product

> **SUMMARY NOTE**
> - Water treatment works make lake and river water fit to drink. They employ filtration followed by chlorination.
> - Sewage works make used water fit to be emptied into rivers or the sea. They use sedimentation followed by air-oxidation.
> - Water consumption in an industrial country totals about 80 000 tonnes per person per year.

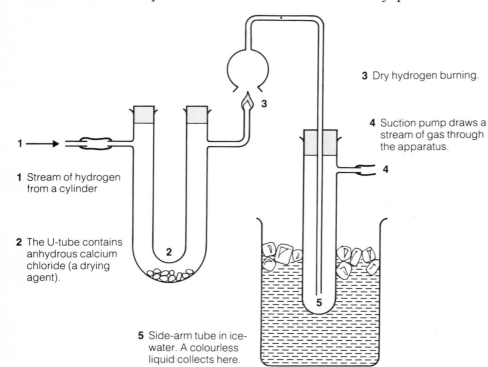

1 Stream of hydrogen from a cylinder

2 The U-tube contains anhydrous calcium chloride (a drying agent).

3 Dry hydrogen burning.

4 Suction pump draws a stream of gas through the apparatus.

5 Side-arm tube in ice-water. A colourless liquid collects here.

Fig. 9.9 What is formed when hydrogen burns in air?

formed when hydrogen burns is a colourless, odourless liquid. You can test it to see whether it is water.

Tests for water

There are two quick tests that will show whether a liquid *contains* water.

Test 1 Water will turn white anhydrous copper(II) sulphate blue:

$$\text{copper(II) sulphate} + \text{water} \rightarrow \text{copper(II) sulphate-5-water}$$
$$\text{CuSO}_4(s) + 5\text{H}_2\text{O}(l) \rightarrow \text{CuSO}_4\cdot 5\text{H}_2\text{O}(s)$$
white solid   blue solid

Test 2 Water will turn blue anhydrous cobalt(II) chloride pink:

$$\text{cobalt(II) chloride} + \text{water} \rightarrow \text{cobalt(II) chloride-6-water}$$
$$\text{CoCl}_2(s) + 6\text{H}_2\text{O}(l) \rightarrow \text{CoCl}_2\cdot 6\text{H}_2\text{O}(s)$$
blue solid   pink solid

These two tests will show that the liquid *contains* water. To test whether it is *pure* water, you can find its boiling and freezing points (see Figs. 2.23 and 2.24). Pure water boils at 100 °C and freezes at 0 °C. The tests show that the liquid is in fact water. Hydrogen burns to form water:

$$\text{hydrogen} + \text{oxygen} \rightarrow \text{water}$$
$$2\text{H}_2(g) + \text{O}_2(g) \rightarrow 2\text{H}_2\text{O}(l)$$

9.9 Water as a solvent

Almost all substances dissolve in water to some extent. Solubility was covered in Section 1.5. For most solids, solubility *increases* with temperature. The solubility of gases often *decreases* with temperature (see Thermal pollution, Section 9.16).

Aqueous solutions freeze below 0 °C and boil above 100 °C (see Section 2.8. A saturated solution of sodium chloride (common salt) freezes at −18 °C.

Since water is such a good solvent, it is difficult to obtain pure water. Distillation is one method of purifying water (see Fig. 2.9). Ion exchange is another method (see Section 9.13). By **pure** water, a chemist means water that contains no dissolved material. This is different from what the public health authority means by **pure** water. Safe drinking water contains dissolved salts: it is not pure water in the chemical sense. Water which contains substances that are harmful to life is **polluted** water.

9.10 Water in limestone regions

In limestone regions, rain water trickles over rocks composed of calcium carbonate (limestone) and magnesium carbonate. These minerals do not dissolve in pure water. Since rain water contains dissolved carbon dioxide, it is weakly acidic. It reacts with the carbonate rocks to form the soluble salts, calcium hydrogencarbonate and magnesium hydrogencarbonate:

SUMMARY NOTE

Water is formed when hydrogen burns in air. Tests for water are:
- it turns anhydrous copper(II) sulphate from white to blue
- it turns anhydrous cobalt(II) chloride from blue to pink
- pure water boils at 100 °C and freezes at 0 °C.

Pure water is obtained by distillation.

calcium carbonate + water + carbon dioxide → calcium hydrogencarbonate solution

$CaCO_3(s) + H_2O(l) + CO_2(g) \rightarrow Ca(HCO_3)_2(aq)$

Gradually, over thousands of years, underground caverns and potholes have been hollowed out as rocks have been converted by acidic rain water into soluble salts (Fig. 9.10).

The process can be reversed. Sometimes, in an underground cavern, a drop of water becomes isolated. With air all round it, water will evaporate. The dissolved calcium hydrogencarbonate changes into a grain of solid calcium carbonate:

calcium hydrogencarbonate → calcium carbonate + water + carbon dioxide

$Ca(HCO_3)_2(aq) \rightarrow CaCO_3(s) + H_2O(l) + CO_2(g)$

Day by day, more grains of calcium carbonate are deposited. Over a century or two, enough calcium carbonate can build up to form a cone hanging from the roof of the cavern. This is called a **stalactite**. The same process can lead to the formation of a **stalagmite** on the floor of the cavern (see Fig. 9.10).

9.11 Soaps and detergents

Oil and grease do not dissolve in water. To wash oil and grease off your hands or clothes, you need a **detergent** (cleaning agent). There are two kinds of detergent: **soaps**, which are made from animal fats and vegetable oils, and **soapless detergents**, which are made from petroleum oil. Both are able to **emulsify** oil and grease (make them miscible with water). Soaps and soapless detergents are sodium salts.

One soap is sodium hexadecanoate, $C_{15}H_{31}CO_2Na$. A model of the soap is shown in Fig. 9.11. It consists of a sodium ion and a hexadecanoate ion, which we will call a *soap* ion for short. The *soap* ion has two parts, shown in Fig. 9.12 as a head and a tail.

Fig. 9.10 Underground cavern with stalactites and stalagmites

SUMMARY NOTE

In limestone regions, acidic rain water reacts with calcium carbonate to form soluble calcium hydrogencarbonate. The reverse process leads to the formation of stalactites and stalagmites.

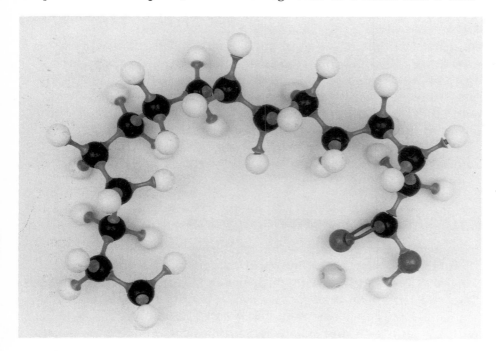

Fig. 9.11 Model of the soap, sodium hexadecanoate

Head is water-loving and fat-hating.

Tail is water-hating and fat-loving.

Fig. 9.12 The soap ion

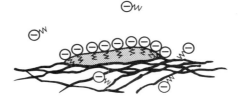
(a) The tails of the **soap** ions are attracted to grease.

(b) The heads of the **soap** ions are attracted to water.

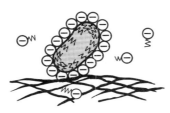
(c) As the heads of the soap ions become surrounded by water molecules, the blob of grease is squeezed away from the cloth.

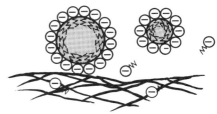
(d) Droplets of grease do not join because their surfaces are negatively charged. The droplets spread through the water. The dirty water must be removed by thorough rinsing.

Fig. 9.13 Soap removes grease from cloth ▷

The tail is a long chain of —CH_2— groups. This is water-hating and fat-loving. The head is a —CO_2^- group. This group is called a **carboxylate** group (see Chapter 17). It is water-loving and fat-hating. Figure 9.13 shows how *soap* ions wash grease from cloth.

The head of each *soap* ion is attracted to the water. The tail of each *soap* ion is attracted to grease. Thus, the *soap* ions form a bridge between the blob of grease and the water. As the dirty clothes are swirled around during washing, the blob of grease surrounded by *soap* ions is dislodged from the cloth to float off into the water. Repulsion between the charges on the —CO_2^- groups keeps the droplets apart. They remain spread through the water — **emulsified**.

Soapless detergents are sodium salts of sulphonic acids. The tail of the detergent ion is a chain of fat-loving —CH_2— groups. The water-loving head is an —SO_3^- group.

9.12 Hard and soft water

Soap lathers easily in distilled water. With some types of tap water, it is *hard* to get a lather, and such water is described as **hard**. Water in which soap lathers easily is **soft** water. Hard water contains substances which combine with *soap* ions to prevent them from doing their job. Experiment shows (see Experiment 9.2) that it is the calcium and magnesium ions in water that make it hard. They combine with *soap* ions to form insoluble calcium and magnesium compounds. These compounds are the **scum** that you see floating on your bath water. The *soap* ions are no longer free, and the soap has little effect.

soap ions + calcium ions → scum
2 $soap^-$(aq) + Ca^{2+}(aq) → $Ca(soap)_2$(s)

Soapless detergents will still work in hard water because their calcium and magnesium salts are soluble. Soapless detergents have taken over from soaps to a large extent. Of the cleaning market, 80 per cent is held by soapless detergents.

9.13 Methods of softening hard water

If you go on adding soap to hard water for long enough, all the calcium and magnesium ions will be precipitated as scum. After that, the soap will be able to work. This is an expensive method of softening water!

Temporary hardness

If the cause of hardness is the presence of dissolved calcium hydrogencarbonate and magnesium hydrogencarbonate, then the water can be softened simply by boiling. These compounds decompose when water is boiled:

calcium hydrogencarbonate → calcium carbonate + carbon dioxide + water

$Ca(HCO_3)_2(aq) \rightarrow CaCO_3(s) + CO_2(g) + H_2O(l)$

The scale which you see on the inside of kettles is calcium carbonate (see Fig. 9.14 and Experiment 9.3). Hardness which can be removed by boiling is called **temporary hardness**.

Permanent hardness

Hardness which cannot be removed by boiling is called **permanent hardness**. It is due to the presence of dissolved calcium and magnesium chlorides and sulphates. These compounds are not decomposed by heat.

Washing soda

One method of softening both temporary and permanent hardness is to add washing soda, sodium carbonate-10-water. Calcium ions and magnesium ions are precipitated as insoluble carbonates:

calcium ions + carbonate ions → calcium carbonate
$Ca^{2+}(aq) + CO_3^{2-}(aq) \rightarrow CaCO_3(s)$

*Exchange resins

Ion exchange resins will take ions of one element out of aqueous solution and replace them with ions of another element. Permutits are manufactured ion exchange resins which will take calcium and magnesium ions out of water and replace them with sodium ions:

calcium ions + sodium permutit → sodium ions + calcium permutit

Figure 9.15 shows how a permutit water softener works.

A permutit cannot go on softening water indefinitely. When the sodium ions in the permutit have been replaced by calcium and magnesium ions, the column must be regenerated. Brine (sodium chloride solution) is passed through the column. It converts calcium permutit back to sodium permutit.

> **SUMMARY NOTE**
>
> Soaps and detergents have ions which form a bridge between grease and water. Hard water contains calcium ions and magnesium ions. They combine to form an insoluble **scum**. Detergents work well even in hard water because their calcium and magnesium salts are soluble.

Fig. 9.14 Scale in a hot water cylinder after only 5 years of use

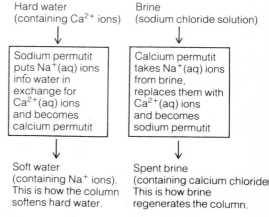

Fig. 9.15 A permutit water softener

SUMMARY NOTE

Methods of softening hard water:
- Temporary hardness is removed by boiling.
- Permanent hardness is removed by adding sodium carbonate (washing soda) or by running water through an exchange resin.

Hard water is better than soft water for drinking.

Pure water is obtained by distillation.

9.14 Advantages of hard water

For health reasons, hard water is preferable to soft water. The calcium content strengthens bones and teeth. It is also beneficial to people with a tendency to develop heart disease.

A number of industries prefer hard water. Brewers like the taste of the dissolved salts in their beer. Tanners find that leather cures better in hard water.

9.15 Pure water

All types of water can be distilled to give pure water. The apparatus shown in Fig. 2.9 could be used.

9.16 Pollution of water

Industrial pollution

Minamata is a fishing village on the shores of Minamata Bay in Japan. In 1953, a mysterious disease began to strike the villagers. Some were crippled, some became mentally deranged, some became blind and some died. A team of scientists traced the cause of the disease to the fish in the bay. The fish contained a high level of mercury compounds. This was a mystery. Although a plastics factory was discharging mercury compounds into the bay, the level was so low at 2 p.p.b. (parts per billion) that the

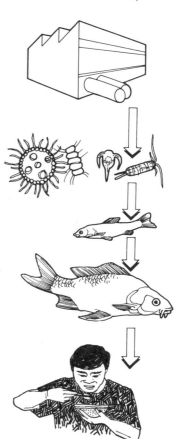

Plastics factory discharges mercury compounds into Minamata Bay. The water in the bay contains 2 p.p.b. of mercury, and is still safe to drink.

Plankton take in mercury compounds from the water. They cannot excrete mercury compounds.

Small fish feed on plankton. Fish cannot excrete mercury compounds. The level of mercury in their flesh builds up to 100 times that in the bay water.

Large fish feed on small fish. Mercury compounds are stored in their flesh.

At the end of this food chain is the human family. They eat fish containing 2000 – 10 000 times the concentration of mercury in the water.

Fig. 9.16 The Minamata food chain ▷

water could be drunk. The scientists found out that when fish and shellfish took in mercury compounds they did not excrete them. The level of mercury in their tissues built up to 5–20 p.p.m. (parts per million). Fish is the staple diet of the villagers. They were eating fish in which the mercury level was 4000 times that of the bay water. Mercury levels built up in their bodies. The toll was 50 dead, 100 crippled and 900 seriously affected.

The Minamata tragedy was an accident. The manufacturers were discharging a low level of pollutant, and they expected the bay to dilute it to a safe level. They did not know how the mercury could be concentrated through a **food chain** (Fig. 9.16).

Unfortunately, other countries were slow to learn from the Japanese example. In 1967, many lakes and rivers in Sweden were found to be so contaminated by mercury that fishing had to stop.

In 1970, high mercury levels were found in hundreds of lakes in Canada and the USA. Now that the danger is known, chemical plants have been able to reduce spillage of mercury. This is not the end of the danger. Mercury from years of careless practice lies in the sediment at the bottom of lakes. Slowly it is converted by bacteria into soluble mercury compounds.

Thermal pollution

Water is used as a coolant, especially by nuclear power stations. River water is circulated round the plant and returned to the river at a higher temperature. If the temperature of the river is raised by many degrees, this can have serious effects. The solubility of oxygen decreases as the temperature rises. At the same time, demands on oxygen increase. Fish are more active and need more oxygen. The bacteria which feed on decaying organic matter become more active and use oxygen faster.

Agricultural pollution

Plants need a number of elements (about 20) for their growth. The elements carbon, hydrogen, oxygen, nitrogen and phosphorus are needed in substantial amounts. Water always provides enough carbon, hydrogen and oxygen; nitrogen and phosphorus are scarcer. Lake water, being poor in the nutrients nitrogen and phosphorus, supports little plant life. Since there is no decaying vegetation to use up dissolved oxygen, lake water is saturated with oxygen. Fish can thrive.

Farmers use compounds of phosphorus and nitrogen (phosphates and nitrates) to increase their crop yields (see Section 14.3). Often, fertilisers are used in large quantities. Plants can absorb only a limited amount through their roots. If land surrounding a lake receives too much fertiliser, unabsorbed fertiliser washes out of the soil into the lake water.

There the fertiliser nourishes the growth of water plants. Weeds flourish, and thick mats of algae cover the surface (Fig. 9.17). The plant growth can snag the propellers of boats. If water is taken from the lake for use, water filters become clogged with

SUMMARY NOTE

Mercury compounds are poisonous. If mercury gets into a lake or river, it remains there to pollute the water for a long time. Slowly, it is converted into soluble compounds. If these pass through a food chain, they may be eaten by humans.

A large nuclear power station needs 4000 tonnes of water a minute for cooling.

That's like having a river flowing through its pipes.

The water is 10 °C hotter when it comes out than when it goes in.

SUMMARY NOTE

Thermal pollution means warming lakes and rivers. It decreases the solubility of oxygen in the water.

Fig. 9.17 Lake with a strong growth of algae

weeds. The name given to this **unintentional enrichment** of lake water and river water is **eutrophication**.

When algae die, they are decomposed by aerobic bacteria. When these bacteria run out of oxygen, they die. Anaerobic bacteria take over, and convert part of the dead matter into smelly decay products. The rest of the debris falls to the bottom. Slowly a layer of dead plant material builds up on the bottom of the lake. Deprived of oxygen, fish die. The lake becomes a dead lake. Many lakes in the USA and Canada are now empty of fish. The tourist industry centred on the lakes has suffered. If the farmers could cut back on fertilisers, they would save money and the fishermen would benefit too.

Excessive fertiliser can also find its way into the **ground water**. This is the water held underground in porous layers of rock. A third of Britain's drinking water comes from ground water. The World Health Organisation recommends that the level of nitrogen in the form of nitrates should not exceed 11 p.p.m. The level in Britain is far short of this value, but it is increasing, and people are worried that in 20 years' time it may be dangerously high.

Nitrates affect human health in two ways.

1 Nitrates are converted into nitrites (salts containing the NO_2^- ion). Nitrites oxidise the iron in haemoglobin from iron(II) ions to iron(III) ions. In its oxidised form, haemoglobin can no longer transport oxygen round the body. Babies are more at risk than adults. The lower acidity of babies' stomachs encourages the conversion of nitrates to nitrites. The extreme case of nitrite poisoning is the 'blue baby' syndrome, in which the baby turns blue from lack of oxygen.

2 Nitrosoamines have been found in a number of foods. These compounds cause cancer. Some chemists think that nitrites formed from nitrates may be responsible for the formation of nitrosoamines.

Research is being done on ways of reducing the nitrate level. Of course, farmers must use fertilisers, but more care can be taken to match the input of fertiliser to the crop's ability to use it. The correct amount applied in the right season will be used up by the plants. Agricultural chemists are advising farmers how they can make the most efficient use of fertilisers. This will save the farmers money as well as making drinking water safer.

Fertilisers are not the only source of phosphates in water. Soapless detergents contain phosphates as **brighteners**. Detergents are discharged into rivers. This source of pollution could be avoided. People might be persuaded to use detergents which left their shirts just a shade less white. Then phosphates could be omitted from the detergents and stop polluting our lakes and rivers.

SUMMARY NOTE

When nitrates and phosphates from fertilisers wash into lakes and rivers, they stimulate the growth of weeds and algae. Decaying plant material uses oxygen. This leads to a shortage of dissolved oxygen, and fish die.

Nitrates are also washing into the **ground water** from which we obtain much of our drinking water. More sparing use of fertilisers is recommended.

Detergents containing phosphates are discharged into rivers and lakes.

Estuaries and coastal waters

Coastal zones and estuaries are only 0.1% of the ocean surface, but they produce 99% of the seafood. The sea water in coastal zones contains nutrients which have been brought in by rivers. Estuaries are among the richest biological regions on Earth. Fish and shellfish abound in them. Of all the parts of the oceans, the estuaries receive the worst pollution. Shipping lanes criss-cross these waters. Collisions and groundings result in accidental discharges of oil (Fig. 9.18). Some oil tankers wash out their tanks at sea, although it is illegal for them to do so (see Chapter 16).

Coastal waters are also favourite dumping grounds for industry's toxic wastes. Nearly all wastes are finally washed out to sea.

The Mersey is Britain's dirtiest estuary. Raw sewage from Liverpool and other towns pours straight into its waters. The Secretary of State for the Environment called the pollution 'an affront to a civilised society'. In 1983, he authorised a £200 million scheme for the construction of sewage works on the banks. These will remove solid sewage and toxic metals, which are discharged in industrial effluent, and turn them into solid sludge. Ships will dump the sludge in the Irish Sea (Fig. 9.19). This solution to the Mersey's problem will not suit everyone. The Ministry of Agriculture, Fisheries and Food is worried that fish in the Irish Sea will be poisoned by the toxic metals in the sludge. This will mean a loss in income for fishermen and workers in the fish processing industry.

Fig. 9.18 Oil from the wrecked Torrey Canyon fouled holiday beaches in Cornwall in 1967

Fig. 9.19 A sludge ship carries digested sewage sludge out to sea

JUST TESTING 47

1 Water has a natural method for disposing of waste that is dumped into it. What is this method? What is wrong with dumping waste in rivers when water can cope with it in this way?
2 A layer of algae on the surface of a pond is described as **algal bloom**. Why is it undesirable?
3 What is **eutrophication**? Who loses money through eutrophication?

SUMMARY NOTE

Seafood is caught in coastal waters and estuaries. These areas receive the worst pollution from our sewage and industrial waste.

> *4 How does lead get into our water supplies? How can contamination from these sources be reduced? (See Chapters 5, 8, 10.)
>
> *5 What factors influence the rate of aquatic plant growth? Why is it so difficult to single out the effect of one nutrient which is present in the water as a pollutant? Design an experiment to enable you to single out the effect of one pollutant.
>
> 6 a Why are people worried about the concentration of nitrates in drinking water?
>
> b Could nitrates get from ground water into your body if you stopped drinking tap water? Explain your answer.
>
> c Why is it difficult to say exactly what concentration of nitrates in drinking water is dangerous?

9.17 Hydrogen

You have studied water, the most important compound of hydrogen. Now let us have a look at the importance of hydrogen.

Uses of hydrogen

1 **Weather balloons**. Since it is the least dense of gases, meteorologists use hydrogen to fill their weather balloons. The balloons, which are about 2 m in diameter, float up into the atmosphere, carrying a load of instruments. The instruments record information about atmospheric conditions.

2 **Making fertilisers**. A much more consuming demand for hydrogen is the manufacture of ammonia, NH_3 (see Section 14.3). Ammonium salts are used as fertilisers.

3 **Extracting metals**. Some metals are obtained from their ores with the help of hydrogen. Tungsten is the metal which is used to make electric light filaments. It is mined as tungsten oxide. Hydrogen converts heated tungsten oxide into tungsten and water:

tungsten oxide + hydrogen → tungsten + water

Hydrogen removes oxygen from tungsten oxide. Another way of saying this is: hydrogen **reduces** tungsten oxide. This reaction is a **reduction** reaction. **Reduction is the addition of hydrogen to a substance or the removal of oxygen from a substance.** Remember **oxidation** (Section 8.9): the removal of hydrogen from or the addition of oxygen to a substance. Reduction is the opposite of oxidation. Hydrogen is oxidised to water. Tungsten oxide is reduced to tungsten:

```
        ┌─── This is reduction ───┐
        ↓                         ↓
tungsten oxide + hydrogen → tungsten + water
                 ↑                         ↑
                 └─── This is oxidation ───┘
```

Hydrogen does not reduce all metal oxides (see Section 10.8). By experiment, you will find that it reduces copper(II) oxide to copper:

$$\underset{\text{CuO(s)}}{\text{copper(II) oxide}} + \underset{\text{H}_2\text{(g)}}{\text{hydrogen}} \rightarrow \underset{\text{Cu(s)}}{\text{copper}} + \underset{\text{H}_2\text{O(l)}}{\text{water}}$$

(Reduction over the left side; Oxidation over the right side)

Copper(II) oxide is an oxidising agent. Hydrogen is a reducing agent.

4 Margarine. Another industry which consumes hydrogen is the manufacture of margarine. Solid fats, such as butter and margarine, can be sold for a higher price than vegetable oils, such as groundnut oil and sunflower seed oil. Since these vegetable oils can be produced in large quantities, there is profit to be made in converting them into solid fats. Hydrogen will do this, with the help of nickel acting as a catalyst. The process is called **hydrogenation**.

$$\underset{\text{(vapour)}}{\text{vegetable oil}} + \underset{\text{(gas)}}{\text{hydrogen}} \xrightarrow{\text{Pass over heated nickel}} \underset{\text{(solid)}}{\text{margarine}}$$

See also Section 17.3 and Fig. 9.20.

Manufacture

Hydrogen is a by-product of the electrolysis of brine (sodium chloride solution); see Fig. 5.11. It is also made by a reaction between methane (natural gas) and water (see Fig. 9.21).

Fig. 9.20 Sunflower seed oil and margarine

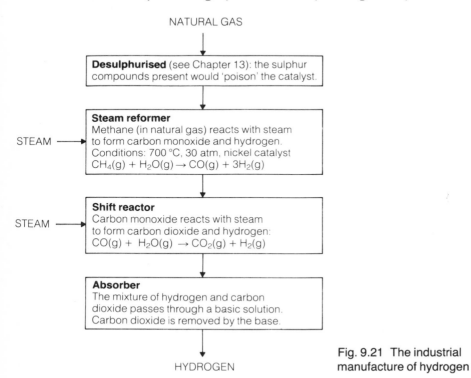

Fig. 9.21 The industrial manufacture of hydrogen

Test for hydrogen

Hydrogen burns in air. If it is pure, it burns quietly and steadily (see Fig. 9.9). If there is oxygen mixed with hydrogen, when you set light to the mixture there is an explosive 'pop' (see

SUMMARY NOTE

Hydrogen is used
- for filling balloons
- in the manufacture of fertilisers
- for reducing metal ores
- in the manufacture of margarine.

Hydrogen is made
- industrially by the reaction between methane and steam
- industrially as a by-product of the electrolysis of brine
- in the laboratory by the action of a metal on an acid.

Hydrogen burns with a squeaky 'pop'.

Experiment 7.1). This reaction can be used as a test for hydrogen.

Many of the reactions of hydrogen have been mentioned in relation to its uses. You will meet hydrogen in many of the topics in this book. Figure 9.22 gives a summary of them.

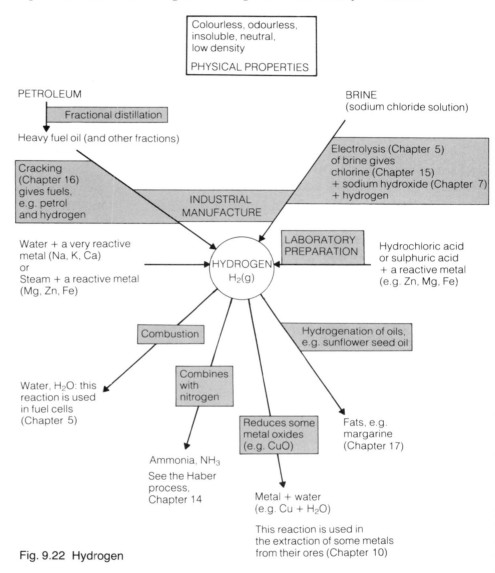

Fig. 9.22 Hydrogen

Exercise 9

1. As sewage plants operate, they produce large quantities of a thick sludge which is rich in plant nutrients. What do you think should be done with it?

2. What is the purest form of water that occurs naturally?
   Which ions should be present in drinking water to promote the growth of strong bones and teeth?

3. Which one of the following tests proves that a liquid is pure water?

   a The liquid is colourless and transparent.
   b It turns anhydrous copper(II) sulphate blue.
   c It boils at 100 °C at a pressure of 1 atmosphere.
   d It has a pH of 7.0.
   e It leaves no solid residue when it is evaporated.

4. Why is heat from industrial processes regarded as **thermal pollution**?

5. Many pollutants are converted into harmless

impurities by bacteria in water. What harm can be done by these pollutants?
What problem arises when plastics are dumped in lakes or rivers?

6 A population of 5000 people discharges raw sewage into a river. The daily sewage from one average person takes 60 g of oxygen from the water. How many tonnes of water, containing 10 p.p.m. of oxygen, are robbed of their oxygen by the sewage from the whole town? (1 tonne = 1000 kg)

7 Explain the importance of chlorine in maintaining a safe supply of drinking water. What used to happen in Britain before water was chlorinated? Does every country now have safe drinking water?

8 A small lake (about 1.5 km across) contains 10 million tonnes of water. The lake water contains the critical level of 0.01 p.p.m. of phosphorus (as compounds). Above this level, algae start to flourish. What mass of phosphorus (as compounds) does the lake contain?

9
- One gram of oxygen will oxidise completely about 0.25 g of petrol or oil.
- Water normally contains 10 p.p.m. of oxygen.
- A drop of oil weighs about 0.030 g.

  a Calculate the mass of oxygen in 1 tonne (1000 kg) of water.
  b How many drops of oil will this mass of oxygen oxidise?
  c What is wrong with Will Chukit's habit of pouring dirty sump oil from his car down the drain?
  d Write a letter to Mr Chukit to explain why he should stop doing this.

10 A woman on a diet decided to eat plenty of tuna fish. It is high in protein and low in fat. She ate about 350 g a day. After some months on this diet, she had difficulty in writing and talking; she had fits of shaking and difficulty in keeping her balance. These are the symptoms of mercury poisoning.
- Tuna fish contains 1 p.p.m. of mercury.
- The lowest level at which mercury has been shown to make people ill is 0.3 mg/day. (1 mg = 1×10^{-3} g)

  a Calculate the woman's intake of mercury.
  b How does it compare with the danger level?

11 a Explain the difference between hard and soft water.
  b What makes water hard? Explain how the substances you mention get into tap water.
  c Give one advantage of soft water and one advantage of hard water.
  d Describe one method of softening hard water. Explain how it works.
  e Compare the action of soaps and detergents in hard water.

12 Explain what hydrogen has to do with
  a the tungsten filaments in electric light bulbs
  b weather balloons
  c margarine.

CHAPTER 10 *Metals*

10.1 Metals and alloys

Metals are important to us. Figure 10.1 shows one of mankind's most highly prized possessions. The car must be reliable in use, attractive in appearance and as inexpensive as possible to manufacture. To enable it to meet this description, it is made largely of **metals** and **alloys**. An alloy is a mixture of metallic elements and, in some cases, non-metallic elements also. Steel is an alloy of iron with other metallic elements and carbon. Figure 10.1 shows how the choice of a metal or alloy for a particular job depends on its properties.

Fig. 10.1 The car

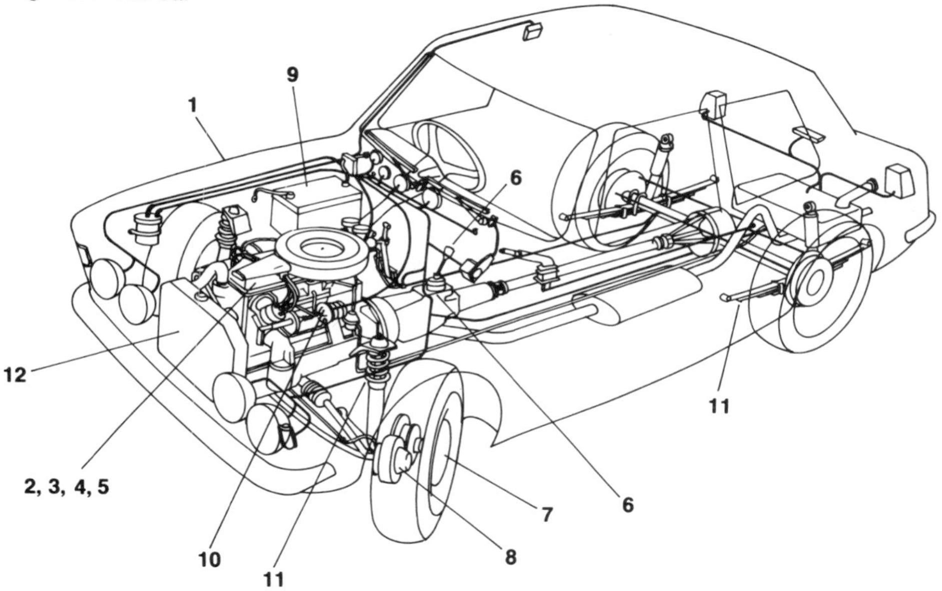

1 The combined body-chassis must be strong. It may need to survive a collision or a roll-over. Steel is chosen because of its strength.
2 The cylinder block is the largest part of the engine. It contains the cylinders in which the combustion of petrol vapour in air takes place. With its complicated shape, it can be made only by pouring molten metal into a mould. Cast iron is used (see Section 10.10).
3 The pistons move up and down inside the cylinders. Pistons are usually made of aluminium, which is less dense than iron. A piston must make a gas-tight fit inside its cylinder. Each piston is encircled by a piston ring of chromium-plated cast iron. Cast iron slides easily inside the cylinder; chromium plating makes it resistant to corrosion.
4 Inlet valves allow petrol vapour and air to enter the cylinders. Exhaust valves let the combustion products escape. Valves must keep their shape at high temperatures, and must not corrode. A steel which contains chromium and nickel is chosen.
5 As each cylinder 'fires' in turn, its piston moves and gives the crankshaft a powerful jerk. To withstand a succession of powerful jerks, a crankshaft must be very tough. It is made of steel which has received special toughening treatments.
6 The gears are made of steel; the gear-box is made of cast iron.
7 The wheels are made of pressed steel or aluminium.
8 The brake discs and drums are made of cast iron.
9 The batteries are lead–acid accumulators.
10 The starter motor is wound with copper wire.
11 The springs need an unusual quality. They must be compressible. A steel containing silicon and manganese was developed for this purpose.
12 The radiator must not become corroded. Aluminium or stainless steel is used.

SUMMARY NOTE

A large number of different metals and alloys are used in making a car. No other materials have the necessary strength.

You find it fairly easy to recognise a metal when you see one (Fig. 10.2). In Experiment 3.1, you looked at the differences between metallic elements and non-metallic elements. Table 3.2 lists the differences.

How much silver is there in a silver coin?

Not a lot! They are made of **cupronickel**, which is 75% copper and 25% nickel.

◁ Fig. 10.2 Metals in use

Alloys

How does alloying alter the properties of metals? Many metallic elements do not have enough strength to allow them to be used for the manufacture of machines and vehicles which will have to withstand stress. Aluminium is in many ways a suitable metal from which to make aeroplanes. It has a low density and is not corroded by the atmosphere. Its drawback is that it is not strong enough. The addition of copper and a little magnesium to aluminium makes an alloy called **duralumin**. This alloy is much stronger than pure aluminium and is used for aircraft manufacture (Fig. 10.3).

Fig. 10.3 Concorde

Alloys have different properties from the metals (and in some cases non-metallic elements) of which they are composed. Brass differs from copper and zinc. It has a lower melting point than either of these metals. This makes it easier to *cast*. Brass makes a pleasanter, more sonorous sound than either of the two elements. It is used for making musical instruments.

Which metal or alloy would you use for electrical wiring? for making saucepans? for making drains? Before you can answer these questions, you need to study the chemical reactions of the metals.

10.2 The metallic bond

What is it that gives metals their strength, enables them to conduct heat and electricity and allows them to be worked into different shapes without breaking? It is the type of chemical bond between the atoms: the **metallic bond**.

SUMMARY NOTE

An alloy is a mixture of metals, with in some cases non-metallic elements also. Alloys have different properties from the elements of which they are composed.

Fig. 10.4 The metallic bond ▷

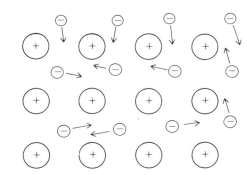

Valency electrons have separated from the atoms to leave metal cations.

Electrons move between the metal cations like a cloud of negative charge. They cancel the repulsion that would drive the cations apart.

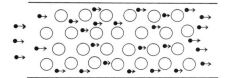

| Electrons are supplied at the LHS. | The sea of electrons flows towards the RHS. | Electrons are removed at the RHS. |

Fig. 10.5 How a metal conducts electricity

A block of metal consists of positive metal ions and free valency electrons (Fig. 10.4). The valency electrons no longer belong to any one atom. They move between the metal ions like a cloud of negative charge. If it were not for this mobile negative charge, the metal cations would be driven apart by repulsion between their positive charges.

This structure explains the properties of metals. They conduct electricity (Fig. 10.5). Unlike ionic crystals, metals can change their shape without breaking (see Figs. 10.6 and 10.7).

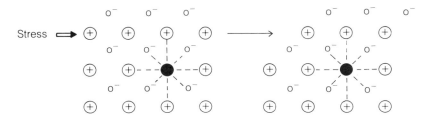

This is a metallic structure. Notice the atom shaded black. It is still bonded to the same number of other atoms after the shape of the metal has been altered.

Fig. 10.6 Metals can bend without breaking

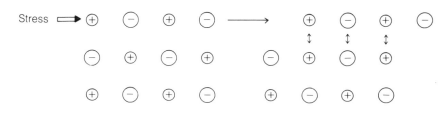

This is an ionic structure. When a force is applied, it alters the shape of the crystal. As a result, cation ↔ cation repulsion and anion ↔ anion repulsion is set up. The crystal fractures.

Fig. 10.7 Ionic compounds shatter easily

> **SUMMARY NOTE**
>
> The nature of the metallic bond is responsible for the strength of metals and their ability to conduct electricity and heat.

10.3 The chemical reactions of metals

Reactions of metals with air

Many metals react with oxygen in the air. You studied this reaction in Experiment 8.1 (see Section 8.9). The results are summarised in Table 10.1.

Table 10.1 The reactions between metals and oxygen

Metal	Symbol	Reaction when heated in air	Reaction with cold air
Potassium	K	⎫	⎫
Sodium	Na		
Lithium	Li		React slowly with air to form a surface film of the metal oxide. This reaction is called **tarnishing**.
Calcium	Ca	Burn in air to form oxides	
Magnesium	Mg		
Aluminium	Al		
Zinc	Zn		
Iron	Fe	⎭	
Tin	Sn	⎫	
Lead	Pb	When heated in air these metals form oxides without burning.	
Copper	Cu	⎭	
Silver	Ag	⎫	Silver tarnishes in air.
Gold	Au	Do not react.	Do not react.
Platinum	Pt	⎭	

JUST TESTING 48

1. State whether the oxides of the following elements are acidic or basic or neutral.
   a sulphur, **b** iron, **c** magnesium, **d** carbon, **e** calcium, **f** phosphorus.
2. Write (i) word equations, (ii) balanced chemical equations for the combustion of the following elements in oxygen to form oxides:
   a magnesium, **b** calcium, **c** sodium.

Reactions of metals with water

You may like to start this topic by doing Experiments 10.1–10.4. The results of the experiments are summarised in Table 10.2.

Table 10.2 Reactions of metals with water and steam

Metal	Reaction with water
Potassium	Reacts violently with cold water (Fig. 10.8). Hydrogen and potassium hydroxide solution, which is a strong alkali, are formed. So much heat is given out (the reaction is so **exothermic**) that the hydrogen burns. The flame is coloured lilac by potassium atoms which have vaporised:

$$\text{potassium} + \text{water} \rightarrow \text{hydrogen} + \text{potassium hydroxide}$$

$$2K(s) + 2H_2O(l) \rightarrow H_2(g) + 2KOH(aq)$$

Potassium is kept under oil to prevent water vapour in the air from attacking it.

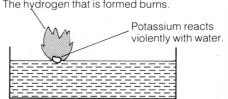

Fig. 10.8 The reaction of potassium with water (NOTE: this reaction must be done as a demonstration behind a safety screen.)

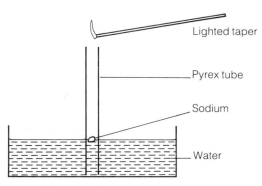

Fig. 10.9 The reaction of sodium with water (NOTE: This violent reaction must be done as a demonstration behind a safety screen.)

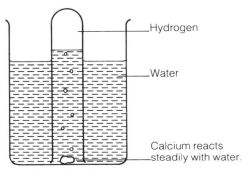

Fig. 10.10 The reaction of calcium with water

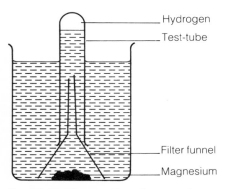

Fig. 10.11 The reaction of magnesium with water

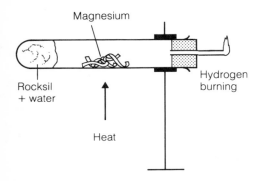

Fig. 10.12 The reaction of magnesium with steam

Sodium	Reacts with slightly less violence than potassium. Hydrogen and sodium hydroxide solution, which is a strong alkali, are formed. The hydrogen formed burns (Fig. 10.9) with a yellow flame. The flame colour is due to the presence of sodium vapour:

$$\text{sodium} + \text{water} \rightarrow \text{hydrogen} + \text{sodium hydroxide}$$
$$2\text{Na}(s) + 2\text{H}_2\text{O}(l) \rightarrow \text{H}_2(g) + 2\text{NaOH}(aq)$$

Like potassium, sodium must be kept under oil.

Lithium	Reacts similarly to sodium but less vigorously. Lithium must be kept under oil.
Calcium	Reacts readily but not violently with cold water to form hydrogen and calcium hydroxide solution, the alkali **limewater** (see Experiment 10.2 and Fig. 10.10):

$$\text{calcium} + \text{water} \rightarrow \text{hydrogen} + \text{calcium hydroxide}$$
$$\text{Ca}(s) + 2\text{H}_2\text{O}(l) \rightarrow \text{H}_2(g) + \text{Ca(OH)}_2(aq)$$

Magnesium	Reacts slowly with cold water (Fig. 10.11) to form hydrogen and magnesium hydroxide:

$$\text{magnesium} + \text{water} \rightarrow \text{hydrogen} + \text{magnesium hydroxide}$$
$$\text{Mg}(s) + 2\text{H}_2\text{O}(l) \rightarrow \text{H}_2(g) + \text{Mg(OH)}_2(aq)$$

Burns in steam (Experiment 10.3 and Fig. 10.12) to form hydrogen and magnesium oxide:

$$\text{magnesium} + \text{steam} \rightarrow \text{hydrogen} + \text{magnesium oxide}$$
$$\text{Mg}(s) + \text{H}_2\text{O}(g) \rightarrow \text{H}_2(g) + \text{MgO}(s)$$

Aluminium	Only if the surface layer of aluminium oxide is removed will aluminium show its true reactivity (Experiment 10.12). Then it reacts to form hydrogen and aluminium oxide.
Zinc	Reacts with steam to form hydrogen and zinc oxide (Experiment 10.4 and Fig. 10.13 overleaf):

$$\text{zinc} + \text{steam} \rightarrow \text{hydrogen} + \text{zinc oxide}$$
$$\text{Zn}(s) + \text{H}_2\text{O}(g) \rightarrow \text{H}_2(g) + \text{ZnO}(s)$$

Iron	Reacts with steam to form hydrogen and the oxide, Fe_3O_4, tri-iron tetraoxide, which is blue–black in colour (Experiment 10.4 and Fig. 10.13):

$$\text{iron} + \text{steam} \rightarrow \text{hydrogen} + \text{tri-iron tetraoxide}$$
$$3\text{Fe}(s) + 4\text{H}_2\text{O}(g) \rightarrow 4\text{H}_2(g) + \text{Fe}_3\text{O}_4(s)$$

Tin Lead Copper Silver Gold Platinum	Do not react.

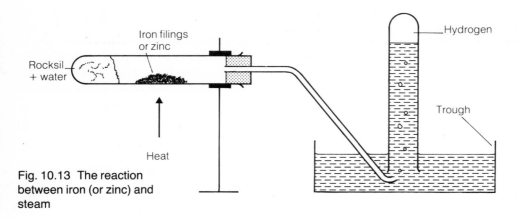

Fig. 10.13 The reaction between iron (or zinc) and steam

Reactions of metals with dilute acids

You have had an introduction to this topic in Section 7.2. Experiments 7.1 and 10.1 will help you. The results are summarised in Table 10.3.

Table 10.3 The reactions of metals with acids

Metal	Reaction
Potassium Sodium Lithium	The reaction is too dangerously violent for you to carry out
Calcium Magnesium Aluminium Zinc Iron Tin Lead	Most metals react with dilute hydrochloric acid to give hydrogen and a solution of the metal chloride, e.g. zinc + hydrochloric acid → hydrogen + zinc chloride $Zn(s) + 2HCl(aq) → H_2(g) + ZnCl_2(aq)$ The speed of the reaction decreases from calcium to lead. Lead reacts very slowly. The reactions of metals with dilute sulphuric acid give hydrogen and sulphates.
Copper Silver Gold Platinum	These metals do not react with dilute hydrochloric acid and dilute sulphuric acid. (Copper reacts with dilute nitric acid to form copper(II) nitrate.)

JUST TESTING 49

1. Fruit juices are acidic. Why are food cans not made simply from iron? What metal could be used to coat iron cans to make them suitable for preserving fruit?
2. Write word equations and symbol equations for the reactions between:
   a magnesium and hydrochloric acid
   b aluminium and hydrochloric acid
   c zinc and sulphuric acid
   d iron and sulphuric acid to form iron(II) sulphate.

Table 10.4 Reactions of metals

Metal	Reaction when heated in oxygen	Reaction with cold water	Reaction with dilute hydrochloric acid
Potassium	Burn to form the oxides.	Displace hydrogen; form alkaline hydroxides.	React dangerously fast.
Sodium			
Lithium			
Calcium			Displace hydrogen; form metal chlorides.
Magnesium		Slow reaction	
Aluminium		No reaction, except for slow rusting of iron; all react with steam.	
Zinc			
Iron			
Tin	Oxides form slowly without burning.	No reaction, even with steam	React very slowly.
Lead			
Copper			No reaction
Silver	No reaction		
Gold			
Platinum			

SUMMARY NOTE

You have studied the reactions of metals with oxygen, with water and with dilute acids.

1 Many metals react with oxygen; some metals burn in oxygen. The oxides of metals are bases.
2 The very reactive metals (e.g. sodium) react with water to give a metal hydroxide and hydrogen. Some metals react with steam to give a metal oxide and hydrogen.
3 Many metals react with dilute acids to give hydrogen and the salt of the metal.

When you look at reactions 1, 2 and 3, you find that the metals follow the same order of reactivity in all the reactions. This order is called the **reactivity series** of the metals. It is one of the important ways of classifying metals.

10.4 The reactivity series

Look at the combined results of the experiments on the reactions of metals with oxygen, water and acids. Table 10.4 summarises them. You will see that a sort of league table of metals emerges. The same metals are the most reactive in the different reactions.

Table 10.4 places the metals in order of **reactivity**, in order of their readiness to take part in chemical reactions. The table is part of the **reactivity series** of the metals. There are 70 metals in the Earth's crust. You may be surprised to see aluminium placed above zinc. Aluminium is so reactive that, when a fresh surface of aluminium meets the air, it immediately reacts with oxygen to form aluminium oxide. The surface layer of aluminium oxide prevents the metal from showing its true reactivity. If you remove the layer of aluminium oxide to expose the metal beneath it, you will see how reactive aluminium really is (Experiment 10.12).

JUST TESTING 50

1 Why does gold occur 'native' whereas zinc does not?
2 Why were silver and copper used as coinage metals?
3 Archaeologists dig up gold and silver objects made by people thousands of years ago. Why are they still in good condition? What has happened to iron objects of the same age?
4 The following metals are listed in order of reactivity:
   sodium > magnesium > zinc > copper
   Explain how this order follows from the reactions of these metals with **a** water, **b** dilute hydrochloric acid.

10.5 Metals in the Periodic Table

The reactivity series (Table 10.5) is one way of classifying metals. You have already met another important classification, the Periodic Table of the elements (see Chapter 4).

You can see from Table 10.6 that it is possible to draw a line across the Periodic Table to separate the metallic elements from the non-metallic elements. The non-metallic elements, including chlorine, oxygen and the noble gases, are on the right-hand side of the line. The metallic elements are on the left-hand side of the line.

On the far left are the very reactive metals lithium, sodium and potassium. These metals, in Group 1 of the Periodic Table, are called **the alkali metals** because their hydroxides are strong alkalis. They are at the top of the reactivity series. The metals in Group 2, **the alkaline earths**, are lower in the reactivity series than Group 1. Aluminium is in Group 3. The less reactive metals, tin and lead, are in Group 4.

The block of metals in between Group 2 and Group 3 are called **the transition metals**. They include iron, copper, silver, gold and zinc. Transition metals can use more than one valency. Iron has valencies 2 and 3: it forms the ions Fe^{2+} and Fe^{3+}. The ions of transition metals are often coloured (see Table 10.7). Transition metals and their ions often act as catalysts. Vanadium(V) oxide is used as a catalyst in the manufacture of sulphuric acid (see Section 13.7). Iron and molybdenum catalyse the manufacture of ammonia (see Section 14.3).

Table 10.5 A section of the reactivity series

Potassium	K
Sodium	Na
Lithium	Li
Calcium	Ca
Magnesium	Mg
Aluminium	Al
Zinc	Zn
Iron	Fe
Tin	Sn
Lead	Pb
Copper	Cu
Silver	Ag
Gold	Au
Platinum	Pt

Table 10.6 The Periodic Table

1	2											3	4	5	6	7	0
		H															He
Li	Be											B	C	N	O	F	Ne
Na	Mg				Transition metals							Al	Si	P	S	Cl	Ar
K	Ca	Sc	Ti	V	Cr	Mn	Fe	Co	Ni	Cu	Zn	Ga	Ge	As	Se	Br	Kr
Rb	Sr	Y	Zr	Nb	Mo	Tc	Ru	Rh	Pd	Ag	Cd	In	Sn	Sb	Te	I	Xe
Cs	Ba	La	Hf	Ta	W	Re	Os	Ir	Pt	Au	Hg	Ti	Pb	Bi	Po	At	Rn

(For symbols, see p. 316)

Career sketch: Medical laboratory technician

Fiona works in a medical laboratory. She analyses blood samples. She measures cholesterol levels and examines blood under a microscope to detect the presence of diseases. She makes cultures of body fluids to find out whether bacteria or other micro-organisms are present.

Fiona has A-levels in chemistry and biology. Senior people in the same laboratory have degrees in biochemistry.

Table 10.7 Colours of transition metal ions

Iron(II) ions	green
Iron(III) ions	rust-coloured
Copper(II) ions	blue
Nickel(II) ions	green
Cobalt(II) ions	pink

> **SUMMARY NOTE**
>
> Another important classification is the Periodic Table. The most reactive metals, the **alkali metals**, are at the left-hand side, in Group 1. Other reactive metals, the **alkaline earths**, are in Group 2, and aluminium is in Group 3. Less reactive metals are in Group 4 and in the block of transition metals.
>
> The non-metallic elements are at the right-hand side of the Periodic Table in Groups 5, 6, 7 and 0.

> **SUMMARY NOTE**
>
> The chemical reactions of metals and non-metallic elements are different. Many metals react with acids. Metals form cations. Their oxides are bases, and their chlorides are crystalline solids.

Silicon and germanium are on the borderline between metals and non-metals. These elements are important in the computer industry (see Section 3.4). They are **semiconductors**, intermediate between metals, which are electrical conductors, and non-metals, which are non-conductors of electricity.

Table 3.2 summarises the differences between the physical properties of metallic and non-metallic elements. Table 10.8 summarises the differences between the chemical properties of metallic and non-metallic elements.

Table 10.8 Chemical properties of metallic and non-metallic elements

Metallic elements	*Non-metallic elements*
1 Metals which are high in the reactivity series react with dilute acids to give hydrogen and a salt of the metal.	Non-metallic elements do not react with dilute acids.
2 Metallic elements form cations, e.g. Na^+, Mg^{2+}, Al^{3+}.	Non-metallic elements form anions, e.g. Cl^-, O^{2-}.
3 The oxides and hydroxides are bases (Section 7.3), e.g. Na_2O, CaO. If they dissolve in water, they give alkaline solutions, e.g. NaOH.	Many oxides are acids and dissolve in water to give acidic solutions, e.g. CO_2, SO_2. Some oxides are neutral and insoluble, e.g. CO.
4 The chlorides of metals are ionic crystalline solids, e.g. NaCl.	The chlorides of non-metals are covalent liquids or gases, e.g. HCl(g), CCl_4(l).

10.6 What predictions can be made from the reactivity series?

Competition between metals for oxygen

You see from the reactivity series that aluminium is more reactive than iron. This means that aluminium combines with oxygen more readily than iron does. What will happen when aluminium is heated with iron(III) oxide? The two metals will be in competition for oxygen. Which will win? Figure 10.14 shows an experiment to find out.

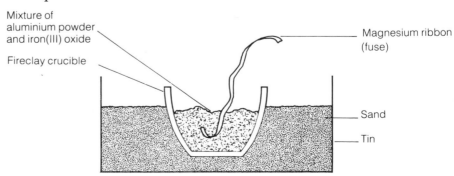

Fig. 10.14 A competition between aluminium and iron (NOTE: this experiment is a demonstration experiment. It should be done behind a safety screen.)

When the magnesium ribbon is lit, it provides enough heat to start the reaction. The reaction that follows is very exothermic. After the crucible has cooled, you can take a lump of solid iron out of the bottom:

aluminium + iron(III) oxide → iron + aluminium oxide
$2Al(s) + Fe_2O_3(s) → 2Fe(s) + Al_2O_3(s)$

- What do you see that tells you that the reaction is exothermic?
- How can you prove that the metal formed is iron, not aluminium?

This reaction is called the **thermit reaction** (therm = heat). Since the iron forms in a molten state, it can be used to weld pieces of metal together (see Fig. 10.16).

Competition between metals for anions

Metals can compete for anions. Experiment 10.5 will show you what happens. Metals can displace other metals from their salts. A metal which is higher in the reactivity series will displace a metal which is lower in the reactivity series from a salt. Copper is above silver in the reactivity series. Figure 10.17 shows what happens when a piece of copper wire is left to stand in a solution of silver nitrate.

Fig. 10.15 Competition for oxygen

JUST TESTING 51

1. Which metal would you use for making saucepans: magnesium, iron or copper? Explain your choice.
2. Which metal would you use for making electrical contacts in a space capsule? (Expense is no object, and it is essential that the contacts do not tarnish.)
3. What would you expect to see if you drop a piece of magnesium ribbon into a test-tube of copper(II) sulphate solution?
   Try it, and see if you are right.
   Write a word equation and a chemical equation for what happens.
4. What would you expect to see if you drop a piece of zinc into a test-tube of lead(II) nitrate solution?
   Try it and see if you are right.
   Why is the reaction slower than in 3?
   Write a word equation and a chemical equation for what happens.
5. What would you expect to happen if you drop a piece of iron into a solution of zinc sulphate? Again, check your prediction.
6. A metal, **X**, is displaced from a solution of one of its salts by a metal, **Y**. A metal, **Z**, displaces **Y** from a solution of one of its salts. Place the metals in order of reactivity.
7. The following metals are listed in order of reactivity, with the most reactive first:
   Na, Mg, Al, Zn, Fe, Pb, Cu, Hg, Au

Fig. 10.16 Mending the line. The thermit mixture is packed into a break in a railway line. The molten iron produced will fill the gap and weld the broken ends together

Fig. 10.17 A silver tree

> **SUMMARY NOTE**
>
> Reactive metals displace metals lower down the reactivity series from their oxides and from their salts.

> List the metals which will
> **a** occur 'native'
> **b** react with dilute acids
> **c** displace lead from lead(II) nitrate solution
> **d** react with cold water
> **e** react with steam.

10.7 Uses of metals and alloys

We find plenty of uses for metals and alloys. Sometimes an industrialist wants a material to do a particular job and there is no metal or alloy which fits the bill. Then metallurgists set to work to invent a new alloy with the right characteristics. The uses of a metal are related to its physical properties and its position in the reactivity series. Table 10.9 gives some examples.

Table 10.9 What are metals and alloys used for?

Metal/Alloy	Characteristics	Uses
Aluminium (**Duralumin** is an important alloy)	Low density Never corroded Good electrical conductor Good thermal conductor Reflector of light	Aircraft manufacture (Duralumin) Food wrapping Electrical cable Saucepans Car headlamps
Brass (alloy of copper and zinc)	Not corroded Easy to work with Sonorous Yellow colour	Ships' propellers Taps, screws Trumpets Ornaments
Bronze (alloy of copper and tin)	Harder than copper Not corroded Sonorous	Coins, medals Statues, springs Church bells
Copper	Good electrical conductor Not corroded	Electrical circuits Water pipes and tanks
Gold	Beautiful colour Never tarnishes	Jewellery Filling teeth Electrical contacts
Iron	Hard, strong Inexpensive, rusts	Construction, transport See Tables 10.10 and 10.11
Lead	Dense Unreactive	Protection from radioactivity Lead is no longer used for water pipes now that we know that it reacts very slowly with water.
Magnesium	Bright flame	Flares and flash bulbs
Mercury	Liquid at room temperature	Thermometers Electrical contacts Dental amalgam for filling teeth

How many miles of copper wire are there in your school buildings?

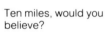
Ten miles, would you believe?

Gold is so **malleable** that it can be hammered into very thin gold leaf.

One cm³ of gold will make 10 square metres of gold leaf!

Silver	Beautiful colour and shine	Jewellery
	Good electrical conductor (Tarnishes in city air)	Electrical contacts in telephones, computers etc.
	Good reflector of light	Mirrors
Sodium	High thermal capacity	Coolant in nuclear reactors
Solder (alloy of tin and lead)	Low melting point	Joining metals, e.g. in an electrical circuit
Steel (alloy of iron etc.)	Strong	See Tables 10.10 and 10.11
Tin	Low in reactivity series	Coating 'tin cans'
Zinc	High in reactivity series	Protection of iron and steel; see Table 10.12

SUMMARY NOTE

Metals and alloys have thousands of uses. Metals and alloys are chosen for particular uses on the basis of their physical properties and their chemical reactions. You should know a few examples of the uses of metals and alloys.

Why do they call titanium 'the aerospace metal'?

It's low in density. And it stays strong at the high and low temperatures that supersonic aircraft have to stand up to.

10.8 Compounds and the reactivity series

How stable are metal oxides?

- The more reactive a metal is, the more readily it forms compounds.
- The more reactive a metal is, the more difficult it is to split up its compounds.

Copper is an unreactive metal. Copper(II) oxide is easily reduced to copper (reduction; see Section 9.17). Hydrogen will reduce hot copper(II) oxide to copper (see Experiment 10.7):

$$\text{copper(II) oxide} + \text{hydrogen} \xrightarrow{\text{heat}} \text{copper} + \text{water}$$
$$CuO(s) + H_2(g) \rightarrow Cu(s) + H_2O(l)$$

Carbon is another reducing agent. When heated, it will reduce the oxides of metals which are low in the reactivity series (see Experiment 10.14):

$$\text{lead(II) oxide} + \text{carbon} \xrightarrow{\text{heat}} \text{lead} + \text{carbon monoxide}$$
$$PbO(s) + C(s) \rightarrow Pb(s) + CO(g)$$

Metals which are low in the reactivity series are often obtained from their ores by reduction with carbon:

$$\text{zinc oxide} + \text{carbon} \xrightarrow{\text{heat}} \text{zinc} + \text{carbon monoxide}$$
$$ZnO(s) + C(s) \rightarrow Zn(s) + CO(g)$$

The oxides of metals which are high in the reactivity series are not reduced by hydrogen or carbon. Aluminium is high in the reactivity series; its oxide is difficult to reduce. The method used to obtain aluminium from aluminium oxide is electrolysis (see Section 10.12).

If a metal is very low in the reactivity series, its oxide will decompose when heated. The oxides of silver and mercury decompose when heated (see Fig. 3.9):

$$\text{mercury(II) oxide} \xrightarrow{\text{heat}} \text{mercury} + \text{oxygen}$$
$$2HgO(s) \rightarrow 2Hg(l) + O_2(g)$$

The stability of other compounds

Some compounds are more **stable** than others. This means that they are more difficult to decompose (split up) by heat than others. Table 10.10 shows how the stability of the compounds of a metal ties in with the position of the metal in the reactivity series.

Table 10.10 Action of heat on compounds

Cation	Anion				
	Oxide	Chloride	Sulphate	Carbonate	Hydroxide
Potassium Sodium Calcium	No decomposition	No decomposition	No decomposition		
Magnesium Aluminium Zinc Iron Lead Copper			Oxide + sulphur trioxide, $MO + SO_3$	Oxide + carbon dioxide, $MO + CO_2$	Oxide + water, $MO + H_2O$
Silver Gold	Metal + oxygen		Metal + $O_2 + SO_3$	Metal + $O_2 + CO_2$	Metal + $O_2 + H_2O$

SUMMARY NOTE

The oxides of the less reactive metals (e.g. Cu) are easily reduced (e.g. by carbon or hydrogen). The oxides of reactive metals at the top of the reactivity series are difficult to reduce.

Compounds of the less reactive metals (e.g. hydroxides, sulphates and carbonates) are decomposed by heat. Compounds of the very reactive metals do not undergo thermal decomposition.

The sulphates and carbonates and hydroxides of the most reactive metals are not decomposed by heat. Those of other metals decompose to give oxides, e.g.

calcium carbonate $\xrightarrow{heat}$ calcium oxide + carbon dioxide
$CaCO_3(s) \rightarrow CaO(s) + CO_2(g)$

copper(II) hydroxide $\xrightarrow{heat}$ copper(II) oxide + water
$Cu(OH)_2(s) \rightarrow CuO(s) + H_2O(l)$

lead(II) sulphate $\xrightarrow{heat}$ lead(II) oxide + sulphur trioxide
$PbSO_4(s) \rightarrow PbO(s) + SO_3(g)$

10.9 Extraction of metals from their ores

Metals occur in rocks in the Earth's crust. Some metals, such as gold, occur as the free metal, uncombined. They are said to occur 'native'. This is unusual: only metals which are very unreactive can withstand the action of air and water for thousands of years without being converted into compounds. Most metals are mined as compounds.

The business of obtaining metals from deposits of their compounds is expensive. Rock containing the metal compound is mined. Lumps of rock are crushed and ground by heavy machinery. Then a chemical method must be found for extracting the metal. All these stages cost money. If the rock contains enough of the metal compound to make it pay to extract the metal, the rock is called an **ore**.

Fig. 10.18 Mining: you can see the rock samples which have been drilled out

The method used to extract a metal from its ore depends on the position of the metal in the reactivity series. The compounds of reactive metals are difficult to reduce to the metals. (See Table 10.11.)

Table 10.11 Methods used for the extraction of metals

Potassium Sodium Calcium Magnesium	The anhydrous chloride is melted and electrolysed.
Aluminium	The molten anhydrous oxide is electrolysed.
Zinc Iron Copper Lead	Found as sulphides and oxides. Sulphides are roasted to give oxides; oxides are reduced with carbon.
Silver Gold	Found 'native' (as the free metals).

Sodium

Sodium is mined as sodium chloride (rock salt) in Cheshire (see Section 15.1). Electrolysis is the method used to obtain sodium. Molten anhydrous sodium chloride (not brine, which is aqueous sodium chloride) is electrolysed to give sodium at the negative electrode and chlorine at the positive electrode.

Cathode: sodium ion + electron → sodium atom
$$Na^+(l) + e^- \rightarrow Na(s)$$

The electrolysis of the molten anhydrous chloride is used to obtain other reactive metals, e.g. potassium, calcium and magnesium. It is an expensive method of obtaining metals because of the cost of the electricity consumed.

Aluminium

Aluminium is mined as **bauxite**, an ore which contains aluminium oxide, $Al_2O_3 \cdot 2H_2O$. This ore is very plentiful, yet aluminium was not extracted from it until 1827. In 1886, a commercial method of obtaining aluminium was invented. First, anhydrous pure aluminium oxide is obtained from the ore. The oxide must be melted before it can be electrolysed. This is difficult because aluminium oxide melts at 2050 °C. The big breakthrough in 1886 came when Charles Martin Hall and Paul Héroult discovered that the electrolysis of aluminium oxide could be carried out in molten **cryolite**, Na_3AlF_6, at 700 °C. Their invention, the Hall–Héroult cell, is shown in Fig. 10.19.

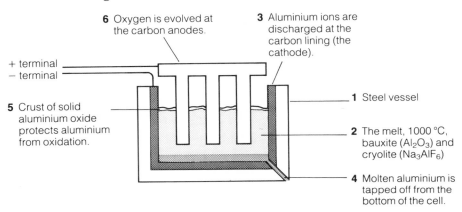

Fig. 10.19 Electrolysis of aluminium oxide

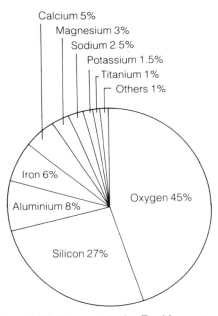

Fig. 10.20 Elements in the Earth's crust

Aluminium is the third commonest element in the Earth's crust, coming after oxygen and silicon (Fig. 10.20). It occurs mainly in clays. No economic method of extraction from clays has yet been found, although chemists are working on the problem.

Iron

Iron is mined as the oxides **haematite**, Fe_2O_3, and **magnetite**, Fe_3O_4, and as the sulphide **iron pyrites**, FeS_2. The sulphide ore is roasted in air to convert it into an oxide. The oxide ores are reduced to iron in a **blast furnace** (Fig. 10.21). A blast furnace is about 50 metres high. It is a tower made of steel plates and lined with heat-resisting bricks. It costs about £1 million. A blast of hot air is sent in near the bottom. Iron ore, coke and limestone are fed in at the top. Iron ores and limestone are plentiful resources. Coke is made by heating coal.

A number of chemical reactions take place in the blast furnace.

1 Coke burns in the blast of air at the bottom of the furnace. Carbon dioxide is formed:

$$\text{carbon (coke)} + \text{oxygen} \xrightarrow{\text{heat}} \text{carbon dioxide}$$
$$C(s) + O_2(g) \rightarrow CO_2(g)$$

2 Higher up the furnace, carbon dioxide reacts with coke to form carbon monoxide:

$$\text{carbon dioxide} + \text{carbon (coke)} \xrightarrow{\text{heat}} \text{carbon monoxide}$$
$$CO_2(g) + C(s) \rightarrow 2CO(g)$$

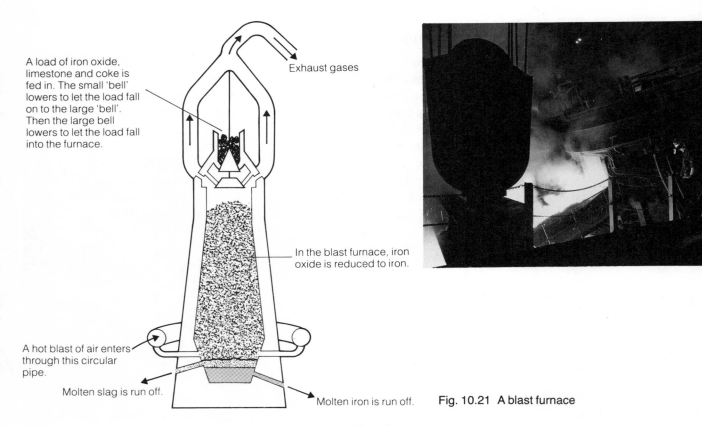

Fig. 10.21 A blast furnace

3 Carbon monoxide is the reducing agent in the blast furnace. It reduces iron oxides to iron:

iron(III) oxide + carbon monoxide $\xrightarrow{\text{heat}}$ iron + carbon dioxide
$Fe_2O_3(s) + 3CO(g) \rightarrow 2Fe(s) + 3CO_2(g)$

As a result, molten iron trickles to the bottom of the furnace.

4 Limestone decomposes in the blast furnace:

calcium carbonate (limestone) $\xrightarrow{\text{heat}}$ calcium oxide + carbon dioxide
$CaCO_3(s) \rightarrow CaO(s) + CO_2(g)$

The calcium oxide produced combines with acidic impurities, e.g. silicon(IV) oxide (sand), SiO_2, in the iron ore. A molten mixture of compounds called a **slag** is formed:

calcium oxide + silicon(IV) oxide (sand) $\xrightarrow{\text{heat}}$ calcium silicate (slag)
$CaO(s) + SiO_2(s) \rightarrow CaSiO_3(l)$

The process runs continuously. The raw materials are fed in at the top, and molten iron and molten slag are run off separately at the bottom. The slag goes to builders and road-makers, who use it to lay foundations. The process is much cheaper to run than an electrolytic method. The availability of the raw materials and the low cost of extraction make iron cheaper than other metals.

Copper

Copper is far down the reactivity series. It is found 'native' (uncombined) in some parts of the world. It is also mined as the sulphide. This is roasted in air to give impure copper. Pure copper is obtained from this by the electrolytic method shown in Figs. 10.22 and 10.23.

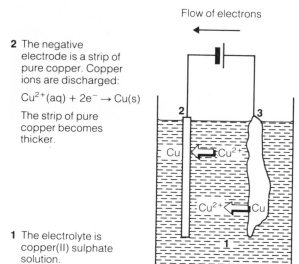

2 The negative electrode is a strip of pure copper. Copper ions are discharged:

$$Cu^{2+}(aq) + 2e^- \rightarrow Cu(s)$$

The strip of pure copper becomes thicker.

1 The electrolyte is copper(II) sulphate solution.

3 The positive electrode is a lump of impure copper. When a current flows, this electrode takes electrons from copper atoms. These go into solution as copper ions:

$$Cu(s) \rightarrow Cu^{2+}(aq) + 2e^-$$

The electrons flow through the external circuit from the positive electrode to the negative electrode.

Fig. 10.22 Purifying copper by electrolysis

Fig. 10.23 'Electrowinning' of copper

After the cell has operated for a week, the negative electrode becomes very thick. It is lifted out of the cell, and a new thin sheet of copper is put in its place. When all the copper has been removed from the impure slab of metal which forms the positive electrode (the anode), a new piece of impure copper is substituted. Undissolved matter from the old anode lies on the bottom of the cell as **anode sludge**.

Any impurities in the copper are likely to be metals such as iron and zinc, which are more reactive than copper. Their ions stay in solution while copper ions are discharged. Metals which are less reactive than copper, that is, silver and gold, accumulate in the anode sludge. They can be extracted from this material.

Silver and gold

The unreactive metals, silver and gold, are found 'native'. Figure 10.24 shows a prospector 'panning' for gold in Northern Ireland, where a new deposit was discovered in 1982.

Fig. 10.24 Panning for gold

JUST TESTING 52

1 Refer to the electrolytic method of purifying copper in Fig. 10.22.
   Impure copper contains other metals, such as iron. Why do iron ions remain in solution while copper ions are discharged?
   Write the equations for the processes which occur at the electrodes. (See Section 5.10 for help.)

2 Why has the human race been using copper, silver and gold for longer than other metals?

3 Suggest what A, B, C and D might be.
   A is a metal which reacts violently with water to form an alkaline solution.
   B is a metal which does not react with steam or with dilute hydrochloric acid.
   C is a metal which, when exposed to air, immediately becomes coated with a layer of oxide.
   D is a metal which is easily worked and which reacts only very very slowly with water.

4 Copy and complete the following equations. Be careful! Some of the suggested reactions do not happen. In these cases write 'no reaction'.
   a calcium + dilute hydrochloric acid →
   b copper + dilute sulphuric acid →
   c copper + oxygen →
   d gold + dilute sulphuric acid →
   e aluminium + iron(III) oxide →
   f tin + iron(III) oxide →
   g carbon monoxide + iron(III) oxide →
   h carbon monoxide + aluminium oxide →
   i zinc + copper(II) sulphate solution →
   j lead + dilute sulphuric acid →

SUMMARY NOTE

The method used for extracting a metal from its ores depends on the position of the metal in the reactivity series. For the very reactive metals (e.g. sodium), no chemical method of reduction can be used. Electrolysis of a molten compound is employed. Less reactive metals (e.g. iron and zinc) are obtained by the action of a chemical reducing agent (e.g. carbon or hydrogen) on the oxide. The metals at the bottom of the reactivity series occur 'native'.

10.10 Focus on iron and steel

The iron that comes out of the blast furnace is called **pig iron** or **cast iron** (see Section 10.1). It contains 3% to 4% carbon. The carbon content lowers the melting point and makes cast iron easier to melt and mould than pure iron. Cast iron expands slightly as it solidifies. This makes it flow into all the corners of a mould and reproduce the shape exactly. By casting, objects with complicated shapes can be made, e.g. engine blocks (see Fig. 10.1).

The carbon atoms in cast iron interfere with the metallic bonding (see Section 10.2) and make the metal brittle. Cast iron cannot be bent without snapping. It cannot be used for the bodywork of a car (Fig. 10.1).

Iron which contains less than 0.25% carbon is called **wrought iron**. Wrought iron is tough, is easily worked and resists corrosion well (Fig. 10.25). Ornamental gates and horse shoes used to be made from wrought iron. Nowadays, mild steel is used.

Fig. 10.25 Cannon from the Mary Rose: the cannon is made of bronze; cannonballs are made of iron

Steel

The high carbon content of cast iron makes it brittle. Fortunately, carbon can be burnt off as its oxides, the gases carbon monoxide, CO, and carbon dioxide, CO_2. Iron is less easily oxidised. Other impurities (e.g. sulphur, phosphorus and silicon) are also converted into acidic oxides. These are not gases, and something must be added to remove them. A base, such as calcium oxide, will do this. It combines with acidic oxides to form a mixture of compounds of

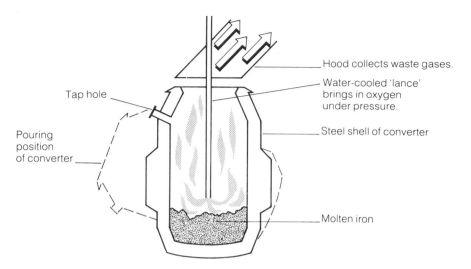

Fig. 10.26 A basic oxygen furnace holding 150–300 tonnes of steel

low melting point called a **slag**. When the percentage of carbon is less than 1%, the iron–carbon alloy is a **steel**. Other elements may be alloyed with steel to give alloys with different properties.

Figure 10.26 shows a **basic oxygen furnace**. In it, cast iron is converted into steel. One converter can produce 150–300 tonnes of steel in an hour.

There are various types of steel. They differ in their carbon content and are used for different purposes.

Table 10.12 Cast iron and types of steel and their uses

Type of steel	Properties	Uses
Mild steel (<0.25% carbon)	Pliable (can be bent without breaking).	Chains and pylons
Medium steel (0.25–0.45% carbon)	Tougher than mild steel, more springy than high carbon steel	Nuts and bolts Car springs and axles Bridges
High carbon steel (0.45–1.5% carbon)	The carbon content makes it both tough and brittle.	Chisels, files, razor blades, saws, cutting tools
Cast iron (2.5–4.5% carbon)	Cheaper than steel, easily moulded into complicated shapes	Drain pipes, fire grates, engine blocks (these articles will break if they are hammered or dropped)

Alloy steels

Metallurgists are constantly designing new alloys suited to different uses. Many elements are used for alloying with steel. The different alloy steels have different properties which fit them for different uses.

Table 10.13 Different alloy steels

Element	Properties of alloy	Uses
Chromium	Prevents rusting if > 10%.	Stainless steels, acid-resisting steels, cutlery, car accessories, tools
Cobalt	Takes a sharp cutting edge. Can be strongly magnetised.	High-speed cutting tools Permanent magnets
Manganese	Increases strength and toughness.	Some is used in all steels; steel in railway points and safes contains a high percentage of manganese.
Molybdenum	Strong even at high temperatures	Rifle barrels, propeller shafts
Nickel	Resists heat and acids.	Stainless steel cutlery, industrial plants which must withstand acidic conditions
Tungsten	Stays hard and tough at high temperatures.	High-speed cutting tools
Vanadium	Increases springiness.	Springs, machinery

Figure 10.27 illustrates the relationship between iron and steel.

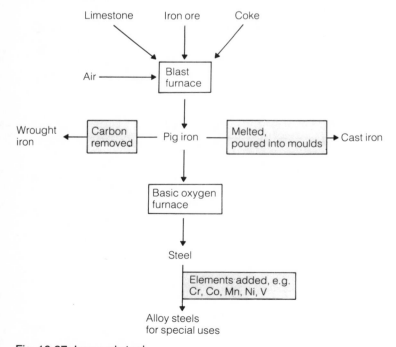

Fig. 10.27 Iron and steel

SUMMARY NOTE

- Cast iron contains up to 4% carbon. It is easy to mould, but it is brittle.
- Wrought iron is the purest form of iron. It is easily worked without breaking.
- Steels contain carbon. Mild (low carbon) steel is pliable; high carbon steel is hard; medium steel is intermediate. The steels are suited to different uses.
- Alloy steels contain other elements. The other elements are added to give a steel with the right characteristics for a particular job. You should know a few examples of steels and their uses.

JUST TESTING 53

1. Think what is the most important characteristic of the steel that must be used for making each of the articles listed below. Should the steel be pliable, or springy or hard? Then say whether you would use mild steel or medium steel or high carbon steel or cast iron for making these articles.
   a car axles, **b** axes, **c** car springs, **d** ornamental gates, **e** drain pipes, **f** chisels, **g** sewing needles, **h** picks, **i** saws, **j** food cans.
2. What is the difference between cast iron and wrought iron? Name two objects made from cast iron and two objects made from wrought iron.

10.11 Rusting of iron and steel

Experiments 8.2, 8.3, 10.9 and 10.10 will introduce you to this topic. Iron and steel rust. Rust is hydrated iron(III) oxide, $Fe_2O_3 \cdot nH_2O$. The number n of water molecules in the formula varies. The conditions iron needs to rust are water and air and acidity. (See Fig. 10.28 and Question **8** of Just Testing 54.) The carbon dioxide present in air provides sufficient acidity. The presence of salts increases the speed at which iron rusts.

The rusting of iron is a serious problem. Replacing rusted iron and steel structures costs the UK £500 million a year. Car engines have a very long life, but the average British car lasts only ten years.

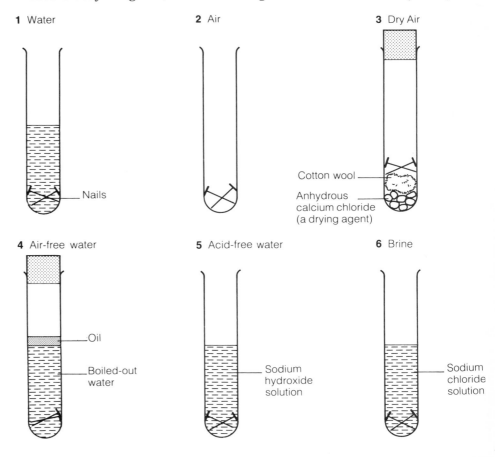

Fig. 10.28 Experiments on rusting

Table 10.14 Rust prevention

Method	Where it is used	Comment
1 A coat of paint	Large objects, e.g. ships and bridges	If the paint is scratched, the iron beneath it starts to rust. Corrosion can spread under parts of the paintwork which are still sound.
2 A film of oil or grease	Moving parts of machinery, e.g. a bicycle chain	The protective film must be renewed.
3 A coat of metal		
Chromium plating	Car bumpers, cycle handlebars	Applied by electroplating (see Chapter 5). Gives a decorative finish as well as protection.
Galvanising (zinc plating)	Galvanised steel girders are used in the construction of buildings and bridges.	Even if the layer of zinc is scratched, the iron underneath does not rust. Zinc is above iron in the reactivity series: as long as some zinc remains, zinc will corrode in preference to iron.
Tin plating	Food cans	Steel cans are coated with tin by electroplating them or by dipping them into molten tin. Tin is below iron in the reactivity series. If the layer of tin is scratched, the iron beneath it rusts. Zinc cannot be used for food cans because zinc and its compounds are poisonous.
4 Stainless steel	Cutlery, car accessories (e.g. radiator grilles)	Steels containing chromium (10%–25%) or nickel (10%–20%) do not rust.
5 Sacrificial protection	Ships	Bars of zinc are attached to the hull of the ship. Being above iron in the reactivity series, zinc is corroded. As long as the zinc bars remain, they protect the ship from rusting. When they have corroded, they must be replaced.
	Underground pipes	Bags of magnesium scrap are attached to underground pipes. Magnesium corrodes in preference to the pipes. From time to time, the magnesium must be replaced.

SUMMARY NOTE

Iron and steel rust when exposed to air and water in slightly acidic conditions. Many different treatments are used to protect iron and steel from rusting:
- a protective coat of oil or grease or another metal (e.g. chromium, zinc or tin)
- alloying with nickel and chromium
- the sacrifice of a more reactive metal (e.g. magnesium or zinc).

Corrosion of the bodywork is the factor that limits the life of the car. Table 10.14 lists some of the methods which are used to protect iron and steel against rusting. Some of them use a protective coating to exclude water and air, while others work by sacrificing a more reactive metal.

The Earth contains vast deposits of metal ores. During the twentieth century we have used more metal than in all the previous centuries together. We need our machinery and our means of transport. The Earth's resources are not inexhaustible. If we keep on mining iron ore and allowing tonnes of iron and steel to rust every year, we may one day run short of iron. We throw tonnes of used iron and steel objects on the scrap heap (Fig. 10.29). It makes better sense to collect scrap iron and steel and recycle it, that is, melt it down and re-use it.

Fig. 10.29 Car dump

JUST TESTING 54

1. In what ways is steel better than iron?
   What chemical reaction occurs when iron is converted into steel?
   Briefly describe one industrial method of converting iron into steel.
2. What are **a** cast iron, **b** wrought iron, **c** steel?
   Give one use for each metal.
3. Describe how the rusting of iron is prevented
   a on the moving parts of a bicycle
   b on a food can
   c on parts of a ship above the water-line
   d on parts of a ship below the water-line.
4. Galvanised iron is used for roofing. Explain why this is a weatherproof material.

5 Why did the Bronze Age come before the Iron Age? Give two examples of things which were made during the Industrial Revolution and which needed iron or steel for their manufacture.

6 Cars are spray painted. The paint reaches some areas more easily than others. On which parts of a car body is the paint layer likely to be thinnest? Which parts of a car body are the most likely to be affected by the salt that is spread on roads in winter?

*7 A number of metals are used in making a car. A solution of salt is an electrolyte. What happens when two different metals are in contact with an electrolyte solution? (See Section 5.13 if you need help.)

8 Refer to Fig. 10.28.
   a In which tubes do the iron nails rust?
   b Explain the difference between tubes **2** and **3**. Are the times taken for the nails to rust different?
   c What has been boiled out of the water in tube **4**? What does the layer of oil do?
   d The nails in tube **5** do not rust. What does this show about rusting?
   e Why is experiment **6** important? Where does iron come into contact with sodium chloride solution?

10.12 Focus on aluminium

Uses of aluminium

Aluminium is a versatile metal with thousands of applications. It is not a very strong metal. Its alloys, such as duralumin (which contains copper and magnesium), are used when strength is needed (see Table 10.15).

Austin Rover's car of tomorrow

The MG EX-E was unveiled at the Frankfurt Motor Show in 1985. The aluminium 'space frame' chassis is held together with glue instead of being welded. It is covered with a plastic body made of injection-moulded panels (see Section 17.4). It is powered by a three litre all-aluminium engine. The streamlined body reduces 'drag'. The car can go from 0 to 60 m.p.h. in 5 seconds.

Fig. 10.30 The MG EX-E

Aluminium wire can be pulled out *really* fine. Guess how much aluminium you'd need to make a wire that would go right round the world!

Only 10 kilograms. Am I right?

Table 10.15 Some uses of aluminium and its alloys

Property	Use for aluminium which depends on this property
1 Low density	Aircraft manufacture: the alloy **duralumin** is used because it is stronger than aluminium.
2 Good electrical conductor	Used for overhead cables. It is replacing copper because cables of aluminium are lighter and need less massive pylons to carry them.
3 Never corroded (except by bases)	Door-frames and window-frames are often made of aluminium. **Anodised aluminium** is used. The thickness of the protective layer of aluminium oxide has been increased by **anodising** (see Experiment 5.10).
4 Non-toxic	Packaging food: milk bottle tops, cases for frozen foods etc., baking foil
5 Reflects light when polished	Car headlamp reflectors
6 Good thermal conductor	Saucepans etc.
7 When highly polished, insulates by reflecting heat	Aluminium can also be used as a thermal insulator. It reflects heat when it is highly polished. Aluminium blankets are used to wrap premature babies. They keep the baby warm by reflecting heat back to the body. Firefighters wear aluminium fabric suits to reflect heat away from their bodies.

Some problems

Bauxite is found near the surface in Australia, Guinea, Brazil and other countries. It is extracted by open-cast mining (Fig. 10.31). A surface layer 1 m to 60 m thick is excavated, and an ugly scar is left on the landscape. In some places, mining companies have taken measures to restore the landscape after a mine has been exhausted.

Often bauxite is converted into pure aluminium oxide at the mine. An impurity which is always found in bauxite is iron(III) oxide, which is red. After it has been separated from the ore, iron(III) oxide is pumped into vast ponds which become the unsightly **red mud ponds**. Near Quebec in Canada is a vast area (many square kilometres) of red mud ponds. The mud cannot be pumped into the sea because it would harm the fish.

After purification, aluminium oxide is shipped to an aluminium **smelter** (a plant for extracting aluminium). The high consumption of electricity makes the electrolytic method of obtaining aluminium expensive. Aluminium smelters are often built in areas which have **hydroelectric power** (electricity from water-driven generators). This is relatively cheap electricity. Waterfalls and fast-flowing rivers provide hydroelectric power. They are found in areas of outstanding

Fig. 10.31 Open-cast mining

natural beauty, such as the highlands of Scotland and Norwegian fiords. Conservationists often object to the siting of aluminium smelters in such beauty spots.

Sometimes there are other objections. There is the cost of carrying ore to a remote area and then transporting aluminium away for sale. There is the problem of either finding a workforce locally or bringing workers into the area and providing housing. It is often more cost-effective to build an aluminium plant in an area which has a bigger population and better transport, and to pay more for electricity from a coal-fired power station.

The exhaust gases from aluminium smelters are 'scrubbed' with water. The waste water contains fluoride from the electrolyte. It is discharged into rivers. The remaining exhaust gases are discharged into the atmosphere through tall chimneys. Fluoride emissions have been known to kill grass and cause tooth decay and lameness in cattle. In the past, farmers in the USA successfully sued aluminium smelters for damage to their cattle. Smelters now take more care to control fluoride emission. Chemists are working on the possibility of replacing the fluoride in the electrolysis cell with a chloride.

SUMMARY NOTE

Aluminium is a metal with a multitude of uses. With its low density and its resistance to corrosion, it has advantages over other metals.

Both the mining and the smelting of aluminium create environmental problems. Open-cast mines and red mud pools spoil the landscape. Fluoride emission has harmed cattle.

Before a decision is made on a site for a new aluminium smelter, various factors are weighed up. They include
- the cost of electricity
- transport costs
- the availability of a workforce.

Fig. 10.32 One way of saving resources: recycle aluminium

JUST TESTING 55

1 The needs of an aluminium plant are
- raw materials, purified bauxite and cryolite imported by sea
- electric power (from coal-fired or gas-fired or hydroelectric power stations)
- transport to bring raw materials to the site and take away the product by rail and sea
- a workforce.

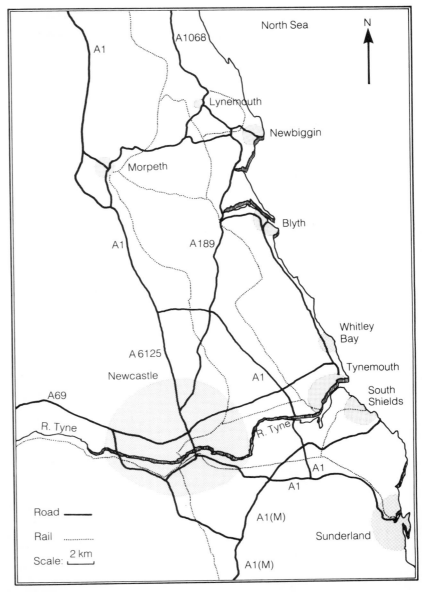

Fig. 10.33 Alcan Aluminium (UK) Ltd., Lynemouth (opened 1972)

Study the map in Fig. 10.33. Explain how the choice of the Lynemouth site satisfies these needs.

2 Why was aluminium smelting carried out at Niagara Falls? What is the disadvantage of siting a plant there?

3 Emperor Napoleon III gave money to finance research on the extraction of aluminium. He was interested in the possibility of aluminium armour for his troops. What advantages over iron would aluminium have for armour?

4 Copy this passage, and fill in the gaps.
Aluminium is extracted from the ore _____. The ore is purified by removing _____, which is pumped away to form ugly _____. After purification, _____ aluminium oxide is dissolved in molten _____ and electrolysed. The electrodes are made of _____. Aluminium is deposited at the _____. At the other electrode, _____ is formed.

> Some of this reacts with the electrode to form the gas _____. The exhaust gas from an aluminium smelter is scrubbed to remove the pollutant _____. This pollutant is discharged into _____.
>
> *5 Find the percentage of aluminium in bauxite, $Al_2O_3 \cdot 2H_2O$ (after studying Chapter 11).
>
> 6 What advantages does aluminium have over steel? Give three examples of objects which are made from aluminium. Point out why aluminium is chosen for their manufacture.

10.13 Some problems which metallurgists have solved

Which metal can be used as the filament in an electric light bulb?

The filament of an electric light bulb glows when it is heated. The filament must be made of metal because metals are the only good electrical conductors. Cost is not of great importance in selecting a metal for the job because the mass of metal needed is small. The colour of light emitted depends on the temperature: at 2500 °C, white light is emitted. To heat the filament to this temperature, an electric current is passed through it. More heat is generated when an electric current passes through a thin wire than through a thick wire. A thin filament must be used.

The filament metal must fit this description:
- It must not melt or evaporate at 2500 °C.
- It must be able to withstand the thermal shock of being switched on and off.
- It must not break when it is drawn into very fine wire: it must be ductile.

The only metal which combines the properties of high melting point (3410 °C) and high ductility is tungsten.

If tungsten came into contact with air at 2500 °C, it would burn. The filament is therefore enclosed in a glass bulb filled with a noble gas, usually argon (Fig. 10.34).

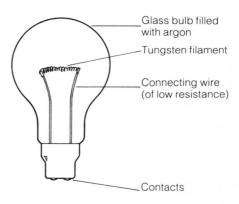

Fig. 10.34 An electric light bulb

Which metal can be used as the 'element' in an electric fire?

Heat is radiated by a wire glowing red hot at 800 °C. The wire must have a high resistance if it is to give out much heat when an electric current is passed through it. It must not be oxidised in air at 800 °C. An alloy of nickel and chromium (80% Ni, 20% Cr) is used. It can operate at the working temperature without melting or becoming deformed or corroded.

The wire for the 'element' is wound on a ceramic rod. Ceramics are thermal and electrical insulators. A reflecting surface is placed behind the element. Stainless steel is used for this purpose. It can be polished to give a shiny surface, and it does not tarnish (Fig. 10.35).

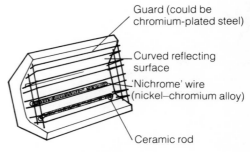

Fig. 10.35 An electric fire

Which metal can be used for water pipes?

Water pipes are needed in such large quantities that they must be made from a reasonably inexpensive material. The material must be strong and hard wearing. It must be easily worked when the pipes are being installed and during later repairs. Only metals have the strength and the ease of working needed for the purpose. It is essential that the metal used does not become corroded by constant contact with water, and that it is completely non-toxic. The metals available in quantity are iron, copper, lead, aluminium and stainless steel.

Lead is easy to work, even when cold. It was the traditional metal to use for water pipes. It gave its name to plumbing (plumbum = lead in Latin; plwm = lead in Welsh). Over a period of many years, lead reacts very slowly with water to form lead compounds. These are poisonous, and lead is no longer used for water pipes.

Copper is the next easiest metal to work with at room temperature. Joints are easy to make by soldering.

Stainless steel pipes are an alternative. Joining stainless steel pipes is more difficult. Stainless steel contains chromium. A layer of chromium oxide forms on the surface of the metal. This layer protects the metal beneath it from attack by air or water. It does, however, make soldering difficult. The oxide layer must be removed before a soldered joint can be made. A **flux**, containing phosphoric acid, is used to remove the chromium oxide so that the metal beneath it can alloy with the solder, and soldered joints can be made.

Aluminium also has a surface film of the oxide. This layer of oxide must be removed before the metal can alloy with solder. It is more difficult to remove the surface layer from aluminium than from stainless steel.

Stainless steel and copper are the main choices for water pipes.

Mild steel pipes are used in central heating radiators. They are less expensive than stainless steel. The same water circulates round and round through the pipes and radiators. A layer of calcium carbonate forms inside the pipes and radiators (see Section 9.13 on hard water). This layer protects the mild steel pipes from attack by water. Mild steel cannot be used for hot water tanks because the water is constantly being changed.

Cast iron is used for the large diameter pipes needed for water mains. These pipes must be very strong to withstand changes in the pressure of the ground around them.

Exercise 10

1 Which of the following are electrical conductors?
sulphur, iron, zinc, mercury, charcoal, steel, diamond, copper, brass, graphite.

2 Name
   a a metal which is liquid at room temperature
   b a metal widely used in electrical wiring
   c a metal which is soft enough to leave marks on paper
   d a metal used for reflecting light
   e a metal used in aircraft construction
   f a metal which is alloyed to form solder
   g the metal present in both brass and bronze
   h the metal other than tin in 'tin' cans.

3 Name two gases which react with heated magnesium to give magnesium oxide. Write word equations for the reactions.
  *Write chemical equations.

4 Explain why silver is found 'native', but calcium is found as its compounds.

5 The following is a list of metals in order of decreasing reactivity. **X** and **Y** are two unknown metals:
  K **X** Ca Mg Al Zn **Y** Fe Cu
  a Will **X** react with cold water?
  b Will **Y** react with cold water?
  c Will **Y** react with dilute hydrochloric acid?
  Explain how you arrive at your answers.
  d What reaction would you expect between **X** and copper(II) sulphate solution?
  e Which is more easily decomposed, **X** sulphate or **Y** sulphate?

6 Here are the characteristics of three types of solder:
  Solder 1: 40% lead, 60% tin, melts at 180 °C
  Solder 2: 70% lead, 30% tin, melts at 80–260 °C
  Solder 3: 5% tin, melts at 310 °C
  Which solder would you use for each of the following jobs? Explain your choice.
  a Filling a gap in a car body (after removing rust, before repainting)
  b Mending a leaking car radiator
  c Joining wires in an electric circuit.

7 Look at the following pairs of chemicals. If a reaction happens, copy the word equation and complete the right-hand side.
  a carbon + lead(II) oxide →
  b hydrogen + calcium oxide →
  c aluminium + tin(II) oxide →
  d magnesium + sulphuric acid →
  e carbon + magnesium oxide →
  f hydrogen + potassium oxide →
  g carbon monoxide + iron(III) oxide →
  h carbon monoxide + aluminium oxide →
  i lead + sulphuric acid →
  j copper + zinc sulphate →

*8 Look at the following pairs of chemicals. If a reaction takes place, copy the equation and complete the right-hand side. Balance the equation.
  a $C(s) + CuO(s) \rightarrow$
  b $Al(s) + PbO(s) \rightarrow$
  c $Au(s) + O_2(g) \rightarrow$
  d $H_2(g) + AgO(s) \rightarrow$
  e $CO(g) + Fe_2O_3(s) \rightarrow$
  f $Ca(s) + HCl(aq) \rightarrow$
  g $Zn(s) + CuSO_4(aq) \rightarrow$
  h $Cu(s) + H_2SO_4(aq) \rightarrow$
  i $Na(s) + O_2(g) \rightarrow$
  j $Cu(s) + O_2(g) \rightarrow$
  k $Pb(s) + CuSO_4(aq) \rightarrow$
  l $Al(s) + CaO(s) \rightarrow$

9 Aluminium is obtained by electrolysing aluminium oxide dissolved in molten cryolite at 700 °C. Carbon electrodes are used.
  a At which electrode is aluminium discharged?
  b What is discharged at the other electrode?
  c Why are carbon electrodes used? Why are they gradually eaten away?
  d When a solution of an aluminium salt is electrolysed, aluminium is not obtained. Why is this? What is formed?
  e Where would you expect an aluminium refinery to be situated?
  *f Write the equations for the electrode processes in **a** and **b**.

*10 For centuries, alchemists tried to change other metals into gold. Why did they value gold more than other metals? If the alchemists had succeeded, and gold were a cheap metal, what uses do you think we would make of gold today? For what purposes is gold unsuitable?
  (Density = 19.3 g/cm^3; m.p. = 1063 °C)

11 From which metal is each of the following objects made? Explain why that metal is chosen.
  a a kettle, b telephone wire, c a post box, d a pen nib, e a horse shoe, f a church bell, g a trumpet, h baking foil, i a ship's propeller.

12 Aluminium is used for the manufacture of a saucepans, b overhead cables, c small boats, d aeroplanes, e hospital blankets. Explain what physical property of aluminium makes it suitable for each of these uses.

13 a • The thermit reaction shows that aluminium is a reactive metal.
     • Aluminium is used for door-frames and window-frames.
     How can both these statements be true?
   b • Aluminium conducts heat.
     • Aluminium is used as a thermal insulator.
     How can both these statements be true?

14 a Aluminium oxide is a common mineral. Seven thousand years elapsed between the discovery of copper and the day when

aluminium was first seen. What was the reason for this gap in time?
b Explain why recycling aluminium is easier than recycling scrap iron.

15 Draw a blast furnace. Name the raw materials used in it. Name the two substances produced.

16 Explain how steel is made from cast iron. What advantages does steel have over cast iron? Draw a diagram of the process used to convert cast iron into steel.

17 Explain how rusting is prevented by
a galvanising, b tin plating, c sacrificial protection, d painting.

18 Explain why
a saucepans are made from aluminium
b aeroplanes are made from aluminium alloys
c bells are made from bronze
d bridges are made from steel
e solder is made from lead
f dental amalgams are made from mercury
g teeth can be capped with gold caps.

19 Write short notes to explain the meanings of the following words:

a element, b malleable, c ductile, d abundant, e alloy.

20 a Copy this table. In the gaps write a tick if a chemical reaction occurs and a cross if no reaction happens.

Metal	Reaction with cold water	Reaction with hot water	Reaction with dilute hydrochloric acid
Sodium			
Zinc			
Magnesium			
Tin			
Calcium			
Copper			

b Write a list of the metals in order of increasing reactivity.

21 The map below shows Port Talbot and its surroundings. Explain why a big iron and steel industry developed here. List the needs of the industry, and explain how Port Talbot meets these needs.

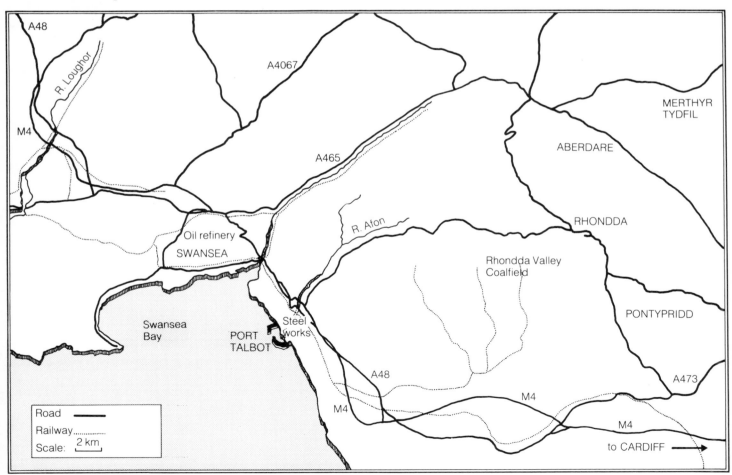

22 Some of the metals and alloys used in spacecraft are listed below.

Metal	Density in g/cm^{-3}	Melting point in °C	Useful properties
Stainless steel	7.8	1400	Very strong
Aluminium	2.7	660	Resists corrosion. Can be strengthened by alloying
Beryllium	1.8	1280	Keeps its strength at high and low temperatures
Duralumin	2.8	600	Hard and strong
Gold	19.3	1060	Highly reflecting. Very malleable
Magnesium	1.7	650	Strong
Titanium	4.5	1670	Keeps its strength at high and low temperatures

a A Skylark research rocket is made of stainless steel and duralumin. Explain why these metals are chosen.

b A satellite carries sensitive electronic equipment. Very high temperatures could damage this equipment. The danger is avoided by covering the satellite with metal foil. Which metal do you think is used? How does it avoid high temperatures inside the satellite?

c Military aircraft reach speeds higher than that of Concorde. The engine must withstand high temperatures without being deformed. Which metal or metals in the list could be used?

d Space rockets are jet-propelled. The fuel burns to form a stream of gas at high temperature and high pressure, which passes out of a nozzle. Fuel does not burn continuously; the nozzle experiences the low temperature of space alternating with the high temperature of the exhaust gases. The metal chosen for a space rocket motor must also be of low density and high melting point. Which of the metals listed is suitable?

e The chief structural material in the Prospero spacecraft is magnesium. The exterior panels are made of glass-reinforced plastics. What is the chief advantage of magnesium over the other metals listed? Magnesium is chosen for satellites which have to orbit the Earth for many years because it can withstand irradiation by cosmic rays. What kind of space vehicle could *not* be made of magnesium? Why?

CHAPTER 11 Chemical calculations

11.1 Relative atomic mass

The masses of atoms are very small. Some examples are
- mass of hydrogen atom, H = 1.4×10^{-24} g
- mass of lead atom, Pb = 2.9×10^{-22} g
- mass of carbon atom, C = 1.7×10^{-23} g.

These masses are so uncomfortable to deal with that we use **relative atomic masses**. The hydrogen atom is the lightest of atoms. Chemists had the idea of comparing the masses of other atoms with that of the hydrogen atom. On the relative atomic mass scale

- the relative atomic mass of hydrogen is 1
- the relative atomic mass of lead is 207 (a lead atom is 207 times heavier than a hydrogen atom)
- the relative atomic mass of carbon is 12 (a carbon atom is 12 times heavier than a hydrogen atom).

The symbol for relative atomic mass is A_r. Since relative atomic mass is a ratio of two masses, the mass units cancel and relative atomic mass is a number without a unit.

A complete list of relative atomic masses is given on p. 316. Those of some common elements are listed in Table 11.1.

*For more advanced work, chemists take the mass of one atom of carbon-12 as the reference point for the relative atomic mass scale. On this scale

$$\text{Relative atomic mass of element} = \frac{\text{Mass of one atom of the element}}{(1/12) \text{ Mass of one atom of carbon-12}}$$

Table 11.1 Some relative atomic masses

Element	A_r	Element	A_r
Aluminium	27	Magnesium	24
Barium	137	Mercury	200
Bromine	80	Nitrogen	14
Calcium	40	Oxygen	16
Chlorine	35.5	Phosphorus	31
Copper	63.5	Potassium	39
Hydrogen	1	Sodium	23
Iron	56	Sulphur	32
Lead	207	Zinc	65

SUMMARY NOTE

The relative atomic mass A_r of an element is the mass of one atom of the element compared with the mass of one atom of hydrogen. On the relative atomic mass scale, $A_r(H) = 1$.

JUST TESTING 56

1. Refer to Table 11.1.
   a. How many times heavier is one atom of barium than one atom of hydrogen?
   b. What is the ratio
      mass of one atom of mercury to mass of one atom of calcium?
   c. How many atoms of nitrogen are needed to equal the mass of one atom of iron?
   d. How many atoms of sodium are needed to equal the mass of one atom of lead?

11.2 Relative molecular mass

The mass of a molecule is the sum of the masses of all the atoms in it. The relative molecular mass M_r of a compound is the sum of the relative atomic masses of all the atoms in a molecule of the compound (Fig. 11.1).

You can find the relative molecular mass of sulphuric acid in this way:

Formula of compound is H_2SO_4
2 atoms of H ($A_r = 1$) = 2
1 atom of S ($A_r = 32$) = 32
4 atoms of O ($A_r = 16$) = 64
Total   = 98
Relative molecular mass M_r of H_2SO_4 = 98

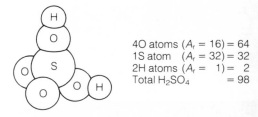

4O atoms ($A_r = 16$) = 64
1S atom ($A_r = 32$) = 32
2H atoms ($A_r = 1$) = 2
Total H_2SO_4 = 98

Fig. 11.1 Atoms in a molecule of sulphuric acid

11.3 Relative formula mass

Many compounds consist of ions, not molecules (see Chapter 5). For ionic compounds, the formula represents a **formula unit**, rather than a **molecule** of the compound. A formula unit of calcium chloride is $CaCl_2$. You cannot call it a molecule of calcium chloride because it consists of ions. The term **relative formula mass** is used for ionic compounds, and it can be used for covalent compounds as well. The symbol M_r can be used for both covalent compounds and ionic compounds.

SUMMARY NOTE

The relative molecular mass of a covalent compound equals the sum of the relative atomic masses of all the atoms in one molecule of the compound.

M_r = Sum of A_r values

The relative formula mass of an ionic compound equals the sum of the relative atomic masses of all the atoms in a formula unit of the compound.

JUST TESTING 57

1 Work out the relative formula masses of these compounds: CO_2, SO_2, NaOH, KCl, $Ca(OH)_2$, HNO_3, $CuCO_3$, NH_4NO_3, $CuSO_4$, $CuSO_4 \cdot 5H_2O$, $Mg(HCO_3)_2$.

11.4 Percentage composition

From the formula of a compound you can find the percentage by mass of the elements in the compound.

Example 1 Find the percentage by mass of (a) barium and (b) oxygen in barium oxide.

Method First find the relative formula mass.
The formula is BaO.
Relative formula mass = $A_r(Ba) + A_r(O)$
= 137 + 16 = 153

Percentage of barium = $\frac{137}{153} \times 100 = 89.5\%$

Percentage of oxygen = $\frac{16}{153} \times 100 = 10.5\%$

You can see that the two percentages add up to 100%.

*Example 2 Find the percentage of water in crystals of magnesium sulphate-7-water.

Method Find the relative formula mass.
The formula is $MgSO_4 \cdot 7H_2O$.

Relative formula mass = Sum of relative atomic masses

1 atom of magnesium ($A_r = 24$)	= 24
1 atom of sulphur ($A_r = 32$)	= 32
4 atoms of oxygen ($A_r = 16$)	= 64
7 molecules of water, $7 \times [(2 \times 1) + 16]$	= 126
Total = Relative formula mass	= 246

Percentage of water = $\dfrac{\text{Mass of water in formula}}{\text{Relative formula mass}} \times 100\%$

$= \dfrac{126}{246} \times 100\% = 51.2\%$

The percentage of water in magnesium sulphate crystals is 51%.

> **SUMMARY NOTE**
>
> You can calculate the percentage by mass composition of a compound from its formula.

JUST TESTING 58

You do not need calculators for these problems.

1. Find the percentage by mass of beryllium, Be, in beryllium oxide, BeO.

2. Find the percentage by mass of iron in iron(III) oxide, Fe_2O_3.

3. Calculate the percentages by mass of
   a carbon and hydrogen in ethane, C_2H_6
   b nitrogen and oxygen in nitrogen monoxide, NO
   c magnesium and nitrogen in magnesium nitride, Mg_3N_2
   d sodium and iodine in sodium iodide, NaI.

4. Calculate the percentage by mass of water in
   a copper(II) sulphate-5-water, $CuSO_4 \cdot 5H_2O$ (take $A_r(Cu) = 64$)
   b sodium sulphide-9-water, $Na_2S \cdot 9H_2O$.

*11.5 The mole

Chemical equations tell us what products are formed when substances react. Equations can also be used to tell us *how much* product as well as *which* product, that is, *what mass* of solid product or *what volume* of gaseous product is formed. A reaction of industrial importance is

calcium carbonate → calcium oxide + carbon dioxide
$CaCO_3(s)$ → $CaO(s)$ + $CO_2(g)$

Cement manufacturers use this reaction to make calcium oxide (lime) from calcium carbonate (limestone). They sell the carbon dioxide gas produced to fizzy drink manufacturers. It is possible to calculate what mass of calcium oxide and what volume of carbon dioxide are formed when a certain mass of calcium carbonate dissociates. The key to the calculation is the idea which chemists call the **mole concept**.

We owe the reasoning behind the mole concept to a nineteenth century Italian chemist called Avogadro. This is how he argued:

We know from their relative atomic masses that one atom of magnesium is twice as heavy as one atom of carbon:
$A_r(Mg) = 24$, $A_r(C) = 12$. *Therefore we can say:*

If 1 atom of magnesium is twice as heavy as 1 atom of carbon, then 1 hundred Mg atoms are twice as heavy as 1 hundred C atoms,
and 5 million Mg atoms are twice as heavy as 5 million C atoms,
and it follows that, if we have a piece of magnesium which has twice the mass of a piece of carbon, the two masses must contain equal numbers of atoms.

2 grams of magnesium and 1 gram of carbon contain the same number of atoms;

10 tonnes of magnesium and 5 tonnes of carbon contain the same number of atoms.

The same argument applies to the other elements. Take the relative atomic mass in grams of any element:

12 g carbon	24 g magnesium	56 g iron	40 g calcium	108 g silver	238 g uranium	207 g lead

All these masses contain the same number of atoms. The number is 6.022×10^{23}.

> The amount of an element that contains 6×10^{23} atoms (the same number of atoms as 12 g of carbon-12) is called one **mole** of that element.

The symbol for **mole** is **mol**. The ratio 6.022×10^{23}/mol is called the **Avogadro constant**. When you weigh out 12 g of carbon, you are counting out 6×10^{23} atoms of carbon. This amount of carbon is one mole (1 mol) of carbon atoms. Similarly, 80 g of calcium is two moles (2 mol) of calcium atoms. You can say that the **amount** of calcium is two moles (2 mol).

You can have a mole of carbon atoms, C, a mole of sodium ions, Na^+, a mole of sulphuric acid molecules, H_2SO_4. One mole of sulphuric acid contains 6×10^{23} molecules of H_2SO_4, that is, 98 g of H_2SO_4 (the molar mass in grams). To write 'a mole of oxygen' is imprecise: one mole of oxygen atoms, O, has a mass of 16 grams; one mole of oxygen molecules, O_2, has a mass of 32 grams.

Molar mass

The mass of one mole of a substance is called the **molar mass**, symbol M. The molar mass of carbon is 12 g/mol; the molar mass of calcium is 40 g/mol. The term molar mass applies to compounds as well as elements. The molar mass of an element is the relative atomic mass expressed in grams per mole; the

If you had 6×10^{23} pound coins and you distributed them equally between all the people on the Earth, how much would each person receive?

There are 4 billion people, so each one gets £150 million million!

molar mass of a compound is the relative formula mass expressed in grams per mole. Sulphuric acid, H_2SO_4, has a relative molecular mass of 98; its molar mass is 98 g/mol. Notice the units: relative molecular mass has no unit; molar mass has the unit g/mol.

> Amount (in moles) of substance = $\dfrac{\text{Mass of substance}}{\text{Molar mass of substance}}$
>
> Molar mass of element = Relative atomic mass in grams per mole
>
> Molar mass of compound = Relative formula mass in grams per mole

SUMMARY NOTE

The number of atoms in 12.000 g of carbon-12 is 6.022×10^{23}. This is the number of atoms in a mass of any element equal to its relative atomic mass expressed in grams. This number of atoms is called one **mole** (1 mol) of atoms. The ratio 6.022×10^{23}/mol is called the Avogadro constant. The number of moles of a substance is called simply the **amount** of that substance. The mass of one mole of an element or compound is the **molar mass** M of that substance.

M of an element = A_r expressed in g/mol

M of a compound = M_r expressed in g/mol

Example What is the amount of lead present in 414 g of lead?

Method

A_r of lead = 207
Molar mass of lead = 207 g/mol
Amount of lead = $\dfrac{\text{Mass of lead}}{\text{Molar mass of lead}}$
= $\dfrac{414 \text{ g}}{207 \text{ g/mol}}$
= 0.5 mol

The amount (number of moles) of lead is 0.5 mol.

Example If you need 2.5 mol of sodium chloride, what mass of sodium chloride do you have to weigh out?

Method

Relative formula mass of NaCl = 23 + 35.5 = 58.5
Molar mass of NaCl = 58.5 g/mol
Amount of substance = $\dfrac{\text{Mass of substance}}{\text{Molar mass of substance}}$
2.5 mol = $\dfrac{\text{Mass}}{58.5 \text{ g/mol}}$
Mass = 58.5 g/mol × 2.5 mol
= 147 g

You need to weigh out 147 g of sodium chloride.

JUST TESTING 59

1 State the mass of
   a 1 mol of sodium atoms
   b 3 mol of bromine molecules, Br_2
   c 0.25 mol of nitrogen atoms, N
   d 0.25 mol of nitrogen molecules, N_2
   e 0.5 mol of sulphur atoms, S
   f 0.5 mol of sulphur molecules, S_8.

2 Find the amount (moles) of each element present in
   a 46 g of sodium, Na
   b 0.6 g of magnesium, Mg
   c 32 g of oxygen, O_2
   d 127 g of copper, Cu.

3 State the mass of
  a 1.0 mol of sulphur dioxide molecules, SO_2
  b 10 mol of sulphuric acid, H_2SO_4
  c 0.50 mol of sodium hydroxide, NaOH
  d 2.0 mol of calcium carbonate, $CaCO_3$.

4 The Avogadro constant is 6×10^{23}/mol.
  a What mass of aluminium contains (i) 6×10^{23} atoms, (ii) 3×10^{25} atoms?
  b What mass of aluminium contains the same number of atoms as (i) 62 g of phosphorus, (ii) 4.0 g of carbon?

*11.6 Calculating the mass of reactant or mass of product

The reason for studying the mole concept is that it enables us to work out useful information, such as the mass of product that can be obtained from a given mass of starting material.

Example 1 What mass of quicklime (calcium oxide) is obtained when 10 tonnes of limestone (calcium carbonate) are decomposed?

Method First write the equation for the reaction:

  calcium carbonate → calcium oxide + carbon dioxide
  $CaCO_3(s)$ → $CaO(s)$ + $CO_2(g)$

The equation tells us that

  1 formula unit of calcium carbonate forms 1 formula unit of calcium oxide

  1 mole of calcium carbonate therefore forms 1 mole of calcium oxide.

Using the molar masses $M(CaCO_3) = 100$ g/mol, $M(CaO) = 56$ g/mol,

  100 g of calcium carbonate forms 56 g of calcium oxide.

The mass of calcium oxide from 10 tonnes of calcium carbonate is therefore given by

  Mass of CaO $= \dfrac{56}{100} \times$ (Mass of $CaCO_3$)

  $= \dfrac{56}{100} \times 10$ tonnes $= 5.6$ tonnes

Example 2 What mass of tungsten(III) oxide, W_2O_3, must be reduced with hydrogen to yield 10 kilograms of tungsten?

Method The equation comes first:

  tungsten(III) oxide + hydrogen → tungsten + water
  $W_2O_3(s)$ + $3H_2(g)$ → $2W(s)$ + $3H_2O(l)$

From the equation you can see that

  1 mole of tungsten(III) oxide forms 2 moles of tungsten.

SUMMARY NOTE

The balanced equation for a chemical reaction tells you how many moles of product are formed from one mole of reactant. You can use the equation with the molar masses of the reactant and the product to find out what mass of product is formed from a certain mass of reactant.

Remember: The balanced equation for the reaction is the key to each of these calculations.

Using the molar masses $M(W_2O_3) = 416$ g/mol, $M(W) = 184$ g/mol,

416 g of tungsten(III) oxide form 368 g of tungsten.

The mass of tungsten(III) oxide required to yield 10 kg of tungsten is therefore given by

$$\text{Mass of tungsten(III) oxide} = \frac{416}{368} \times \text{Mass of tungsten}$$
$$= \frac{416}{368} \times 10 \text{ kg}$$
$$= 11.3 \text{ kg}$$

JUST TESTING 60

You do not need calculators for these problems.

1. What mass of copper(II) oxide, CuO, is formed when 127 g of copper are completely oxidised?

2. Mercury(II) oxide, HgO, decomposes to give mercury and oxygen when heated. What mass of mercury is obtained from 54 g of mercury(II) oxide?

3. What mass of carbon dioxide can be made by decomposing 200 g of calcium carbonate?

4. When hydrochloric acid neutralises potassium hydroxide, the products of the reaction are potassium chloride and water:
    $HCl(aq) + KOH(aq) \rightarrow KCl(aq) + H_2O(l)$
    Find what mass of potassium chloride is formed when 14 g of potassium hydroxide are completely neutralised by hydrochloric acid.

5. Sodium sulphate can be made by neutralising sodium hydroxide with sulphuric acid:
    $2NaOH(aq) + H_2SO_4(aq) \rightarrow Na_2SO_4(aq) + 2H_2O(l)$
    Calculate the mass of sodium sulphate that can be made from 80 g of sodium hydroxide and an excess of sulphuric acid.

*11.7 Finding the equation for a reaction

If you know the mass of each reactant taking part in a reaction, you can calculate the number of moles of each reactant taking part in the reaction. This will give you the equation for the reaction.

Example Under certain conditions, nitrogen combines with oxygen to form an oxide. Given that 0.56 g of nitrogen form 1.20 g of the oxide of nitrogen, find the equation for the reaction.

Method Relative atomic masses are $A_r(N) = 14$, $A_r(O) = 16$

$$\begin{aligned}
\text{Amount of nitrogen, } N_2 &= \text{Mass/Molar mass} \\
&= 0.56 \text{ g} / 28 \text{ g/mol} = 2.0 \times 10^{-2} \text{ mol} \\
\text{Mass of oxygen} &= 1.20 \text{ g} - 0.56 \text{ g} = 0.64 \text{ g} \\
\text{Amount of oxygen, } O_2 &= \text{Mass/Molar mass} \\
&= 0.64 \text{ g} / 32 \text{ g/mol} \\
&= 2.0 \times 10^{-2} \text{ mol}
\end{aligned}$$

Therefore

2.0×10^{-2} mol N_2 reacts with 2.0×10^{-2} mol O_2

It follows that

1 mol N_2 reacts with 1 mol O_2.

The equation must be

$N_2(g) + O_2(g) \rightarrow$

Balancing the equation gives

$N_2(g) + O_2(g) \rightarrow 2NO(g)$

> **SUMMARY NOTE**
>
> If you know the mass of product that is formed from a certain mass of reactant, and you know their molar masses, you can work out the equation for the reaction.

*JUST TESTING 61

1. One gram of mercury reacts with iodine to form 1.635 g of an iodide of mercury.
   a. What amount (how many moles) of mercury reacts?
   b. What amount (how many moles) of iodine reacts?
   c. What is the ratio: moles of mercury/moles of iodine?
   d. Find the equation for the reaction.

2. A sample of powdered copper weighs 1.27 g. It reacts with oxygen to form 1.43 g of an oxide of copper.
   a. What is the amount (mol) of copper that reacts?
   b. What is the amount (mol) of oxygen that reacts?
   c. Find the ratio: amount (mol) copper/amount (mol) oxygen.
   d. Work out the equation for the reaction.

*11.8 Finding formulas

The formula of a compound is worked out from the percentage composition by mass of the compound.

Example 1: Finding the formula of magnesium oxide

Experiment 11.1 can be done to find the mass of oxygen that combines with a weighed amount of magnesium. A certain mass of magnesium is heated as shown in Fig. 11.2.

When all the magnesium has been converted into magnesium oxide, the mass of magnesium oxide is found. The mass of oxygen that has combined with the magnesium is found by subtraction. A typical set of results is given below.

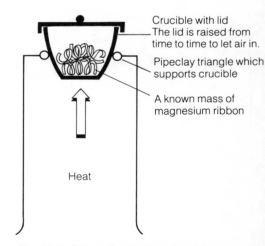

Fig. 11.2 Converting magnesium into magnesium oxide

(1) Mass of crucible = 16.18 g
(2) Mass of crucible + magnesium = 17.24 g
(3) **Mass of magnesium** = (2) − (1) = 1.06 g
(4) Mass of crucible + magnesium oxide = 17.95 g
(5) **Mass of oxygen combined** = (4) − (2) = 0.71 g

The results are treated in this way:

Element	Magnesium	Oxygen
Mass	1.06 g	0.71 g
A_r	24	16
Amount in moles	1.06/24 = 0.044	0.71/16 = 0.044
Divide through by 0.044	1 mole Mg to	1 mole O
Formula	MgO	

The formula MgO is the simplest formula which fits the results. Other formulas (Mg_2O_2, Mg_3O_3 etc.) also fit the results. MgO is the **empirical formula** for magnesium oxide.

The empirical formula of a compound is the simplest formula which represents the composition by mass of the compound.

Example 2: Finding the empirical formula of copper oxide

Figure 11.3 shows an apparatus for reducing copper oxide to copper (see Experiment 11.2). A set of results might look like this:

(1) Mass of 'boat' = 21.04 g
(2) Mass of 'boat' + copper oxide = 23.76 g
(3) Mass of 'boat' + copper = 23.21 g
(4) Mass of copper = (3) − (1) = 2.17 g
(5) Mass of oxygen = (2) − (3) = 0.55 g

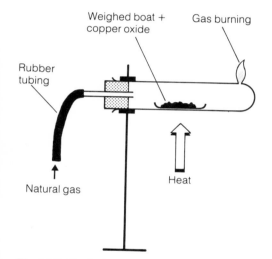

Fig. 11.3 Reducing copper oxide to copper

The results are treated in this way:

Element	Copper	Oxygen
Mass	2.17 g	0.55 g
A_r	63.5	16
Amount in moles	2.17/63.5 = 0.034	0.55/16 = 0.034
Ratio	1 mole of Cu to	1 mole of O
Empirical formula	CuO	

The oxide is copper(II) oxide.

Example 3: Finding the formula of copper(II) sulphate crystals

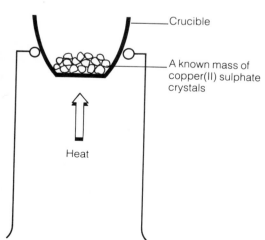

Fig. 11.4 Heating copper(II) sulphate crystals

When copper(II) sulphate crystals are heated gently, they lose water of crystallisation (Fig. 11.4). The result is **anhydrous** copper(II) sulphate. The mass of anhydrous copper(II) sulphate

187

formed from a weighed mass of crystals can be found. The results can be treated in this way:

Suppose that 4.00 g of crystals give 2.56 g of anhydrous copper(II) sulphate. Then the mass of water lost is 1.44 g.

Compound	Copper(II) sulphate		Water
Formula	$CuSO_4$		H_2O
Mass	2.56 g		1.44 g
M_r	159.5		18
Amount in moles	2.56/159.5		1.44/18
	= 0.0160		= 0.080
Divide by 0.0160	= 1 mol	to	5 mol
Empirical formula	$CuSO_4 \cdot 5H_2O$		

SUMMARY NOTE

The empirical formula of a compound shows the symbols of the elements present and the ratio of the number of atoms of each element present in the compound.

*JUST TESTING 62

You do not need calculators for these problems.

1 Find the empirical formulas of the following compounds:
  a a compound of 25.4 g of copper and 3.2 g of oxygen
  b a compound of 24.0 g of calcium and 5.6 g of nitrogen
  c a compound of 0.31 g of phosphorus and 1.07 g of chlorine.

2 Calculate the empirical formulas of the compounds which have the following percentage compositions by mass:
  a 40% calcium, 12% carbon, 48% oxygen
  b 36% beryllium, 64% oxygen
  c 80% carbon, 20% hydrogen.

3 Find the empirical formulas of the compounds formed when
  a 4.00 g of mercury form 4.64 g of a mercury sulphide
  b 0.62 g of phosphorus form 1.10 g of a phosphorus chloride
  c 5.60 g of iron form 10.65 g of an iron chloride.

4 Find n in the empirical formulas of these hydrates:
  a barium chloride crystals, $BaCl_2 \cdot nH_2O$, given that 1.22 g of the crystals lose water of crystallisation on heating to form 1.04 g of anhydrous barium chloride
  b magnesium sulphate crystals, $MgSO_4 \cdot nH_2O$, given that 1.23 g of the crystals form 0.60 g of anhydrous magnesium sulphate.

*11.9 Molecular formula and empirical formula

The empirical formula of a compound is the simplest formula that states the composition of the compound. The empirical formula mass may not be the same as the molar mass. The formula that gives the correct molar mass is called the **molecular**

SUMMARY NOTE

The molecular formula of a compound is a multiple of its empirical formula. The correct molecular formula is the one that gives the correct molar mass.

M of molecular formula
= n × M of empirical formula.

formula. The molecular formula is a multiple of the empirical formula.

Example The empirical formula of a compound is CH. The molar mass is 78 g/mol. What is the molecular formula of the compound?

Method

Empirical formula mass of CH = 12 + 1 = 13 g/mol
Molar mass = 78 g/mol
Molar mass/Empirical formula mass = 6
The molecular formula is 6 × the empirical formula
The molecular formula is C_6H_6.

*JUST TESTING 63

1 The empirical formula of a compound is CHO_2. Its molar mass is 90 g/mol. What is the molecular formula of the compound?

2 The empirical formula of a compound is C_3H_6O. If its molar mass is 58 g/mol, what is its molecular formula?

*11.10 The volumes of reacting gases

The volume of a fixed mass of gas depends on its temperature and pressure (see Chapter 1):

- As the temperature increases, the volume of gas increases.
- As the pressure increases, the volume of gas decreases.

We therefore have to state the temperature and pressure at which the volume of a gas is measured. We shall compare gas volumes at room temperature (20 °C) and 1 atmosphere. These conditions are called room temperature and pressure, **r.t.p.**

It has been found that one mole of gas occupies 24 litres at r.t.p. It does not make any difference what the gas is:

- 2 g of hydrogen
- 32 g of oxygen
- 44 g of carbon dioxide

All these are one mole of gas (the molar mass expressed in grams). All occupy 24 litres at r.t.p.

This volume, 24 litres, is called the **gas molar volume**.

The fact that different gases have the same molar volume makes calculations on the reacting volumes of gases easy. When you look at the ratio of the numbers of moles of gases reacting, you are looking at the ratio of the volumes of gases reacting. If the equation shows

$$2A(g) + 3B(g) \rightarrow C(g) + D(g)$$

you know that

2 moles of A react with 3 moles of B

and (at r.t.p.)

2 × 24 litres of A react with 3 × 24 litres of B

that is,

2 litres of A react with 3 litres of B

and, in general,

the volume of B is 1.5 times the volume of A.

Example 1 A car engine burns octane, C_8H_{18}:

octane + oxygen → carbon dioxide + water
$2C_8H_{18}(g) + 25O_2(g) → 16CO_2(g) + 18H_2O(l)$

What volume of oxygen is needed to ensure the complete combustion of 1 litre of octane vapour?

Method From the equation, you see that

2 moles of octane burn in 25 moles of oxygen

therefore

1 mole of octane burns in 12.5 moles of oxygen

that is,

24 litres of octane vapour burn in 12.5 × 24 litres of oxygen

1 litre of octane vapour burns in 12.5 litres of oxygen.

Example 2 A coal-fired electric power station burns coal which contains sulphur. The sulphur burns to form sulphur dioxide, $SO_2(g)$. If the power station burns 56 tonnes of sulphur in its coal every day, what volume of sulphur dioxide does it send into the air?
(1 tonne = 1000 kg = 1 000 000 g = 10^6 g)

Method The equation for combustion of sulphur is

sulphur + oxygen → sulphur dioxide
$S(s) + O_2(g) → SO_2(g)$

The equation tells you that

1 mol of sulphur forms 1 mol of sulphur dioxide

that is,

32 g of sulphur form 24 litres of sulphur dioxide

therefore

56 tonnes of sulphur form $\dfrac{56 \times 10^6}{32} \times 24$ litres of SO_2

$= 42 \times 10^6$ litres

This power station gives out 42×10^6 litres (42 million litres) of sulphur dioxide in a day.

SUMMARY NOTE

One mole of any gas occupies 24 litres at room temperature and pressure: the **gas molar volume** is 24 l at r.t.p. The volumes of the gaseous reactants taking part in a chemical reaction are in the same ratio as the numbers of molecules of the different gaseous reactants shown in the equation for the reaction.

> ### *JUST TESTING 64
>
> (Use gas molar volume = 24 litres at r.t.p.)
>
> 1. Calculate the volume of oxygen needed for the complete combustion of 75 cm³ of methane, $CH_4(g)$. What volume of carbon dioxide is formed?
>
> 2. What volume of carbon dioxide is formed by the complete combustion of 60 g of carbon?
>
> 3. What volume of hydrogen is formed when 6.0 g of magnesium react with an excess of sulphuric acid? The equation is
> $$Mg(s) + H_2SO_4(aq) \rightarrow H_2(g) + MgSO_4(aq)$$
>
> 4. What mass of calcium carbonate would be needed to provide 120 litres of carbon dioxide? The equation for the thermal decomposition is
> $$CaCO_3(s) \rightarrow CaO(s) + CO_2(g)$$

*11.11 Reactions in solution

Many reactions take place in solution. One way of stating the concentration of a solute in a solution is to state the mass of solute present in 1 litre of solution, e.g. grams per litre, g/l. A more convenient method is to state the amount in moles of a solute present in 1 litre of solution (see Fig. 11.5):

- If 1 mole of solute is present in 1 litre of solution, the concentration of solute is 1 mole per litre, 1 mol/l
- 2 moles of solute in 1 litre of solution:
  concentration = 2 mol/l
- 4 moles of solute in 10 litres of solution:
  concentration = 0.4 mol/l

| 1 litre = 1000 cm³ = 1 dm³ |

One litre is 1000 cubic centimetres. A litre is also known as a cubic decimetre, dm³.

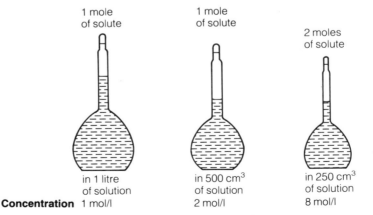

Fig. 11.5 Making solutions of known concentration

> Concentration in moles per litre = $\dfrac{\text{Amount of solute in moles}}{\text{Volume of solution in litres}}$
>
> Rearranging,
>
> Amount of solute (mol) = Volume of solution (l) × Concentration (mol/l)

Example 1 Calculate the concentration of a solution that was made by dissolving 60 g of sodium hydroxide and making the solution up to 1.00 litre.

Method

Molar mass of sodium hydroxide, NaOH = 23 + 16 + 1
= 40 g/mol
Amount in moles = Mass/Molar mass = 60 g / 40 g/mol
= 1.5 mol
Volume of solution = 1.0 litre
Concentration = Amount in moles/Volume in litres
= 1.5 mol / 1.0 l
= 1.5 mol/l

Example 2 Calculate the amount in moles of solute present in 250 cm^3 of a solution of hydrochloric acid which has a concentration of 2.0 mol/l.

Method

Amount (mol) = Volume (l) × Concentration (mol/l)
Amount of solute = 250 × 10^{-3} l × 2.0 mol/l
= 0.50 mol

Note that when you are given the volume in cm^3, you have to change it into litres. This makes the units come right:

Amount (moles) = **Volume (litres)** × **Concentration (moles per litre)**

SUMMARY NOTE

The concentration of a solute in a solution can be stated in grams of solute per litre of solution, g/l.

More often, concentration is stated in moles of solute per litre of solution, mol/l:

Concentration (mol/l) = $\dfrac{\text{Amount of solute (mol)}}{\text{Volume of solution (l)}}$

> ***JUST TESTING 65**
>
> 1 Calculate the concentrations of the following solutions:
>   **a** 5.6 g of potassium hydroxide in 500 cm^3 of solution
>   **b** 8.0 g of sodium hydroxide in 5.0 l of solution
>   **c** 9.8 g of sulphuric acid in 2.5 l of solution
>   **d** 7.3 g of hydrogen chloride in 250 cm^3 of solution.
>
> 2 Find the amount of solute in moles present in the following solutions:
>   **a** 250 cm^3 of a solution of sodium hydroxide of concentration 2.0 mol/l
>   **b** 1.00 l of hydrochloric acid of concentration 0.020 mol/l
>   **c** 500 cm^3 of sulphuric acid of concentration 0.12 mol/l
>   **d** 10 cm^3 of a 0.25 mol/l solution of potassium hydroxide.

Neutralisation

One of the reactions carried out in solution is neutralisation. Take the neutralisation of hydrochloric acid by sodium hydroxide solution:

hydrochloric acid + sodium hydroxide → sodium chloride + water

HCl(aq) + NaOH(aq) → NaCl(aq) + H$_2$O(l)

You see from the equation that

1 mole of HCl neutralises 1 mole of NaOH.

1 mole of HCl is present in 1 litre of solution of concentration 1 mol/l.

1 mole of NaOH is present in 1 litre of solution of concentration 1 mol/l.

Therefore

1 litre of hydrochloric acid of concentration 1 mol/l neutralises 1 litre of sodium hydroxide solution of concentration 1 mol/l.

You cannot assume that one mole of acid always neutralises one mole of base. You always have to look at the equation for the reaction. As in all the other types of calculation, the starting point is the balanced chemical equation for the reaction.

Here are some examples of other acid–base reactions.

1 hydrochloric acid + calcium hydroxide → calcium chloride + water

2HCl(aq) + Ca(OH)$_2$(aq) → CaCl$_2$(aq) + 2H$_2$O(l)

Here, 2 moles of acid neutralise 1 mole of base.

2 sulphuric acid + potassium hydroxide → potassium sulphate + water

H$_2$SO$_4$(aq) + 2KOH(aq) → K$_2$SO$_4$(aq) + 2H$_2$O(l)

Here, 1 mole of acid neutralises 2 moles of base.

Titration

A solution of known concentration is called a **standard** solution. You can use a standard solution of a base to find out the concentration of a solution of an acid. The method of doing this is **titration** (see Fig. 11.6). The idea is to find out what volume of, say, acid of unknown concentration is needed to neutralise a known volume, usually 25 cm^3, of a standard solution of a base.

Example 1 A titration tells you that 18.5 cm^3 of hydrochloric acid neutralise 25.0 cm^3 of a 0.100 mol/l solution of sodium hydroxide. You can find out the concentration of hydrochloric acid.

Method

1 First write the equation

hydrochloric acid + sodium hydroxide → sodium chloride + water

HCl(aq) + NaOH(aq) → NaCl(aq) + H$_2$O(l)

You see that 1 mole of HCl neutralises 1 mole of NaOH.

2 Now work out the amount in moles of base. You must start with the base because you know the concentration of base, and you do not know the concentration of acid.

Remember: Amount (mol) = Volume (l) × Concentration (mol/l)

$$\text{Amount in moles of NaOH} = \text{Volume (25.0 cm}^3\text{)} \times \text{Concentration (0.100 mol/l)}$$
$$= 25.0 \times 10^{-3} \times 0.100$$
$$= 2.50 \times 10^{-3} \text{ mol}$$

3 Now work out the concentration of acid:

Amount in moles of HCl = Amount in moles of NaOH
$$= 2.50 \times 10^{-3} \text{ mol}$$

also

$$\text{Amount (mol) of HCl} = \text{Volume of HCl(aq)} \times \text{Concentration of HCl(aq)}$$

therefore, if c mol/l is the concentration of HCl,

$$2.50 \times 10^{-3} \text{ mol} = 18.5 \times 10^{-3} \text{ l} \times c \text{ mol/l}$$
$$\text{and } c \text{ mol/l} = 2.50 \times 10^{-3} \text{ mol}/18.5 \times 10^{-3} \text{ l}$$
$$= 0.135 \text{ mol/l}$$

The concentration of hydrochloric acid is 0.135 mol/l.

Example 2 In a titration, 25.0 cm³ of sulphuric acid of concentration 0.120 mol/l neutralised 23.0 cm³ of potassium hydroxide solution. Find the concentration of the potassium hydroxide solution.

Method

1 First write the equation

sulphuric acid + potassium hydroxide → potassium sulphate + water

$$H_2SO_4(aq) + 2KOH(aq) \rightarrow K_2SO_4(aq) + 2H_2O(l)$$

You see that 1 mole of H_2SO_4 neutralises 2 moles of KOH.

2 Now work out the amount in moles of acid. You must choose the acid because you do not know the concentration of the base:

$$\text{Amount in moles of acid} = \text{Volume (25.0 cm}^3\text{)} \times \text{Concentration (0.120 mol/l)}$$
$$= 25.0 \times 10^{-3} \times 0.120 \text{ mol}$$
$$= 3.00 \times 10^{-3} \text{ mol}$$

3 Now work out the concentration of base:

Amount in moles of KOH = 2 × Amount in moles of H_2SO_4
$$= 6.00 \times 10^{-3} \text{ mol}$$

also

$$\text{Amount (mol) of KOH} = \text{Volume of KOH(aq)} \times \text{Concentration of KOH(aq)}$$

SUMMARY NOTE

A solution of known concentration is called a **standard solution**.
The concentration of a solution of an acid can be found by **titrating** it against a standard solution of a base. Similarly, a standard solution of an acid can be employed in a titration to find the concentration of a solution of a base.

therefore, if c mol/l is the concentration of KOH

6.00×10^{-3} mol $= 23.0 \times 10^{-3}$ l $\times c$ mol/l
c mol/l $= 6.00 \times 10^{-3}$ mol$/23.0 \times 10^{-3}$ l
$= 0.261$ mol/l

The concentration of potassium hydroxide is 0.261 mol/l.

---

*JUST TESTING 66

1. 25.0 cm³ of hydrochloric acid are neutralised by 30.0 cm³ of a solution of sodium hydroxide of concentration 0.25 mol/l. Find the concentration of the hydrochloric acid.

2. 25.0 cm³ of hydrochloric acid are neutralised by 20.0 cm³ of a solution of 0.15 mol/l sodium carbonate solution. Find the concentration of the hydrochloric acid.

3. A solution of sodium hydroxide contains 20 g/l. What volume of this solution would be needed to neutralise 25.0 cm³ of 0.10 mol/l hydrochloric acid?

4. What is the volume of sulphuric acid of concentration 0.100 mol/l that will neutralise **a** 5.3 g of sodium carbonate, **b** 2.1 g of sodium hydrogencarbonate?

5. The following results were obtained in tests on antacid indigestion tablets:

Brand	Price of 100 tablets	Volume of 0.01 mol/l acid required to neutralise 1 tablet
Neutro	75p	2.5 cm³
Fixit	90p	3.3 cm³
Baso	125p	4.0 cm³
Alko	150p	4.5 cm³

Which antacid tablets offer the best value for money?

Exercise 11

You do not need a calculator for most of these problems. Those that need a calculator say so.

1. Use Avogadro constant = 6×10^{23}/mol. If the price of one billion (10^9) mercury atoms were 1p, what would you have to pay for 1 milligram (10^{-3} g) of mercury?

2. How many carbon atoms are there in a 2-carat diamond? One carat = 0.2 g, Avogadro constant = 6×10^{23}/mol. If the diamond costs £1500, what is the price of one carbon atom?

3. Ethanol, molecular formula C_2H_6O, is found in 'alcoholic' drinks. If you have 9.2 g of ethanol, how many moles do you have of **a** ethanol, **b** carbon atoms, **c** hydrogen atoms, **d** oxygen atoms?

4. (Calculator needed.) Find the empirical formula of the compounds which analyse as
   a 40.0% C, 6.7% H, 53.3% O
   b 19.2% Na, 0.8% H, 26.7% S, 53.3% O.

5. Vitamin C has the empirical formula $C_3H_4O_3$, and a molar mass of 176 g/mol. What is its molecular formula?

6. Melamine is a plastic with the empirical formula CH_2N_2. Its molar mass is 126 g/mol. What is its molecular formula?

7. Aluminium is made by electrolysing aluminium oxide. What mass of pure aluminium oxide, Al_2O_3, must be electrolysed to give 54 tonnes of aluminium?

8. (Calculator needed.) In certain illnesses, the body becomes short of salt (sodium chloride). A nurse must inject a sodium chloride solution into a vein. A 0.85 per cent solution is used (0.85 g sodium chloride per 100 g of water). What is the concentration of sodium chloride in mol/l in this solution?

9. When nitrogen and oxygen combine to form an oxide of nitrogen, under certain conditions, 0.70 g of nitrogen forms 2.30 g of an oxide. By means of a calculation, find the balanced equation for the reaction.

10. The diameter of a silver atom is 2.7×10^{-8} cm. $A_r(Ag) = 207$, Avogadro constant $= 6 \times 10^{23}$/mol. Imagine the atoms in 20.7 g of silver placed in a straight line, touching each other. What distance (km) would they cover?

11. A manufacturer's chimney emits 100 000 kg of nitrogen monoxide, NO, daily. The manufacturer decides to use the clean-up reaction:

    natural gas + nitrogen monoxide → nitrogen + carbon dioxide + water

    $CH_4(g) + 4NO(g) \rightarrow 2N_2(g) + CO_2(g) + 2H_2O(l)$

    The cost of natural gas is 60p per cubic metre.
    a. What is the cost of running the clean-up operation?
    (First convert kg of NO to cubic metres of NO; use the gas molar volume to do this. Then use the reaction equation to find out how many cubic metres of CH_4 are needed.)
    b. The cost of setting up the cleaning plant is high. Then there is the daily running cost. Would you spend money on installing and running a clean-up operation if you were the manufacturer? Explain your answer.

12. The emission of nitrogen monoxide from motor vehicle exhausts can be reduced by injecting a stream of ammonia into the exhaust vapour:

    nitrogen monoxide + ammonia → nitrogen + water vapour

    $6NO(g) + 4NH_3(g) \rightarrow 5N_2(g) + 6H_2O(g)$

    A large car emits about 5 g of nitrogen monoxide per mile. If a car is driven 10 000 miles a year, what mass of ammonia is needed to clean up the exhaust gas?

CHAPTER 12 Limestone, chalk and sand

12.1 Concrete

The Humber Bridge is the largest single-span suspension bridge in the world. It is a triumph of civil engineering — one of the wonders of modern technology (Fig. 12.1). There are two traffic lanes in both directions, as well as lanes for cyclists and pedestrians. The main span is 1410 m, and the overall length is 2220 m. This is the world's first suspension bridge to hang from high towers of reinforced concrete. On all the other large bridges, the towers are of steel. The Humber Bridge contains 500 000 tonnes of concrete. The concrete pillars have foundations that reach 40 m down into the ground beneath the estuary. Steel was also used in vast quantities: 30 000 tonnes.

Fig. 12.1 The Humber Bridge

SUMMARY NOTE

Concrete is a strong construction material. It is made from chalk or limestone and clay or shale. Vast quantities of these raw materials are quarried to meet the demand for concrete.

What is this strong construction material, concrete? How is it made? The first stage is the manufacture of cement. The raw materials used to make cement are

- limestone or chalk
- shale or clay
- a small percentage of calcium sulphate (**gypsum**).

Figure 12.2 shows the method for making cement, and Fig. 12.3 shows how concrete is made from cement.

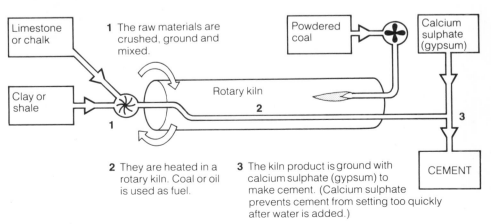

Fig. 12.2 Making cement

Chemical reactions occur between (a) chalk or limestone ($CaCO_3$) and (b) clay or shale, which consist largely of silicon(IV) oxide (SiO_2), aluminium oxide (Al_2O_3), silicates (compounds of silicon, oxygen and a metal) and aluminates (compounds of aluminium, oxygen and another metal). Cement, which is the product of the reactions, consists chiefly of calcium silicate ($CaSiO_3$), and calcium aluminate ($CaAl_2O_4$). A little calcium sulphate is added to slow down the rate at which concrete sets.

Concrete is one of the most versatile of all the materials used in civil engineering. It is used in the construction of all kinds of homes — from cottages to skyscrapers — as well as schools and factories, power stations and reservoirs. Transport by land, sea and air depends on concrete used in roads and bridges, docks and harbours, runways and hangars. **Reinforced** concrete is strengthened by steel supports.

There are large deposits of both limestone and chalk in the UK. The UK quarries 50 million tonnes of limestone and chalk every year. Limestone is quarried by blasting a hillside with explosive charges. Chalk can be dug out by mechanical excavators. Limestone, chalk and also marble are all forms of calcium carbonate, $CaCO_3$. They differ in the way the ions are arranged in their crystals. The quarrying of vast quantities of limestone and chalk can create difficulties. These minerals often occur in regions of outstanding natural beauty (Fig. 12.4). We do not want to despoil our countryside, but we do want the useful materials which chemists can obtain from limestone and chalk. Mining companies have to take steps to restore the countryside after they finish working a deposit of limestone or chalk (Fig. 12.5). As well as the manufacture of concrete and cement, limestone is used in the manufacture of iron in blast furnaces (see Chapter 17), in the manufacture of glass (see Section 12.9) and in the manufacture of lime (see below).

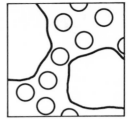
1 Sand and gravel are mixed.

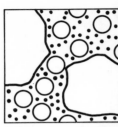
2 Cement powder is added.

3 Water is added. Crystals start to grow from the cement.

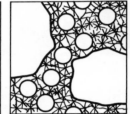
4 The crystals grow and interlock with the surfaces of sand and gravel, binding the mass together.

Fig. 12.3 Making concrete

Fig. 12.4 A limestone quarry

SUMMARY NOTE

Concrete is a strong construction material. It is made from chalk or limestone and clay or shale. Vast quantities of these raw materials are quarried to meet the demand for concrete.

1 Before quarrying

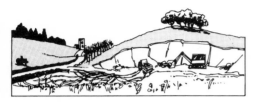
2 Topsoil is removed and stored. Trees are planted to shield the village from the sight of the quarry. The deposit of limestone is then worked.

3 The hole left by removal of the limestone is filled. Topsoil is spread over the area.

4 The land is restored to agricultural use

Fig. 12.5 Restoring the scenery

12.2 Limestone and lime

You may like to do Experiments 12.1 and 12.3 on heating limestone and other carbonates.

When calcium carbonate is heated strongly, it dissociates (splits up) to give calcium oxide and carbon dioxide:

calcium carbonate → calcium oxide + carbon dioxide

$$CaCO_3(s) \rightarrow CaO(s) + CO_2(g)$$

This reaction is an important industrial process. The products, calcium oxide (called **lime**) and carbon dioxide, are both important chemicals. They are made by heating limestone in towers called **lime kilns** (Fig. 12.6).

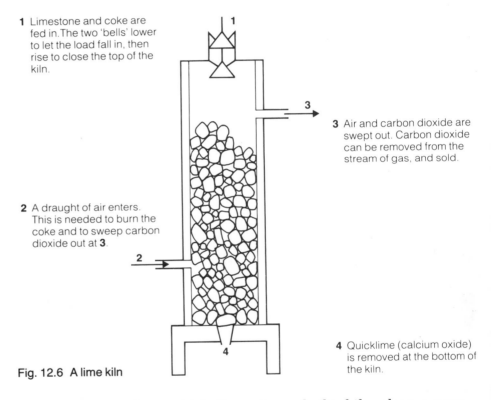

1 Limestone and coke are fed in. The two 'bells' lower to let the load fall in, then rise to close the top of the kiln.

2 A draught of air enters. This is needed to burn the coke and to sweep carbon dioxide out at 3.

3 Air and carbon dioxide are swept out. Carbon dioxide can be removed from the stream of gas, and sold.

4 Quicklime (calcium oxide) is removed at the bottom of the kiln.

Fig. 12.6 A lime kiln

The draught of air which flows through the kiln plays a very important part in the process. For one thing, it enables coke to burn and heat the limestone. Also, it removes the carbon dioxide formed. If the carbon dioxide (an acid gas) were not removed, it would combine with the base, calcium oxide:

carbon dioxide + calcium oxide → calcium carbonate

$$CO_2(g) + CaO(s) \rightarrow CaCO_3(s)$$

(**acid** + **base** → **salt**)

If the carbon dioxide is carried off as it is formed, it cannot linger to combine with calcium oxide and reform calcium carbonate.

The reaction

calcium carbonate ⇌ calcium oxide + carbon dioxide

$CaCO_3(s) \rightleftharpoons CaO(s) + CO_2(g)$

is a **reversible** reaction: it can go from left to right and also from right to left, depending on the temperature and the pressure.

Calcium oxide is called **lime** or **quicklime**. It is used on building sites. It is first allowed to react with water (see Experiment 12.3). This converts it into calcium hydroxide, which is called **slaked lime**:

calcium oxide + water → calcium hydroxide

$CaO(s) + H_2O(l) \rightarrow Ca(OH)_2(s)$

Calcium oxide, slaked lime, is mixed with sand to give **mortar**. Mortar is used to hold bricks together. As it is exposed to the air, it gradually hardens as it reacts with carbon dioxide in the air to form calcium carbonate (see Experiment 12.2).

Calcium oxide, lime, is also used in large quantities as a base. Farmers spread it on fields to neutralise excessively acidic soils (Fig. 12.7).

SUMMARY NOTE

Calcium carbonate (limestone) is heated in lime kilns to give calcium oxide (called both lime and quicklime) and carbon dioxide. Calcium oxide is mixed with sand to make mortar. It is also used to neutralise acidic soils.

Fig. 12.7 Applying lime to fields

JUST TESTING 67

1 Can you remember from Experiment 12.3 why calcium oxide is called **quicklime** and why calcium hydroxide is called **slaked lime**? What is the name commonly used for a solution of calcium hydroxide in water?

12.3 Carbonates

You have studied carbonates in Experiments 12.1 and 12.3. Figure 12.8 gives a summary.

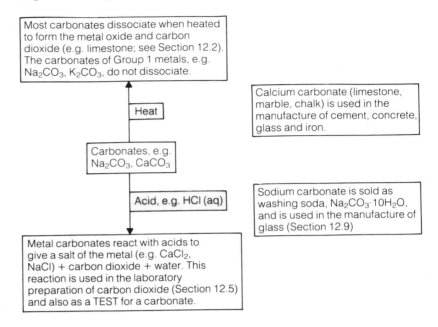

Fig. 12.8 Some reactions of carbonates

12.4 Hydrogencarbonates

Figure 12.9 summarises some of the reactions of hydrogencarbonates.

Hydrogencarbonates

| The hydrogencarbonates of Group 1, e.g. sodium hydrogencarbonate, $NaHCO_3$, decompose when warmed to give carbon dioxide and water and the metal carbonate. This is why sodium hydrogencarbonate is used in baking powder. The carbon dioxide which it provides makes bread and cakes rise in the oven. | Calcium hydrogencarbonate, $Ca(HCO_3)_2$, and magnesium hydrogencarbonate, $Mg(HCO_3)_2$, are present in tap water. They cause **temporary hardness**. When temporarily hard water is boiled, the hydrogencarbonates decompose to form the carbonates, and the water becomes **soft**. |

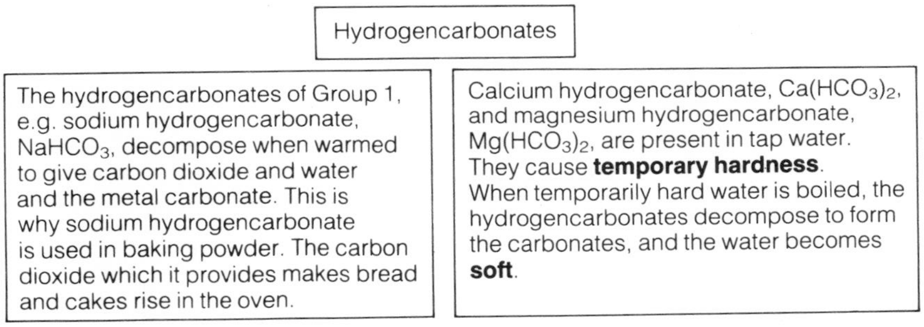

Fig. 12.9 Some reactions of hydrogencarbonates

SUMMARY NOTE

- Carbonates and hydrogencarbonates react with dilute acids to give carbon dioxide and a salt.
- Hydrogencarbonates decompose when heated to give carbon dioxide and a carbonate.

JUST TESTING 68

1. What are the salts present in temporarily hard water?
2. How do these salts make the water hard?
3. What is the 'scale' which forms gradually in a kettle which is used for boiling hard water?
4. Sodium hydrogencarbonate is an ingredient of baking powder. When it dissociates in the oven, the reaction which occurs is

   sodium hydrogencarbonate → sodium carbonate + carbon dioxide + water

   Write a balanced chemical equation for this reaction.

12.5 Carbon dioxide

Carbon dioxide is formed when many carbonates and hydrogencarbonates are heated and when carbonates and hydrogencarbonates react with acids (see Experiments 12.3 and 12.4). Figure 12.10 shows the laboratory preparation and collection of carbon dioxide.

Fig. 12.11 Enjoying a solution of CO_2

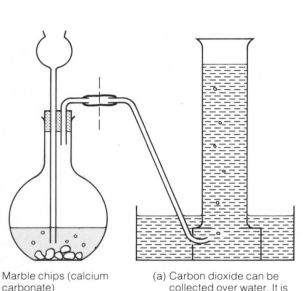
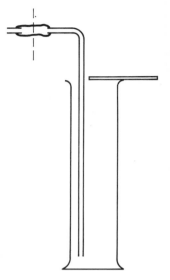

Marble chips (calcium carbonate)

+ dilute hydrochloric acid

(a) Carbon dioxide can be collected over water. It is only slightly soluble in water.

(b) Carbon dioxide, being denser than air, can be collected by downward delivery. As it is an invisible gas, it is difficult to tell when the gas jar is full.

Fig. 12.10 The preparation and collection of carbon dioxide

Properties of carbon dioxide

You have investigated the properties of carbon dioxide in Experiment 12.4. Carbon dioxide is colourless and odourless. Although it can be collected over water, it dissolves to a slight extent to give a solution of the weak acid, carbonic acid, H_2CO_3. Under pressure, the solubility of carbon dioxide increases. Dissolving carbon dioxide in water under pressure, and adding flavourings, has been turned into big business by the soft drinks industry. Many people enjoy these 'carbonated' drinks (Fig. 12.11). When the cap is taken off a bottle, the pressure is decreased, and carbon dioxide comes out of solution.

Carbon dioxide is denser than air, so it can be collected by downward delivery (Fig. 12.10b). It extinguishes a burning splint. The combination of its inability to support combustion and its high density makes carbon dioxide a suitable gas to use in fire-extinguishers (Fig. 12.12).

Some metals can burn in carbon dioxide. As you saw in Experiment 12.4, burning magnesium continues to burn in carbon dioxide. The temperature of the magnesium flame is high enough to bring about the reaction

magnesium + carbon dioxide → magnesium oxide + carbon

$2Mg(s)\quad +\quad CO_2(g)\quad \rightarrow \quad 2MgO(s)\quad +\quad C(s)$

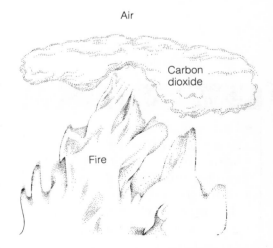

Air is less dense than carbon dioxide; it stays above the blanket of carbon dioxide, and cannot reach the fire.
Being denser than air, carbon dioxide stays at ground level and keeps air away from the fire.

Fig. 12.12 Carbon dioxide as a fire-extinguisher

> **SUMMARY NOTE**
>
> Carbon dioxide is prepared by the reaction of an acid on a carbonate. It is
> - slightly soluble in water
> - denser than air
> - used as a fire-extinguisher.

You can find out more about fire-extinguishers in Section 12.8.

> **JUST TESTING 69**
>
> 1 Why must a bottle of fizzy drink have a well-fitting cap? Why can you not see bubbles inside the closed bottle?

12.6 The carbon cycle

In clean dry air the percentage by volume of carbon dioxide is 0.03%. This small percentage of carbon dioxide is essential to plants. They take in carbon dioxide through their leaves, and use it together with water taken in through their roots to **synthesise** sugars. The reaction will take place only in sunlight and only in green leaves, which possess the green pigment **chlorophyll** that catalyses the reaction. This reaction is called **photosynthesis**. (**Synthesis** means making a compound from its elements or from simpler substances; **photo** means light.)

Photosynthesis:

$$\text{sunlight energy} + \text{carbon dioxide} + \text{water} \xrightarrow{\text{chlorophyll}} \text{glucose (a sugar)} + \text{oxygen}$$

$$\text{energy} + 6CO_2(g) + 6H_2O(l) \longrightarrow C_6H_{12}O_6(aq) + 6O_2(g)$$

The energy of sunlight is transformed into the energy of the chemical bonds in glucose. In the process, oxygen is released. Glucose is converted into starch for storage in plants.

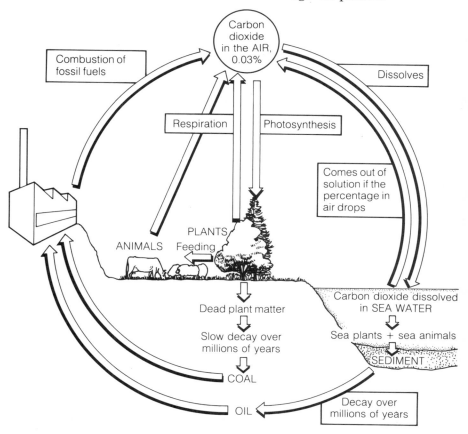

Fig. 12.13 The carbon cycle

The reverse reaction takes place in animals. In **respiration**, oxygen is used to oxidise sugars (and starches) to carbon dioxide and water with the release of energy (see Section 8.11). Plants respire to obtain energy.

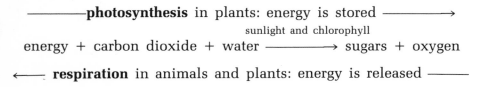

> **SUMMARY NOTE**
>
> Plants take carbon dioxide from the air in photosynthesis. Animals put carbon dioxide into the air in respiration. Nature maintains a balance between these processes called the **carbon cycle**.

The processes which take carbon dioxide out of the air and the processes which put carbon dioxide into the air are delicately balanced. The balance is called the **carbon cycle**, and is shown in Fig. 12.13.

12.7 The greenhouse effect

The Earth's climate is affected by the levels of carbon dioxide and water vapour in the atmosphere. These gases are responsible for the **greenhouse effect**. The Earth is heated by the sun. Heat energy from the Earth is radiated back into space. The balance keeps the temperature of the planet fairly constant. If anything happens to upset the balance, the result will be a change in the world's climate. A rise in temperature could melt the ice caps at the North and South Poles. A fall could decrease the world's food production. Sunlight is radiation of wavelengths in the visible region of the spectrum. It passes unhindered through the atmosphere, unless it meets cloud cover. When the Earth sends heat energy back into space, it changes the wavelength to that of infrared radiation. Unlike sunlight, infrared radiation cannot travel freely through the air surrounding the Earth. Water vapour and carbon dioxide absorb infrared radiation. They act as 'blankets' round the Earth, hindering the escape of heat into space. Without these blankets, the Earth's surface would be −40 °C, instead of a life-supporting 15 °C. If additional water vapour and carbon dioxide were added to the atmosphere, it would be like adding another blanket. The Earth would heat up. This blanketing by water vapour and carbon dioxide is called **the greenhouse effect** (Fig. 12.14).

The greenhouse effect was named, as you can imagine, after the way a greenhouse warms up. Sunlight enters through the glass panes, but infrared radiation has difficulty in escaping from the greenhouse through glass.

There is more water vapour than carbon dioxide in the atmosphere. Most of the greenhouse effect is due to the blanket of water vapour. There is a 'hole' in the water vapour blanket: it does not absorb radiation of certain wavelengths. Carbon dioxide covers part of the 'hole' by absorbing light at some of the wavelengths which water vapour allows to escape.

The natural carbon dioxide balance shown in Fig. 12.13 is being disturbed by the massive quantities of carbon dioxide that we send into the air by the combustion of coal and oil. The 'blanket' is getting thicker. Calculations have shown that the

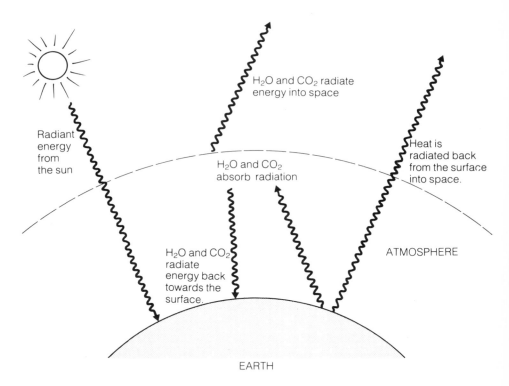

Fig. 12.14 The greenhouse effect

SUMMARY NOTE

A layer of carbon dioxide and water vapour surrounds the Earth. This layer reduces the amount of heat which the Earth radiates into space. By burning hydrocarbon fuels, we pour carbon dioxide and water vapour into the atmosphere. The layer of carbon dioxide and water vapour is growing thicker. If it becomes too thick, the Earth will warm up, and the ice caps could begin to melt.

We also put dust into the atmosphere. This has the effect of scattering some of the sun's radiation and cooling the Earth.

level of carbon dioxide could, at its present rate of increase, double in your lifetime. The result would be an increase of 2 °C in the average temperature of the Earth. This would mean that the massive ice caps of the Arctic and Antarctic regions would slowly begin to melt. The levels of oceans would begin to rise, and coastal areas would eventually be flooded.

Secondary effects would multiply the effect of carbon dioxide. The increase in temperature would make more water vaporise from the oceans. It would also drive out some of the carbon dioxide dissolved in the oceans. Thus, a still thicker greenhouse 'blanket' would be formed.

While some scientists are investigating the greenhouse effect, others are more concerned about the increasing level of particulate matter (dust etc., see Chapter 8). Sunlight which would otherwise reach the Earth is reflected into space by dust. Scientists have calculated that a fourfold increase in particulate matter would decrease the Earth's temperature by about 3 °C. They believe that such a fall in temperature would trigger a new ice age.

JUST TESTING 70

1 In the graph shown in Fig. 12.15, what could have caused the increase in temperature after 1920?

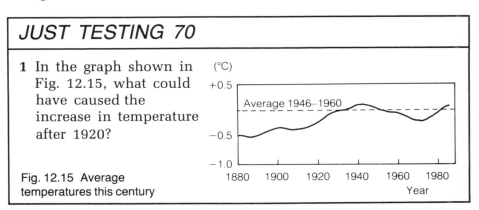

Fig. 12.15 Average temperatures this century

12.8 Fighting fire with carbon dioxide

The fire triangle

A fire needs three things: fuel, oxygen and heat. For every fuel there is an **ignition temperature**, below which the fuel will not burn. Heat is needed to raise the fuel to this temperature. The three essentials are represented by the three sides of the **fire triangle**; see Fig. 12.16. If one of the three sides is removed, the triangle will be destroyed: the fire will go out. Different fire-extinguishers attack different sides of the triangle.

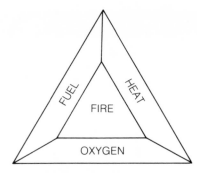

Fig. 12.16 The fire triangle

Fuel. A fire can burn itself out. When fire-fighters are combating a forest fire, they cut down a band of trees in the path of the fire. Deprived of trees to burn, the fire cannot spread and gradually dies out.

Heat. The easiest way to remove heat is to throw water on to a fire (Fig. 12.17). The **soda–acid** type of extinguisher is able to provide a powerful jet of water quickly (see Fig. 12.18). There are some kinds of fires on which water cannot be used. Some metals (e.g. sodium and potassium) react with cold water. Other metals (e.g. magnesium and zinc) react with water when they are hot to produce the flammable gas hydrogen. Throwing water on to a blazing metal may therefore make the fire much worse.

◁ Fig. 12.17 Fire-fighting

1 Strike the knob.

2 The knob smashes a bottle of concentrated sulphuric acid against a wire gauze.

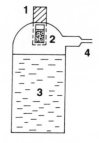

3 Sulphuric acid reacts with the solution of sodium hydrogencarbonate. Carbon dioxide is formed.

4 The pressure of gas forces a jet of the aqueous solution out of the nozzle.

Fig. 12.18 A soda-acid extinguisher

Oil fires cannot be extinguished by water. Oil floats on top of water, and continues to burn. This is why fires at sea are so catastrophic.

Water should not be used on electrical fires. The firefighter could receive a shock from electricity conducted along the stream of water.

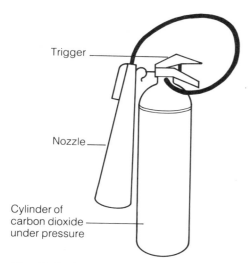

Fig. 12.19 A pressurised carbon dioxide extinguisher

Fig. 12.20 A powder extinguisher

Why do they say you shouldn't throw water on to a blazing chip pan?

It makes burning oil splash out of the pan and spreads the fire.

What should you do then?

Switch off the gas or electricity. Wet a cloth and wring it out. Then cover the pan with the damp cloth. This will keep air out.

Oxygen. A carbon dioxide fire-extinguisher acts by starving a fire of oxygen. You may like to do Experiment 12.5 on fire-extinguishers. Figure 12.19 shows one type of carbon dioxide fire-extinguisher. It is a cylinder of carbon dioxide under pressure. When the pin is removed and the trigger is squeezed, a jet of carbon dioxide comes out of the nozzle. It lasts for about $\frac{1}{2}$ minute. After use, the cylinder must be refilled ready for the next emergency.

Powder extinguishers also release carbon dioxide. They contain compounds which decompose to give carbon dioxide when thrown on to a fire, e.g. sodium hydrogencarbonate (see Experiment 12.5). Powder extinguishers are small and convenient for carrying in cars (Fig. 12.20). They can be used on oil fires and electrical fires.

Foam extinguishers use a combination of the cooling effect of water and the exclusion of air by carbon dioxide. The extinguisher is activated by striking a knob. This releases acid into a solution containing sodium hydrogencarbonate and a foam stabiliser. The acid and the hydrogencarbonate react to form carbon dioxide. The foam stabiliser prevents the bubbles of gas from escaping. A stream of foam — bubbles of carbon dioxide and aqueous solution — is propelled out of the nozzle. A layer of foam blankets the fire. It is denser than carbon dioxide alone. Foam cannot be used on the types of fires for which water is dangerous.

SUMMARY NOTE

The fire triangle:
Fuel: A fire will burn itself out if it runs out of fuel.
Heat: A fire will go out if you cool the fuel below its ignition temperature.
Oxygen: A fire will go out if it has insufficient oxygen.
Many fire-extinguishers cover the fire with a dense layer of carbon dioxide to keep out oxygen.

JUST TESTING 71

1 What type of fire-extinguisher would you use on each of the following types of fire?
   a a wood fire, **b** a fire in a magnesium factory, **c** a fire in a plastics factory, **d** a petrol fire, **e** a person whose clothing has caught fire, **f** a chip-pan fire, **g** a burning gas (e.g. natural gas), **h** an oil fire.

2 On which types of fire would you use each of the following types of extinguisher?
   a soda–acid, **b** foam, **c** dry powder, **d** carbon dioxide under pressure.

12.9 Sand and limestone: a four thousand year old combination

Seeing round corners

Mr Murray believes he has a stomach ulcer. The X-ray photographs show nothing abnormal, and his doctor is unconvinced. The doctor decides to take a look. He slips a thin, flexible plastic tube about 1 metre long down Mr Murray's throat. Mr Murray is mildly sedated but fully conscious. The doctor peers down the tube to view the pinkish-white walls of his patient's oesophagus. How can he see down a bent tube? The reason is that the tube contains a bundle of **glass fibres**.

The individual glass fibres are as thin as a human hair and as pliable. As each glass fibre is drawn out, it is sealed in a protective coating of plastic. A bundle of glass fibres can follow the tortuous turns of the digestive tract without snapping. Light travelling along one bundle of glass fibres illuminates the tissues. A second bundle of fibres carries a colour image back to the doctor's eyes.

The tube, the doctor's **endoscope**, moves further on its way to his patient's stomach. He is looking for ulcers, but suddenly the endoscope shows up a patch of white tissue. It looks like the first sign of a cancer. The X-ray photograph has not picked it up because slight changes in colour are not detected by X-rays. An attachment allows the doctor to take a photograph of the cancer as he plans the operation he will have to do.

Looking through glass

Glass fibres are made from **optical** glass, a very pure kind of glass. Optical glass is made from the same materials as ordinary window glass: sand (silicon(IV) oxide, SiO_2), limestone (calcium carbonate, $CaCO_3$) and sodium carbonate (Na_2CO_3). The recipe for glass is 4500 years old. The ancient Egyptians discovered it. When they melted sand with limestone and sodium carbonate, they obtained a transparent material, glass. Egypt is one of the few places where sodium carbonate occurs naturally.

Silicon(IV) oxide, SiO_2, is a crystalline substance. It has a macromolecular structure (see Fig. 12.21). In glass, many of the Si—O bonds have been broken, and the ordered arrangement has been largely destroyed. Glass is a mixture of calcium silicate, $CaSiO_3$, and sodium silicate, Na_2SiO_3. Glass is neither a liquid nor a crystalline solid. It is a **supercooled liquid**: it appears to be solid, but has no sharp melting point. X-rays show that the regular, orderly packing of atoms found in other solids is not present in glass.

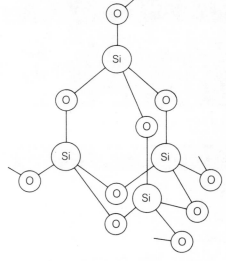

Fig. 12.21 The structure of silicon(IV) oxide

Speaking through glass

High-grade optical glass is made from pure silicon(IV) oxide instead of sand. Optical fibre costs about £1 per metre.

The main future of optical fibres lies in communications. A telephone exchange needs to use thick copper cables to transmit

> **SUMMARY NOTE**
>
> Sand and limestone are used to make glass. Glass has been used for 4500 years. New uses are being found for glass in optical fibre technology. Glass fibres enable people to see round corners. They can carry information in the form of pulses of light.

hundreds of telephone calls simultaneously. A pair of optical glass fibres will replace a thick copper cable. It can carry 10 000 simultaneous calls, transmitting them in the form of pulses of light. Each pattern of sound is converted into a pulse of light. The pulses of light are decoded by the receiver and converted back into sound. Optical fibres will carry a beam of light for a distance of 10 km. Fibres can be joined to cover longer distances.

Through optical fibre cables, two-way television is possible; that is, people can speak and also see each other at the same time. In the future, we shall have more sophisticated services than copper wire can provide. People will be able to communicate with shops, libraries and hospitals by two-way television.

12.10 Silicon

Silicon is the second most abundant element in the Earth's crust (see Fig. 10.20). It occurs as silicon(IV) oxide, SiO_2, in sand and quartz and as silicates (compounds of silicon, oxygen and a metal) in clays and many rocks. Many of the silicates in clays and rocks contain aluminium, and chemists are working on methods of extracting aluminium from them (see Section 10.12). Asbestos is an **aluminosilicate**.

Silicon(IV) oxide or **silica** is our source of the element silicon. Silicon is a dark-coloured shiny solid which melts at 1410 °C. It is in Group 4 of the Periodic Table, beneath carbon. The importance of silicon as a semiconductor has been described in Section 3.4. When it is heated with carbon at a high temperature, silicon(IV) oxide is reduced to silicon:

$$\text{silicon(IV) oxide} + \text{carbon} \xrightarrow{\text{heat}} \text{silicon} + \text{carbon monoxide}$$

$$SiO_2(s) + 2C(s) \rightarrow Si(s) + 2CO(g)$$

When sand is reduced, the silicon formed is not pure. Several stages of purification are needed to transform it into the very pure silicon used in silicon chips.

> **SUMMARY NOTE**
>
> Silicon is obtained from silicon(IV) oxide (sand). Very pure silicon is used in silicon chips in microcomputers.

12.11 Some problems and their solutions

For thousands of years, craftsmen used glass without knowing anything about its chemistry. Since chemists began to study glass, they have invented many new types. Let us look at some of the problems they tackled. Glass is in the forefront of modern technology.

Problem 1 The problem was to make glass which would stand sudden changes in temperature without breaking.

Solution The solution was to add boron oxide during manufacture. The result was Pyrex® glass.

How was the problem solved? A search was made for oxides sufficiently like silicon(IV) oxide to form a glass but with somewhat different properties. From their positions in the Periodic Table, you would expect boron and silicon to have similar oxides.

Problem 2 The problem was to make plates of glass which are the same thickness all over. This is necessary for window glass.

The old solution was to grind a sheet of glass to the required thickness and smoothness. In the process, 30% of the sheet was wasted.

The new solution was found by the Pilkington Glass Company. It is the **float glass** process. A stream of molten glass flows on to a bath of molten tin. As the glass cools and solidifies, both the top and the bottom surfaces are perfectly smooth and planar (Fig. 12.22). The glass is rolled to the required thickness while it is still soft and then cut into sections.

Fig. 12.22 Float glass is washed before being automatically cut and stacked

How was the problem solved? A team of chemists and engineers in St. Helens, Cheshire, worked for 7 years to perfect the process. They had to overcome enormous problems with the revolutionary new process. It is now used all over the world.

Problem 3 The problem was to make sunglasses which darken in bright light and lighten again when the light fades.

Solution The solution was to include silver chloride in the glass. The silver salt darkens on exposure to light in the same way as do the silver salts in photographic film. Photosensitive glass reverts to colourless when the light fades.

How was the problem solved? Chemists applied their knowledge of what happens in photography to a different situation.

Problem 4 Here the solution came before the problem! How? By accident! A researcher left some photosensitive glass in a furnace overnight. He found that the glass had become opaque. Believing that glass you cannot see through is useless, he threw it out, and found to his surprise that the glass would not break. He had made **glass ceramic**, which is almost unbreakable.

The problem of finding uses for the new material was not difficult. Makers of ovenware and electrical insulation immediately seized on glass ceramic. It is now found in the nose cones of space rockets and the tiles of space shuttles.

Problem 5 In many tropical countries, the disease **bilharzia** (or **schistosomiasis**) is a blight. It is carried by water snails. Scientists discovered that these snails can be killed by copper salts in quite low concentrations. How can health officers maintain a steady, low concentration of copper compounds in the streams and ponds where the snails live?

Solution The copper compounds can be incorporated into a **soluble glass**. To make a soluble glass, phosphorus(V) oxide, P_2O_5, is used instead of silicon(IV) oxide. When pellets of the copper-containing glass are put into the water, they dissolve slowly, releasing copper compounds gradually into the snail-infested water. This treatment is more effective than a single large dose of copper compounds.

Problem 6 A herd of sheep is grazing on land which is short of three essential **trace elements**. (These are elements which are needed in very small quantities.) The farmer wants a convenient method of adding these three elements to their diet. Can you solve his problem?

> **SUMMARY NOTE**
>
> You may like to read about these examples of problem-solving in the glass industry:
> - how to avoid breakage
> - how to make plate glass
> - how to make light-sensitive glass
> - how to make glass ceramic
> - how to make a soluble glass.

Exercise 12

1 **a** State the approximate percentages by volume of nitrogen and oxygen in air.
   b Which of these gases is important in respiration?
   c Give the names of the substances which are produced in respiration.
   d Describe a test for each of these substances.
   e Name the reactants and the products involved in photosynthesis.
   f What other substance is needed for photosynthesis to take place?

2 **a** Where are the great reserves of carbon in nature?
   b How does the flow of carbon dioxide into and out of the atmosphere take place (i) over land and (ii) over the sea?
   c How is the burning of fossil fuels affecting the carbon cycle?
   d The air near the ground in a cornfield in summer contains less carbon dioxide by day than by night. Why is this?
   e The concentration of carbon dioxide in the atmosphere near the Earth increases during the winter and decreases during the summer. Why is this?
   f Sea water is slightly alkaline. Fresh water is slightly acidic. In which type of water would you expect the rate of photosynthesis to be greater? Assume that the solubility of carbon dioxide in the two different kinds of water is the important factor.

3 **a** Why is carbon dioxide the best type of extinguisher to use on an electrical fire? Why is water not used?
   b Name one type of fire which would not be extinguished by carbon dioxide. Explain why this fire would continue to burn.

4 a What is the limewater test for carbon dioxide? How does it work? What happens when you bubble carbon dioxide through limewater for a long time?

b Explain how falling on to limestone rocks makes water hard.

c Why does hard water produce 'fur' in kettles?

5 Each year, 8 million tonnes of sulphuric acid are produced in the US as a by-product of the smelting of sulphide ores. The sulphuric acid is discharged into streams.

a Calculate the mass of calcium oxide (quicklime) needed to neutralise it by means of the reaction:

calcium oxide + sulphuric acid → calcium sulphate + water

$CaO(s) + H_2SO_4(aq) \rightarrow CaSO_4(aq) + H_2O(l)$

b Calculate the mass of calcium carbonate (limestone) that must be heated in a lime kiln to make the calcium oxide needed in **a**.

6

Why should I bother buying stuff in returnable bottles and then having to cart the empties back to the shop?

Go on, tell him why!

7 a What property of pure silicon crystals makes them useful as transistors?

b How can sand be reduced to silicon?

c In the manufacture of glass, what materials are added to provide (i) silicon and (ii) sodium?

d In what group of the Periodic Table are carbon and silicon?

e What type of chemical bond is present in $SiCl_4$?

f Why is silicon(IV) oxide, SiO_2, a high melting point solid, whereas carbon dioxide, CO_2, is a gas?

8 At present, scientists do not know how serious the threat from the greenhouse effect really is. Imagine that a group of scientists were to obtain new evidence that convinced everyone that the threat from the greenhouse effect is very serious. They work out that, by the year 2000, the rise in temperature will start to create deserts and to melt ice caps.

The urgent problem would be to reduce the level of carbon dioxide in the atmosphere. Discuss the ways in which nations would have to cooperate to solve this problem. Say how national aims would have to come second to world aims. Which countries would be most affected? What would happen to the developing nations' drive to industrialise? Would nuclear power help to solve the problem?

Draw up a world plan for action to reduce the level of carbon dioxide in the air. Say how you would persuade nations to follow the plan.

This is rather a big question for one person! Perhaps you should form a small group to tackle it. Or a number of groups could work out solutions and then compare notes.

CHAPTER 13 Chemicals from sulphur

13.1 Sulphur

General Cortes ran out of sulphur in 1521. He had set sail from Spain in 1518 to conquer Mexico. When he ran out of ammunition in 1521, he knew he would have to make his own gunpowder. Two of the ingredients were easy to find in Mexico. The third was sulphur, and the only place he could find it was in the crater of a volcano called Popocatepetl. At the bottom, a flame was burning. Yellow clouds of sulphur vapour swirled upwards, and deposited solid sulphur on the sides of the crater. A volunteer was lowered inside a large basket. He filled the basket with sulphur scraped from the sides of the crater, and was hauled up again. He descended again and again until he had collected 150 kilograms of sulphur. From this, the Spaniards were able to make enough gunpowder to keep the army of occupation supplied until a shipment could be sent from Spain.

Fig. 13.1 One way to obtain sulphur — from your neighbourhood volcano

Charles Goodyear was rather careless with sulphur in 1839. He had some lying on his laboratory bench while he was experimenting with rubber. Rubber is obtained by cutting the bark of certain tropical trees. It trickles out in the form of a liquid called latex. Latex solidifies when ethanoic acid is added. Goodyear accidentally spilt some rubber on to a mound of sulphur. When he picked up the rubber, it had turned from a soft, pliable substance into a much harder material. He quickly realised that he had made a tougher, more hard-wearing, more useful material than pure rubber. The process of **vulcanisation** was born (Fig. 13.2). In vulcanisation, rubber is mixed with sulphur, and the mixture is heated. Hard rubber, such as that used for motor tyres, contains 10% to 50% sulphur (Fig. 13.3).

The tyre industry needs enormous quantities of sulphur. What is this useful substance? Where does the industry obtain it?

Fig. 13.2 Vulcanisation of tyres

Fig. 13.3 Vehicle tyres

Sulphur is a non-metallic element. It does not conduct heat or electricity. It is a brittle, yellow solid, which melts at 112 °C and boils at 444 °C.

Sulphur is found **native**, that is, as free sulphur, in Sicily, Poland and the USA.

Some of the uses of sulphur are shown in Fig. 13.4, and the reactions are summarised in Fig. 13.7.

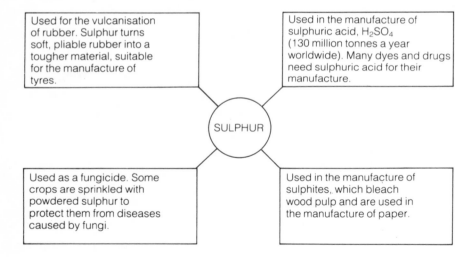
Fig. 13.4 Some of the uses of sulphur

13.2 Allotropes of sulphur

There are two kinds of crystalline sulphur. They are called **rhombic** and **monoclinic** sulphur. Rhombic sulphur is formed when a solution of sulphur is allowed to crystallise below 96 °C (see Experiment 13.1). Methylbenzene can be used as a solvent. Monoclinic sulphur is formed when molten sulphur solidifies.

The two forms of pure sulphur have the same chemical reactions (see Fig. 13.7). They are both pure sulphur. The two forms of the element have crystals of different shapes and are called **allotropes** (see Fig. 13.5).

When solid sulphur is heated, it melts at 112 °C. On further heating, it becomes a **viscous** (treacle-like) liquid. When this

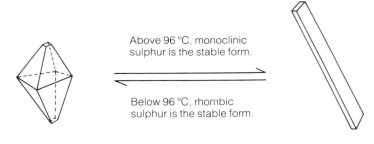

Fig. 13.5 The allotropes of sulphur ▷

Rhombic sulphur crystal (magnified 20×)

Monoclinic sulphur crystal (magnified 20×)

viscous liquid is poured into cold water, it forms **plastic** sulphur (see Fig. 13.6). As plastic sulphur is not a **crystalline** form of the element, it is not an allotrope. Some of the reactions of sulphur are shown in Fig. 13.7.

Solid sulphur consists of S_8 molecules.

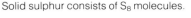

In molten sulphur, long chains of sulphur atoms become entangled. They make the liquid **viscous** (treacle like).
When it is poured into cold water, the liquid solidifies with the atoms in long chains. This is how plastic sulphur is made.

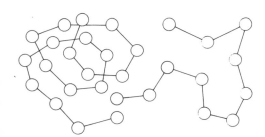

Fig. 13.6 Sulphur molecules

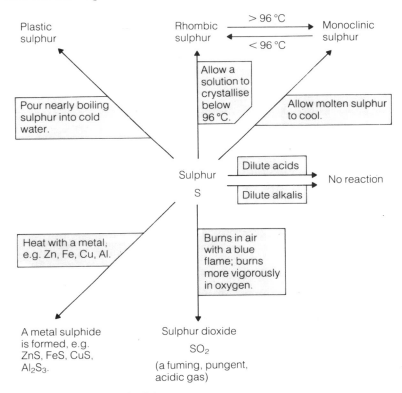

Fig. 13.7 Some reactions of sulphur

SUMMARY NOTE

- Sulphur was used in gunpowder. It is now used in **vulcanisation**, a method of toughening rubber.
- Sulphur is a non-metallic element. It exists in allotropic forms, rhombic and monoclinic. Some reactions of sulphur are shown in Fig. 13.7.

JUST TESTING 72

1 Describe and explain what happens when sulphur is
 a burned in air (see Section 8.9)
 b heated with iron (see Experiment 3.4).

2 Sulphur is a non-metallic element. Give two physical properties and two chemical properties of sulphur which back up this statement.

13.3 Sulphur dioxide

When sulphur burns, sulphur dioxide is formed (Experiment 13.2):

sulphur + oxygen → sulphur dioxide

$S(s) + O_2(g) \rightarrow SO_2(g)$

Sulphur dioxide is a gas with a very unpleasant, penetrating smell. It 'fumes' in moist air. It reacts with water vapour to form a mist. It is very unpleasant to breathe, causing congestion followed by choking.

The food industry uses sulphur dioxide as a preservative. It is added in small quantities to foods such as those shown in Fig. 13.8. It prevents the growth of moulds and bacteria.

*Sulphur dioxide is also a useful bleach (see Experiment 13.2). It is mild in its action and can be used on wool and silk, as well as on paper and straw. These materials are coloured by dyes which contain oxygen. Sulphur dioxide is a **reducing agent**, that is to say, it can remove oxygen from other substances. The **reduced** form of the dye that remains is colourless:

Fig. 13.8

| coloured form of dye | + | solution of sulphur dioxide | → | colourless form of dye | + | oxidation product of sulphur dioxide |

| (dye + oxygen) | + | $SO_2(aq)$ | → | (dye) | + | oxidation product of $SO_2(aq)$ |

The colourless substance formed is gradually oxidised by air to form the coloured dye again. This is why straw, silk and paper turn yellow with age.

Fig. 13.9 Straw boaters

Fig. 13.10 The paper industry

What happens to the solution of sulphur dioxide when it is oxidised? Sulphur dioxide is extremely soluble in water. It reacts with water to form the weak acid, sulphurous acid, H_2SO_3:

sulphur dioxide + water → sulphurous acid

$SO_2(g) + H_2O(l) \rightarrow H_2SO_3(aq)$

Sulphurous acid is slowly oxidised by oxygen in the air to sulphuric acid, H_2SO_4. Sulphuric acid is formed whenever sulphurous acid accepts oxygen from an oxidising agent:

sulphurous acid + oxygen from the air or from an oxidising agent → sulphuric acid

$H_2SO_3(aq) + (O) \rightarrow H_2SO_4(aq)$

This reaction is the basis of two tests for sulphur dioxide. When it reduces the oxidising agents potassium dichromate(VI), $K_2Cr_2O_7$, and potassium manganate(VII), $KMnO_4$, there is a change in colour (see Fig. 13.11):

potassium dichromate(IV), orange + SO_2 → green and then blue

potassium manganate(VII), purple + SO_2 → colourless

The gas jar contains sulphur dioxide. A strip of paper spotted with potassium dichromate(VI) turns from orange to blue. A spot of potassium manganate(VII) turns from purple to colourless.

Fig. 13.11 Testing for sulphur dioxide

*13.4 Sulphites

Being an acid gas, sulphur dioxide reacts with bases. Salts of sulphurous acid called **sulphites** are formed. For example, when sulphur dioxide is bubbled into sodium hydroxide solution, sodium sulphite is formed:

sulphur dioxide + sodium hydroxide → sodium sulphite + water

$SO_2(g) + 2NaOH(aq) \rightarrow Na_2SO_3(aq) + H_2O(l)$

The weak acid, sulphurous acid, is driven out of its salts by strong acids, e.g. hydrochloric acid. This reaction is used as a method of preparing sulphur dioxide in the laboratory:

sodium sulphite + hydrochloric acid → sulphur dioxide + sodium chloride + water

$Na_2SO_3(s) + 2HCl(aq) \rightarrow SO_2(g) + 2NaCl(aq) + H_2O(l)$

It is also used as a **test for a sulphite**.

The reactions of sulphur dioxide are summarised in Fig. 13.12.

SUMMARY NOTE

Sulphur dioxide is a gas with a pungent smell. It is used as a food preservative and as a bleach. Sulphur dioxide is a reducing agent. It is an acidic gas, which reacts with bases to form salts called sulphites. A strong acid drives sulphur dioxide out of sulphites.

JUST TESTING 73

1 Describe two chemical tests for sulphur dioxide.

2 a Name the acid formed when sulphur dioxide dissolves in water.
  b What are the salts of this acid called?
  c What must be added to these salts to make them give sulphur dioxide?

*3 How could you prove that rhombic and monoclinic sulphur are forms of the same element?

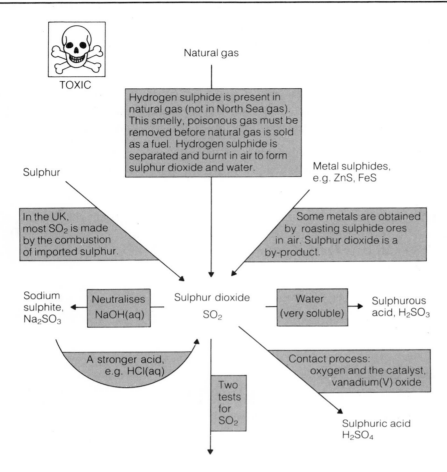

Fig. 13.12 Some reactions of sulphur dioxide

13.5 Sulphuric acid

Figure 13.13 shows a phosphate fertiliser in use. The most plentiful source of phosphate is 'rock phosphate', an ore containing calcium phosphate. It has the disadvantage of being insoluble. The fertiliser industry uses it as the starting point in the manufacture of the soluble fertiliser, ammonium phosphate. Sulphuric acid is essential to the process. Calcium phosphate reacts with sulphuric acid to make phosphoric acid which is neutralised by ammonia to give ammonium phosphate. Sulphuric acid turns something we have plenty of — calcium phosphate — into something we need plenty of — fertiliser. 'Superphosphate' is another soluble phosphate fertiliser made from 'rock phosphate' and sulphuric acid. Figure 13.14 summarises the uses of sulphuric acid.

Fig. 13.13 Applying fertiliser

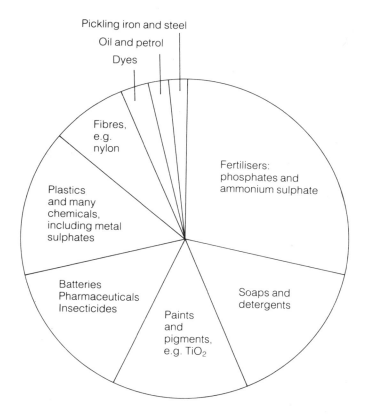

Fig. 13.14 The main uses of sulphuric acid (130 million tonnes a year worldwide)

13.6 Manufacture of sulphuric acid: the Contact Process

The Contact Process for the manufacture of sulphuric acid takes place in stages:

sulphur → sulphur dioxide → sulphur trioxide → sulphuric acid

S → SO_2 → SO_3 → H_2SO_4

Conditions which give a good yield of sulphur trioxide

Sulphur dioxide and oxygen react to form sulphur trioxide:

$$2SO_2(g) + O_2(g) \rightleftharpoons 2SO_3(g)$$

The reaction is reversible: it will take place from right to left as well as from left to right. The proportion of sulphur trioxide present in the reacting mixture depends on the temperature and pressure.

At high temperatures, the molecules of sulphur trioxide formed split up again into sulphur dioxide and oxygen. (The decomposition of sulphur trioxide is an endothermic reaction (see Section 18.4).) High temperatures therefore give a poor yield of sulphur trioxide. The Contact Process uses a temperature of 400 °C to 500 °C. This temperature is high enough to speed up the formation of sulphur trioxide and low enough to prevent its decomposition. A catalyst is used to increase the rate of the reaction. The original Contact Process used platinum as the catalyst. It is easily 'poisoned' (made inactive) by traces of impurities in the gases. Vanadium(V) oxide, V_2O_5, is now used as the catalyst because it is cheaper and less easily poisoned.

The industrial process

A flow diagram of the Contact Process is shown in Fig. 13.15. The sulphur dioxide needed is obtained from

- the combustion of sulphur
- the removal of unpleasant-smelling sulphur compounds from petroleum oil
- the removal of hydrogen sulphide from natural gas (not North Sea gas)
- roasting metal sulphides in order to extract metals from their ores.

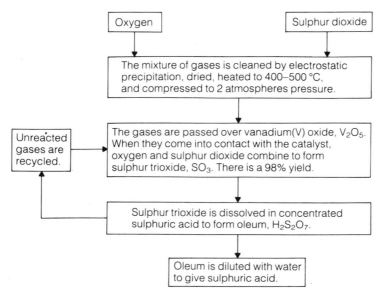

Fig. 13.15 Flow diagram for the Contact Process for the manufacture of sulphuric acid

Sulphur dioxide and air are purified by electrostatic precipitation (see Fig. 8.25).

A sulphuric acid plant must be sited in a place where supplies of sulphur dioxide can reach it, that is, near a port where cargoes of sulphur can arrive by ship, or near a gas refinery or a metal smelter.

Reaction of sulphur trioxide and water

Sulphur trioxide reacts with water to give sulphuric acid:

$$SO_3(g) + H_2O(l) \rightarrow H_2SO_4(l)$$

This reaction is troublesome to carry out on an industrial scale. As soon as sulphur trioxide meets water vapour, it forms a mist of sulphuric acid. Instead of water, sulphuric acid is used to absorb sulphur trioxide. Oleum is formed. On dilution with water, this gives sulphuric acid:

sulphur trioxide + sulphuric acid → oleum
$SO_3(g)$ + $H_2SO_4(l)$ → $H_2S_2O_7(l)$

oleum + water → sulphuric acid
$H_2S_2O_7(l) + H_2O(l) \rightarrow 2H_2SO_4(l)$

> **SUMMARY NOTE**
>
> Sulphuric acid is used to make fertilisers and pesticides, soaps and detergents, paints and plastics. It is made by the Contact Process. Sulphur dioxide and oxygen combine when they come into contact with the catalyst, vanadium(V) oxide. The sulphur trioxide formed is absorbed in sulphuric acid to form oleum. Oleum reacts with water to give sulphuric acid.

13.7 Reactions of sulphuric acid

Dilute sulphuric acid

When concentrated sulphuric acid is mixed with water, a large amount of heat is given out. This is why making dilute sulphuric acid can be a dangerous job if it is not done properly. You must never add water to the concentrated acid. It can reach boiling point in localised regions and boil and splash out of the container. When you add the concentrated acid to water, and stir, the water can absorb the heat given out (see Fig. 13.16).

The reactions of dilute acids, were covered in Chapter 7. The reactions of dilute sulphuric acid are summarised in Fig. 13.17.

> **SUMMARY NOTE**
>
> Figure 13.17 shows the reactions of dilute sulphuric acid.

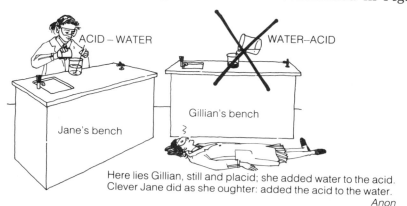

Fig. 13.16

Here lies Gillian, still and placid; she added water to the acid.
Clever Jane did as she oughter: added the acid to the water.
Anon

Concentrated sulphuric acid

When concentrated, sulphuric acid has other reactions (see Fig. 13.18). It acts as a drying agent and as a dehydrating agent.

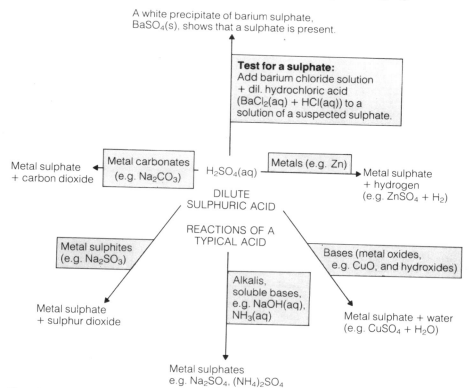

Fig. 13.17 Reactions of dilute sulphuric acid

Reactions as a drying agent. A **drying agent** removes water from mixtures. Gases are bubbled through concentrated sulphuric acid to dry them.

Reactions as a dehydrating agent. A **dehydrating agent** removes the elements H_2O from compounds.

1 Concentrated sulphuric acid dehydrates copper(II) sulphate crystals:

copper(II) sulphate-5-water → copper(II) sulphate + water
(blue crystals) (white, anhydrous)

$$CuSO_4 \cdot 5H_2O(s) \xrightarrow{\text{conc. } H_2SO_4} CuSO_4(s) + 5H_2O(l)$$

Heat is given out as concentrated sulphuric acid reacts with the water of crystallisation, just as it is when you add concentrated sulphuric acid to water.

2 When carbohydrates (sugars and starches) are dehydrated, carbon remains:

glucose (a sugar) → carbon (sugar charcoal) + water

$$C_6H_{12}O_6(s) \xrightarrow{\text{conc. } H_2SO_4(l)} 6C(s) + 6H_2O(l)$$

The reaction is exothermic.

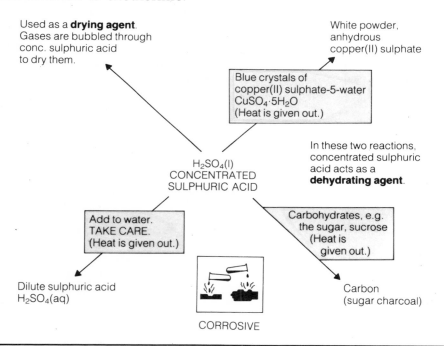

Fig. 13.18 The reactions of concentrated sulphuric acid

SUMMARY NOTE

Concentrated sulphuric acid is a drying agent and a dehydrating agent. Heat is given out when it reacts with water.

JUST TESTING 74

1 A bottle is labelled 'dilute sulphuric acid'. How could you show that it is an acid? How could you prove that it contains sulphate ions?

2 Under what conditions does sulphuric acid react with **a** zinc, **b** copper(II) oxide, **c** copper(II) sulphate-5-water? State what products are formed in each of these reactions.

3 What is the difference between a drying agent and a dehydrating agent?

4 Describe the manufacture of sulphuric acid.
How could a country which has no sulphur deposits economise on the quantity of sulphur which it imports?

13.8 Sulphates

It was the adventurers who explored the continent of South America who discovered the potato. They brought their discovery back to Europe. The potato yields more food more quickly on less land than wheat or corn or rice. After the potato was introduced into Ireland, the health of the people improved, and the population expanded rapidly. In 1845, there were 8 million people living in Ireland. For most of them, the chief item in their diet was potato.

In 1845, a mysterious disease struck the potato crop. The plants wilted, and the potato tubers rotted. Six years of potato famine followed. The potato blight was caused by a fungus which spread throughout Europe. The Irish were the worst affected because they had no other crop to fall back on. A million people died. Another million emigrated to the USA.

The Irish potato harvest was revived by an injection of new stock from South America. Today, Irish potato farmers are still growing potatoes, and they are not afraid of the blight. A simple chemical has come to the rescue. The potato blight fungus can be killed by being sprayed with **Bordeaux Mixture**. This is a very inexpensive fungicide. It is made by dissolving copper(II) sulphate and calcium oxide in water. Bordeaux Mixture is also used to combat fungal infections of apple trees, pear trees and vines. If the Irish had known in 1845 that copper(II) sulphate was a killer — as far as potato blight fungus goes — a million lives would have been saved.

SUMMARY NOTE

Many sulphates have important uses. The methods used for their preparation are shown in Fig. 13.17.

13.9 Methods of preparing sulphates

Methods of making salts were covered in Chapter 7. Reactions which lead to the formation of sulphates are summarised in Fig. 13.17.
Experiments with some interesting sulphates are covered in Experiments 13.3 to 13.5.

JUST TESTING 75

1 What chemicals, in addition to sulphuric acid, would you need in order to make **a** magnesium sulphate, **b** potassium sulphate, **c** copper(II) sulphate, **d** lead(II) sulphate? Describe how you would prepare solid samples of these four sulphates.

Exercise 13

1. An element A burned in air to give a gas B which dissolved in water to give an acid C which changed slowly into an acid D. When heated gently, the element A melted to form a brown liquid.
   - **a** Name the element A and compounds B, C and D.
   - **b** Is A a metallic or a non-metallic element? Give two reasons for your answer.

2. Suggest what the substances A, B, C, D and E, might be.
   - **a** Sulphur burned in the gas A with a very bright blue flame.
   - **b** A solution of B, on addition of barium chloride solution and dilute hydrochloric acid, gave a white precipitate.
   - **c** A white powder C turned blue when water was added.
   - **d** A green solid D effervesced and formed a blue solution when dilute sulphuric acid was added.
   - **e** A black solid E reacted with dilute sulphuric acid to form a blue solution.

3. Do not write in this book. Supply words or phrases to fill the blanks.
   In the ____ process for manufacturing sulphuric acid, the gas sulphur dioxide is obtained by ____. It is mixed with ____ and purified by the method of ____. The mixture of gases is passed over a heated catalyst, ____, and the compound ____ is formed. This compound is absorbed in ____ to form ____. On dilution with water, this gives sulphuric acid.

4. Do not write in this book. State what would be observed in the tests **a** to **f**.

Test	Sodium sulphite (aq)	Sodium sulphate (aq)
Add dilute hydrochloric acid	a	b
Add barium chloride solution	c	d
Add dilute hydrochloric acid and barium chloride solution	e	f

5. Describe how you would prepare samples of
   a rhombic sulphur, **b** monoclinic sulphur, **c** plastic sulphur.

*6. How could you prove that concentrated sulphuric acid is a dehydrating agent?

7. Study this reaction scheme:

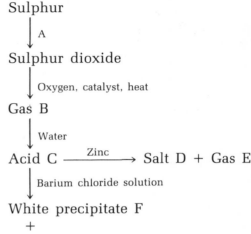

   - **a** Give the conditions for reaction A.
   - **b** Name gas B. Write a word equation and a symbol equation for its formation.
   - **c** Name acid C. Give its chemical formula.
   - **d** Name salt D, gas E and precipitate F.

8. Sulphur dioxide is released into the atmosphere by electricity power stations and sulphuric acid plants. If the sulphur dioxide were recovered from the air, it could be converted into sulphuric acid. This would reduce the need for imported sulphur.
   - **a** Explain briefly why sulphur dioxide is regarded as a pollutant.
   - **b** Explain why coal-fired power stations produce sulphur dioxide.
   - **c** Explain the connection between recovering sulphur dioxide from the air and reducing imports of sulphur.
   - ***d** The process for recovering sulphur dioxide is

     sulphur dioxide + oxygen + water → sulphuric acid

     $2SO_2(g) + O_2(g) + 2H_2O(l) \rightarrow 2H_2SO_4(l)$

     - Take the percentage of sulphur dioxide in the polluted air as 1% by volume.
     - Take the gas molar volume as 25 litres under the conditions of the process.

Calculate the mass of sulphuric acid that could be obtained from 100 000 litres of air. (The steps in the calculation are: volume of SO_2 ... amount (mol) of SO_2 ... amount (mol) of H_2SO_4 ... mass of H_2SO_4. $A_r(O) = 16$, $A_r(S) = 32$, $A_r(H) = 1$.)

Find the fertiliser

To find the name of a fertiliser made from sulphuric acid, fill in the clues across. Then read the answer in 1 down.
1. This non-metallic element burns with a blue flame. (7)
2. Goodyear found out how to ___ rubber. (9)
3. Soldiers needed sulphur to make this. (9)
4. Monoclinic sulphur looks like these. (7)
5. Copper is ___ towards dilute sulphuric acid. (10)
6. The weak acid that contains sulphur. (10)
7. Concentrated sulphuric acid will ___ sugar. (9)
8. The allotrope of sulphur that crystallises at room temperature. (7)
9. The form of sulphur you get by pouring molten sulphur into water. (7)
10. Fungi do not like this sulphate. (6)
11. A measure of acidity. (2 words: 2, 6)
12. The oxide of this metal is used in the Contact Process. (8)
13. Sulphuric acid is a ___ acid. (6)
14. The number of sulphur atoms in a molecule of solid sulphur. (5)

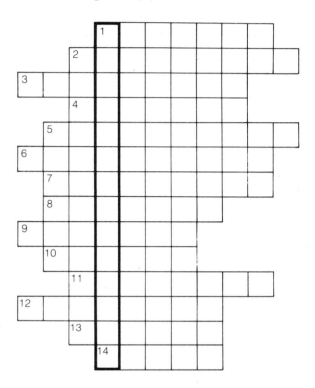

CHAPTER 14 Nitrogen and agriculture

14.1 The nitrogen cycle

Proteins are important nitrogen compounds. They are one of the three main classes of food substances. (The others are fats and carbohydrates.) Plants can synthesise (build) proteins from simpler nitrogen-containing compounds. Animals are unable to synthesise proteins: they obtain protein by eating plants or by eating animals which have eaten plants. Some plants can synthesise proteins from the nitrogen in the air. Beans, peas and clover can do this. They have nodules on their roots which contain **nitrogen-fixing bacteria** (Fig. 14.1). These bacteria **fix** atmospheric nitrogen; that is, turn it into nitrogen compounds. Plants which cannot fix nitrogen synthesise proteins from nitrates. Ammonium salts enter the soil in the excreta of animals and through the decay of plant and animal remains. **Nitrifying bacteria** in the soil convert ammonium salts into nitrates. Another source of nitrates in the soil is rain. Nitrogen and oxygen combine in the atmosphere during a lightning storm to form oxides of nitrogen. These gases react with water to form nitric acid. A rain shower brings nitric acid to Earth, and it reacts with minerals to form nitrates. Plants take in these nitrates through their roots. Some nitrates are converted back into

Fig. 14.1 Nodules on the roots of a leguminous plant

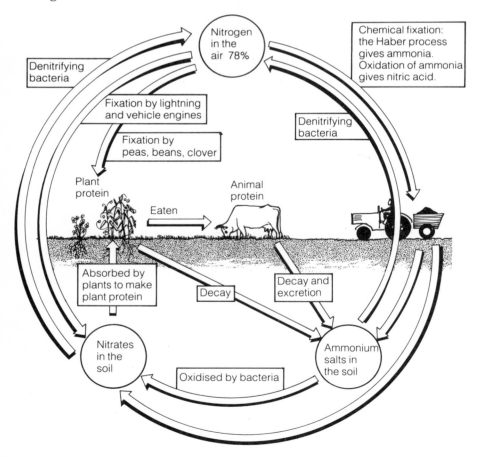

Fig. 14.2 The nitrogen cycle

gaseous nitrogen by **denitrifying bacteria**. Animals obtain proteins by eating plants or by eating other animals.

The way in which nitrogen circulates from air to soil to living things and back again is called the **nitrogen cycle** (see Fig. 14.2).

14.2 Nitrogen

You met nitrogen in Chapter 8. Figure 14.3 summarises its reactions.

> **SUMMARY NOTE**
>
> Proteins contain nitrogen. Plants synthesise proteins from nitrates. Some plants can 'fix' gaseous nitrogen. Plants can utilise the nitrogen in ammonium salts after soil bacteria convert them into nitrates. Animals rely on plants for protein.
>
> **The nitrogen cycle**: This is the way nitrogen circulates from the air to the soil to living things and back again.

NITROGEN

- Colourless, odourless, neutral gas, insoluble in water
- Makes up 78% by volume of air. Obtained by fractional distillation of liquid air; see Section 8.3.
- Used when an inert atmosphere is needed, e.g. for some welding jobs; see Section 8.5.
- Combines with oxygen during lightning storms. Gaseous oxides of nitrogen are formed. These react with water vapour in the air to form nitric acid. Rain containing nitric acid reacts with minerals in the soil to form nitrates. These are fertilisers. These reactions, which convert nitrogen into nitrogen compounds, are called 'atmospheric fixation' of nitrogen.
- Extinguishes a burning splint. Combines with burning magnesium to form magnesium nitride, Mg_3N_2.
- Combines with hydrogen under the conditions of the Haber process to form ammonia. This is the industrial method of 'fixing' nitrogen.

Fig. 14.3 Some reactions of nitrogen

14.3 Fertilisers

As the world population has increased, the demand for food has increased. The world population was 3.5 billion in 1965. By 2000, it will reach 6.5 billion. Many people today are hungry, and many more are under-nourished. Farming has to be intensive, with large crops grown on the land available for agriculture. Crops take nutrients from the soil, and these must be replaced before the next crop is sown.

Before the nineteenth century, farmers relied on the natural fertilisers, manure and compost, to replace nutrients in the soil. When the population began to rise sharply in the nineteenth century, the need for food increased. To grow more food, farmers needed more fertilisers. They asked the chemists to make some artificial fertilisers. Before the chemists could do this, they had to find out which chemicals in manure fertilised the soil. Then they could make chemical fertilisers to do the same job. In 1842, the German chemist Justus von Liebig (of condenser fame) analysed soils to find out the differences between fertile soils (which gave good crops) and infertile soils. He found that fertile soils are richer than poor soils in

- nitrogen compounds
- phosphates
- potassium compounds
- calcium and magnesium compounds.

In Britain, in 1843, John Lawes set up the first agricultural research station in Rothamsted in Hertfordshire. He studied the effects of fertilisers on the growth of crops. His experiments showed that plants need three groups of chemicals.

1 Large quantities of carbon, hydrogen and oxygen compounds; there is no shortage of these elements. Carbon comes from carbon dioxide in the air. Hydrogen and oxygen come from water.

2 Small quantities of the **trace elements** (iron, manganese, boron, copper, cobalt, molybdenum, zinc); the quantities of these elements needed are so small that they do not normally need to be added to the soil.

3 Fairly large quantities of compounds containing nitrogen, phosphorus, potassium, calcium, magnesium, sulphur and, for some crops, sodium

Counting the cost

Most fertilisers concentrate on supplying the necessary nitrogen (N), phosphorus (P) and potassium (K). NPK fertilisers are consumed in huge quantities: in 1980, the world consumption was 19 million tonnes. In the UK, the consumption of NPK fertilisers is about 7 million tonnes a year. The fertilisers cost about £60/tonne. You can see that the fertiliser industry is very big business. The cost of fertilisers is a large item in a farmer's budget. He wants to use them as efficiently as possible. The Ministry of Agriculture, Fisheries and Food has set up an Agricultural Development and Advisory Service to help farmers and market gardeners. They can obtain expert advice on the type and quantity of fertiliser to apply and the best season of the year for applying it. Every farmer and grower has different problems. The agricultural chemists must consider the type of crop and the type of soil before they can recommend the most suitable treatment. The growers want good crops, but they do not want to spend more than they need on fertilisers. It is important not to use too much fertiliser. The excessive use of fertilisers which results in the leaching of nitrates into ground water was mentioned in Section 9.16.

Nitrogen

Nitrogen can be absorbed by plants when it is combined as nitrates. Ammonia and ammonium salts also can be used as fertilisers because ammonium salts are converted into nitrates by organisms which live in soil. All nitrates and all ammonium salts are soluble. This makes them easily absorbed. Nitrogenous fertilisers increase both the size of the crop and the protein content of the plants. Sometimes, concentrated ammonia solution is used as a fertiliser. Usually, solid fertilisers — ammonium nitrate, ammonium sulphate and urea (CON_2H_4) — are applied in pellet form (see Figs. 14.4 and 14.5).

Fig. 14.4 Application of concentrated ammonia solution to a field

Fig. 14.5 Application of pellets of ammonium salts to a field

Phosphorus

Phosphorus is present in phosphates, the salts of phosphoric acid. Phosphates stimulate root development. Healthy roots enable the crop to take in other nutrients. The phosphates which are naturally present in soil are only slightly soluble, and are only slowly absorbed by plants. The soluble fertiliser ammonium phosphate is made from ores containing calcium phosphate (see Section 13.5).

Potassium

Potassium compounds are all soluble, and are therefore readily taken up by plant roots. They help plants to carry on photosynthesis. Potassium chloride is an ingredient of NPK fertilisers.

Sodium

For most crops, there is no need to add extra sodium compounds. Sugar beet is an exception: it grows better if common salt is added as a fertiliser.

The manufacture of nitrogenous fertilisers

Farmers took note of the results of the experiments on fertilisers. From 1850 onwards, European farmers started to import sodium nitrate (**Chile nitre**) from South America to fertilise their land. Vast deposits of sodium nitrate exist in Chile. Nevertheless, Europeans were not completely happy at being dependent on a remote source of fertiliser. They wanted to be able to make their own. Nitrogen was the obvious starting point. Every country has plenty of nitrogen in the air. The problem was that, since nitrogen is so unreactive, it is difficult to convert it into nitrogen compounds. The problem was solved in 1908 by a German chemist called Fritz Haber. He managed to get nitrogen to combine with hydrogen to form ammonia:

$$\text{nitrogen} + \text{hydrogen} \rightleftharpoons \text{ammonia}$$
$$N_2(g) + 3H_2(g) \rightleftharpoons 2NH_3(g)$$

This reaction is reversible: it takes place from right to left as well as from left to right. Some of the ammonia formed dissociates into nitrogen and hydrogen. A mixture of nitrogen, hydrogen and ammonia is formed. To increase the percentage of ammonia in the mixture, a high pressure and a low temperature are used. A low temperature reduces the dissociation of ammonia, but it also makes the reaction very slow. Modern industrial plants use a compromise temperature and speed up the reaction by means of a catalyst. The yield of ammonia is about 10%. Ammonia is condensed out of the mixture. (It is easily condensed because it has a high boiling point.) The nitrogen and hydrogen which have not reacted are recycled through the plant (see Fig. 14.4).

The hydrogen for the Haber process is obtained from natural gas. The method is similar to that shown in Fig. 9.21.

SUMMARY NOTE

Intensive farming is needed to keep the world's population fed. Artificial fertilisers must be used to give big crops. NPK fertilisers contain nitrogen (for protein synthesis), phosphorus (for healthy roots), potassium (to help photosynthesis) and some calcium and magnesium.

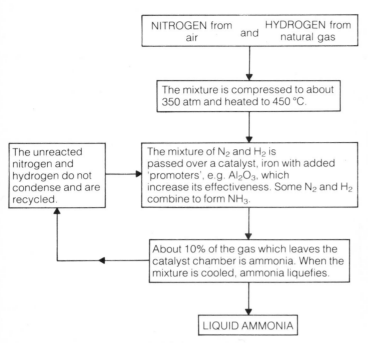

Fig. 14.6 Flow diagram of the Haber Process

Fig. 14.7 ICI, Billingham

JUST TESTING 76

1. Why does every country need fertilisers? Why is nitrogen one of the elements which fertilisers must contain? What is the advantage of obtaining nitrogen from air, instead of from nitrates?
2. Explain how nitrogen is converted into ammonia. Why do ammonium salts make good fertilisers? What other nitrogen compounds are used as fertilisers?
3. Name three routes by which nitrogen from the air finds its way into the soil. How does the amount of nitrogen in the air stay constant?

SUMMARY NOTE

Sodium nitrate is mined in Chile. It was imported and widely used as a fertiliser in Europe. Now nitrogenous fertilisers are made from ammonia. The Haber Process converts nitrogen and hydrogen into ammonia. Hydrogen is obtained from natural gas, and nitrogen is obtained from air.

14.4 Ammonia

Properties

Ammonia has a distinctive smell — very penetrating and not very pleasant. Wet nappies smell of ammonia. Urine contains a nitrogen compound called urea. When bacteria break down urea, ammonia is formed. In large quantities, ammonia is a poison. Ammonia is a very soluble gas.

Basic reactions of ammonia

Ammonia is a basic gas. It turns damp red litmus blue. When it meets the acidic gas hydrogen chloride, they react to form the salt ammonium chloride, a white solid (Fig. 14.8):

ammonia + hydrogen chloride → ammonium chloride

$NH_3(g)$ + $HCl(g)$ → $NH_4Cl(s)$

Fig. 14.8 Reaction between ammonia and hydrogen chloride ▷

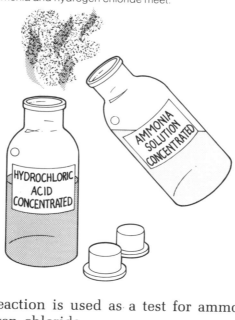

Clouds of ammonium chloride (a white solid) form when ammonia and hydrogen chloride meet.

Fig. 14.9 Ammonia solution as a cleaning fluid

This reaction is used as a test for ammonia and also as a test for hydrogen chloride.

Ammonia is a weak base. Its aqueous solution is alkaline. Some cleaning fluids contain ammonia (see Fig. 14.9). Ammonia solution is used as a degreasing agent. Like other alkalis, it **saponifies** grease, converting it into soluble soaps (see Section 7.3).

The alkaline nature of aqueous ammonia is due to the reaction between ammonia and water:

ammonia + water ⇌ ammonium ions + hydroxide ions
$NH_3(aq)$ + $H_2O(l)$ ⇌ $NH_4^+(aq)$ + $OH^-(aq)$

Ammonia is a weak base. The concentration of hydroxide ions in ammonia solution is low. It is high enough to make ammonia solution act as a typical alkali.

Some typical reactions of aqueous ammonia

1 When ammonia solution is added to a solution containing metal ions, many metals are precipitated as insoluble hydroxides (see Section 7.3). For example,

iron(II) ions + hydroxide ions → iron(II) hydroxide
$Fe^{2+}(aq)$ + $2OH^-(aq)$ → $Fe(OH)_2(s)$

2 Some metal hydroxides are first precipitated and later dissolved when more ammonia solution is added. Such behaviour is due to the formation of soluble **complex ions**. Examples are the tetraamminezinc ion, $(Zn(NH_3)_4)^{2+}$, and the tetraamminecopper(II) ion, $(Cu(NH_3)_4)^{2+}$. The tetraamminecopper(II) ions are deep blue. The formation of this deep blue colour with copper(II) salts is a sensitive test for ammonia.

3 Ammonia solution neutralises acids to give ammonium salts (see Experiments 14.2 and 14.4). For example,

ammonia + sulphuric acid → ammonium sulphate
$2NH_3(aq)$ + $H_2SO_4(aq)$ → $(NH_4)_2SO_4(aq)$

Figure 14.10 gives a summary of the reactions of ammonia.

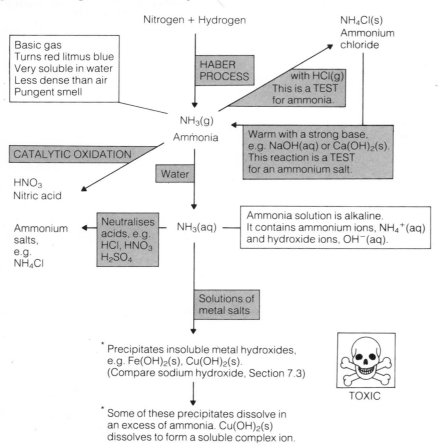

Fig. 14.10 Some reactions of ammonia

SUMMARY NOTE

Ammonia is a gas with a pungent smell. It is basic. It reacts with water to form an alkaline solution, which contains ammonium ions, NH_4^+, and hydroxide ions, OH^-. Ammonia solution reacts as a typical alkali. It precipitates insoluble metal hydroxides from solution. It neutralises acids to form ammonium salts.

Ammonium salts

Ammonium salts are fertilisers (see Section 14.2). Ammonium sulphate and ammonium nitrate are widely used (see Fig. 14.2).

Thermal dissociation

When you heat ammonium chloride (Experiment 14.6), it splits up into ammonia and hydrogen chloride. On cooling, these gases recombine (see Fig. 14.11).

$$\text{ammonium chloride} \underset{\text{cool}}{\overset{\text{heat}}{\rightleftarrows}} \text{ammonia} + \text{hydrogen chloride}$$

$$NH_4Cl(s) \underset{\text{cool}}{\overset{\text{heat}}{\rightleftarrows}} NH_3(g) + HCl(g)$$

This behaviour (decomposition on heating followed by recombination on cooling) is called **thermal dissociation**. Many ammonium salts undergo thermal dissociation (see Section 18.4).

Competition between bases

Being a weak base, ammonia is driven out of its salts by strong bases. To test a substance to see if it is an ammonium salt, you can warm it with sodium hydroxide solution. If ammonia is evolved, this proves that the substance is an ammonium salt (see Fig. 14.12):

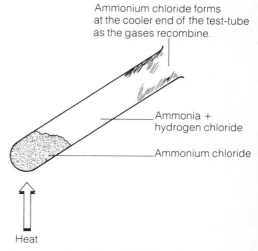

Fig. 14.11 Thermal dissociation of ammonium chloride

SUMMARY NOTE

Ammonium salts are used as fertilisers. A strong base drives ammonia out of its salts. Many ammonium salts undergo thermal dissociation.

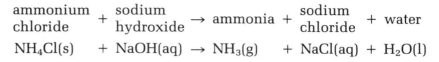

ammonium chloride + sodium hydroxide → ammonia + sodium chloride + water

$NH_4Cl(s) + NaOH(aq) \rightarrow NH_3(g) + NaCl(aq) + H_2O(l)$

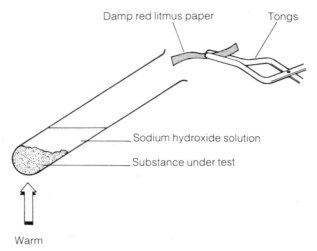

Fig. 14.12 Testing for an ammonium salt

JUST TESTING 77

1. Someone tells you that a bottle labelled 'smelling salts' contains ammonium carbonate. What tests would you do in order to find out whether this is true?
2. Describe how you could separate ammonium chloride and sodium chloride. (See Chapter 2 if you need help.)
3. What is formed when ammonia dissolves in water? Give two uses for the solution.
4. Describe how you would make crystals of ammonium sulphate. Why is this compound sold in large quantities?
*5. Describe what happens when ammonia solution reacts with solutions of **a** iron(III) sulphate, **b** iron(II) sulphate, **c** lead(II) nitrate, **d** copper(II) sulphate. Write equations for the reactions that take place in **a**, **b** and **c**.

14.5 Nitric acid

Dynamite

An important use of nitric acid is the manufacture of an oily liquid called nitroglycerine. This liquid explodes when it is ignited by a lighted fuse. After it was first made in 1846 by an Italian chemist, Professor Sobrero, miners and engineers started to use nitroglycerine for blasting away rock. Unfortunately, nitroglycerine can also explode when it receives a sudden blow. This sometimes happened when it was being transported to the site.

Immanuel Nobel began to manufacture nitroglycerine in Sweden in 1863. A year later, an explosion killed his son, Emil, and three other workers in the factory. Nobel's other son, Alfred, continued manufacturing, but he looked for some way to stabilise nitroglycerine so that it would not be exploded by mechanical shock. He experimented with a type of clay called kieselguhr, which was found near the factory. He found that it would absorb three times its own weight of nitroglycerine. The mixture did not explode when it was struck. It could be detonated only by means of a fuse. Alfred Nobel patented it as **dynamite**.

The new, safer explosive was invaluable for mining, making roads and railways, and blasting tunnels. The industrial revolution was taking place, and Nobel made a fortune out of his discovery. Of course, dynamite was used for armaments too. Concern about the use of his invention for warfare led Nobel to become interested in the peace movement. He decided to finance a prize for the person who had done the most effective work to promote peace between nations. When he died, he left £2 million to fund prizes for work towards peace and work in medicine, physics, chemistry and literature. These prizes have been awarded ever since 1901 by the Nobel Foundation in Stockholm.

Industrial manufacture of nitric acid

An industrial chemist called Ostwald invented a way of making nitric acid from ammonia and air and water. A flow diagram of his process is shown in Fig. 14.14.

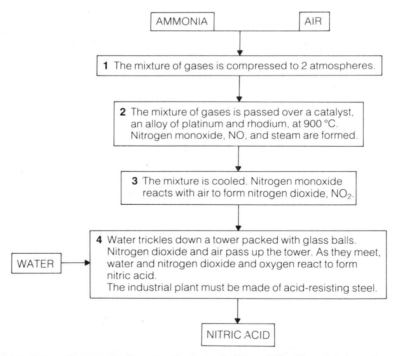

Fig. 14.13 Dynamite in use: blowing up a cooling tower

Fig. 14.14 A flow diagram for the manufacture of nitric acid by the oxidation of ammonia

The reactions that take place are

Stage 2 ammonia + oxygen → nitrogen monoxide, NO + steam
Stage 3 nitrogen monoxide + oxygen → nitrogen dioxide, NO_2
Stage 4 nitrogen dioxide + oxygen + water → nitric acid, HNO_3

Some properties of concentrated nitric acid are shown in Fig. 14.16.

SUMMARY NOTE

Nitric acid is used in the manufacture of dynamite and fertilisers. It is made by the oxidation of ammonia by air in the presence of a catalyst (Pt, Rh).

Concentrated nitric acid is a corrosive liquid. Care must be taken over its storage and transport.

Fig. 14.15 Concentrated nitric acid

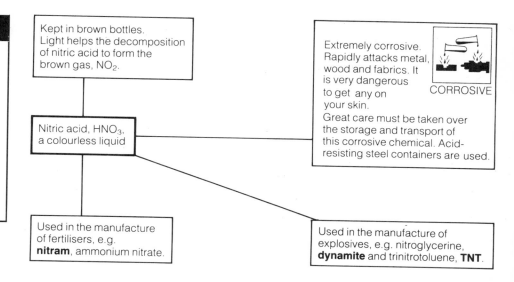

Reactions of dilute nitric acid

Nitric acid reacts as an acid when dissolved in water.

1 Bases. The reactions of dilute nitric acid with metal oxides, metal hydroxides and metal carbonates are summarised in Fig. 14.16.

2 Metals. Dilute nitric acid reacts with metals to give metal nitrates. Hydrogen is not obtained because dilute nitric acid oxidises it to water, and the gas nitrogen monoxide, NO, is formed. This gas forms the brown gas nitrogen dioxide, NO_2, as soon as it meets the air. Dilute nitric acid is a sufficiently powerful oxidising agent to oxidise copper and lead to their nitrates:

$$\text{copper} + \text{nitric acid (dilute)} \rightarrow \text{copper(II) nitrate} + \text{nitrogen monoxide} + \text{nitrogen dioxide} + \text{water}$$

*3 Test for a nitrate: the Devarda's alloy test. Nitric acid and nitrates can be reduced to ammonia. The reducing agent is a mixture of sodium hydroxide and Devarda's alloy (aluminium, copper and zinc). Naturally, if ammonium salts are present,

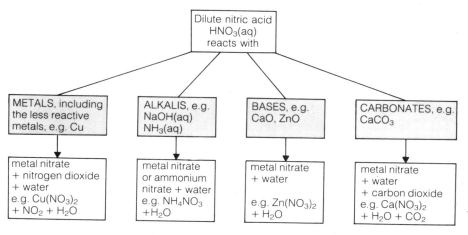

Fig. 14.16 Some reactions of nitric acid

NOTE All nitrates are soluble.

TEST for a nitrate: reduced to ammonia by Devarda's alloy and alkali.

sodium hydroxide will react with them to give ammonia, even in the absence of a nitrate. For this reason, the solution of a suspected nitrate is first boiled with sodium hydroxide solution. When any ammonia present has been driven off, Devarda's alloy is added to the solution. If ammonia is evolved, this proves the presence of a nitrate.

14.6 Nitrates

The preparation of nitrates

The methods for preparing nitrates are summarised in Fig. 14.16. The method of precipitation is *not* used because *all nitrates are soluble*.

> **SUMMARY NOTE**
>
> Dilute nitric acid has many reactions in common with the other mineral acids. In addition, it is an oxidising agent. It oxidises copper and lead to their nitrates.
>
> **Nitrates:** For methods of preparation, see Fig. 14.16.
>
> **Test:** Nitrates give ammonia when they are reduced.

JUST TESTING 78

1. Name three substances that will react with dilute nitric acid to give copper(II) nitrate.
   Describe in detail how you would use *one* of these substances to prepare crystals of copper(II) nitrate.
2. What is formed when dilute nitric acid reacts with **a** lead, **b** zinc, **c** calcium carbonate, **d** sodium hydroxide, **e** copper(II) oxide?
3. Describe how you could obtain crystals of sodium nitrate, starting from sodium hydroxide and nitric acid.
4. Write word equations and symbol equations for the formation of nitrates by the reactions between dilute nitric acid and
   a lead(II) oxide
   b sodium hydroxide
   c calcium carbonate.

14.7 Agricultural chemicals

Fertilisers

Three-quarters of Britain is farmland. Thanks to the use of fertilisers, the land is enormously productive (see Section 14.3).

Herbicides

The well-fertilised soil is a good growing ground for weeds as well as for the crops that the farmers sow. At one time, farm workers used to uproot weeds by hand. Now that agricultural workers are better paid, farmers employ fewer people. They cannot afford hand-weeding and use chemical weedkillers, **herbicides**, instead. **Selective** weedkillers will kill one type of plant and leave others, including the crop, untouched. The commonest, e.g. 2,4-D and 2,4,5-T, kill broad-leaved plants but do not harm grasses and cereals with blade-shaped leaves. They kill dandelions but leave grass. They kill thistles and nettles but leave cereals. These herbicides are organic compounds which contain chlorine. Other weedkillers, e.g. paraquat and sodium chlorate(V), are **non-selective**: they kill all plants.

The benefits of chemical weedkillers have been great. It is difficult to see how scientific farming could continue without them. There have also been some disastrous results of using herbicides.

In 1968, the USA was fighting a war in Vietnam. Their enemy, the Vietcong, found shelter in the jungles. US planes dropped a herbicide called **Agent Orange**, which killed all the trees and other vegetation where it landed. It destroyed the trees which sheltered Vietcong fighters and their supply lines and depots. The herbicide contained impure 2,4,5-T. The impurity, **dioxin**, has appalling **teragenic** effects (effects on unborn babies). Vietnamese mothers gave birth to terribly deformed babies. When US servicemen returned home and started families, some of them were horrified to find that they too had become the parents of seriously defective babies. Agent Orange had affected the fathers' genes. These unforeseen tragedies led to the banning of 2,4,5-T in many countries. It cannot be manufactured without some dioxin appearing as impurity in the product.

Insecticides

If insects were not controlled, they would consume most of the food we grow, and humans would starve. The first chemicals which farmers used to protect their crops from insects were general poisons, e.g. compounds of arsenic, lead and mercury. They kill all animals which swallow them. Farm animals, children and pets can be accidentally poisoned. These poisons are stable compounds: they remain unchanged in water or soil or air for tens of years.

In 1939, the scientist Paul Mueller discovered that a substance called DDT would kill insects and leave other species unharmed. The use of **selective insecticides** took off. DDT is an organic compound which contains chlorine (see Fig. 14.17). It is a petrochemical (see Section 17.8). DDT has saved more lives than any other single substance. By killing insects which eat crops, it preserves crops and fights hunger and starvation. By killing insects which carry disease, it has saved millions of lives and prevented billions of illnesses.

The use of DDT had a brilliant beginning. In 1943, a massive epidemic of typhus broke out in Italy. The disease is carried by lice. By dusting the population with DDT, the epidemic was halted. DDT went on to bring malaria under control in many parts of the world. Malaria is carried by mosquitoes. Typhus and cholera are carried by other insects. By killing the carriers, DDT wiped out the diseases.

In agriculture, DDT was used successfully to control tea parasites in Sri Lanka and cotton pests in the USA. It reduces the damage done by swarms of locusts in the desert regions of North Africa. When locusts consume crops, famine follows and whole populations starve. People predicted that DDT would

By killing locusts, DDT fights famine.

The World Health Organisation says that DDT has saved 5 million lives.

By killing insects, DDT fights disease.

In tropical countries, 550 million people now live free from the threat of malaria.

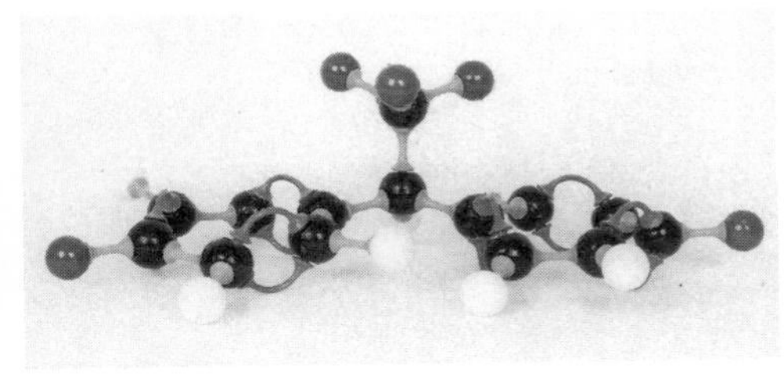

Fig. 14.17 A model of a DDT molecule

exterminate all insect pests. In 1948, Paul Mueller was awarded the Nobel prize for his discovery.

Sometimes lakes are sprayed with DDT to combat mosquitoes. Clear Lake in California is an example. After the lake was sprayed, the level of DDT in the water was 0.02 p.p.m. (parts per million). This is a very low level, and people were surprised when aquatic birds, such as grebes, died. Scientists found out that the deaths were a result of the food chain that exists in the lake (Fig. 14.18). (You saw another food chain illustrated in Fig. 9.16.) Aquatic animals absorb DDT from the water. DDT is more soluble in fats and oils than it is in water. It dissolves in body tissues, and is not excreted. Fish can build up high concentrations of DDT. As well as taking DDT from the water, they eat smaller animals which have absorbed DDT. Aquatic birds which eat fish can build up still higher levels of DDT.

Ten million tonnes of DDT have been applied to the Earth in the last 25 years. DDT is a stable chemical. It does not react with air and water. The amount of DDT in the world is steadily increasing. It kills insects. In sufficiently large amounts, it kills birds. The human family is at the end of the food chain. We eat fish, and we eat birds. What will DDT do to us? There is no direct evidence that DDT produces cancer in humans. There are, however, statistics to show that cancer victims in the USA have twice as much DDT in their fatty tissues as the rest of the population.

DDT is now banned in the USA and Europe because of the dangerous build-up of DDT in wildlife and in humans. DDT is still used for the toughest jobs, e.g. fighting malaria. For agricultural purposes, we use safer but less effective insecticides, e.g. organic compounds of phosphorus, which are broken down in the course of time to safe compounds. At the same time, scientists are looking for other ways of controlling insects.

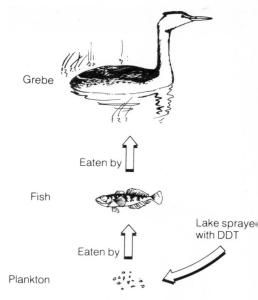

Fig. 14.18 The grebes' food chain

Fungicides

Mildew and potato blight are caused by fungi. To prevent these diseases, crops are sprayed with fungicides. Many copper salts are fungicides, e.g. copper(II) sulphate (see Section 13.9).

Molluscicides

Slugs and snails eat crops and spread diseases. Bilharzia is a tropical disease which is spread by snails. Molluscicides are being used to fight the disease (see Section 12.11).

Did you know British farmers spend £200 million a year on chemicals to kill cereal pests?

An average field of wheat gets 2 herbicides, 2 insecticides and 4 fungicides in one season.

Two herons have been found dead this week on a stretch of the River Avon. This brings the total to 16 since March. Disturbingly high amounts of DDT and dieldrin have been found in the carcases of the birds.
(*The Telegraph*, 8 August 1985)

JUST TESTING 79

1 a Why do farmers use chemical insecticides?
  b What advantage does DDT have over the insecticides which were used previously?
  c No-one has ever been poisoned by DDT. Why has its use on farms been banned?
  d For what purpose is DDT still used? Why is it in such demand?

> **SUMMARY NOTE**
>
> Herbicides and insecticides kill weeds and insects. The use of herbicides ensures that weeds will not rob crops of nutrients. The use of insecticides reduces the fraction of the crops that insects consume. There are dangers in the use of powerful pesticides.

2. Some people think we should give up using chemical insecticides. What would be the result?
3. Birds of prey, such as falcons and eagles, were affected by DDT before the insecticide was banned. These birds do not eat insects. How can they be affected by insecticides?
4. When would copper(II) sulphate have been a welcome fungicide in Ireland? (See Chapter 13 if you need help.)

Exercise 14

1. Suggest what A, B, C, D and E might be.
   a. A is a neutral, rather unreactive gas.
   b. B is a white solid which gives ammonia when it is warmed with sodium hydroxide. A solution of B gives a white precipitate when barium chloride solution and hydrochloric acid are added.
   c. A solution of the blue salt C turns a deep blue when ammonia is present.
   d. The indicator D is turned blue by ammonia.
   e. The gas E reacts with ammonia to form a white solid.

2. a. What class of compounds do plants make from nitrates?
   b. Why can atmospheric nitrogen not be used as a fertiliser by most plants?
   c. Name one natural process that produces nitrogen compounds which can enter the soil and be used by plants.
   d. Explain why natural sources of nitrogen are not sufficient for repeated crop growing on the same land.
   e. Why are ammonium salts used as fertilisers when plants cannot absorb them?
   f. Describe how you could make ammonium sulphate crystals.

3. What reactions occur between ammonia and **a** hydrogen chloride gas, **b** lead(II) nitrate solution, **c** copper(II) sulphate solution? In each case, describe the appearance of the product. Write equations for the reactions.

4. Explain how ammonia is manufactured. What steps are taken to increase the yield?

*5. Describe the industrial method of making nitric acid. Explain why it is possible for every country to manufacture nitric acid.

*6. Calculate the percentage by mass of nitrogen in these fertilisers:
   a. ammonia, b. ammonium nitrate,
   c. ammonium sulphate, d. urea, CON_2H_4.

*7. 100 cm³ of 1.00 mol/l ammonia solution just reacted with a measured volume of 1.00 mol/l sulphuric acid. The solution was heated until it was saturated and then allowed to cool. Crystals of ammonium sulphate were formed. The equation for the reaction is

   $$2NH_3(aq) + H_2SO_4(aq) \rightarrow (NH_4)_2SO_4(aq)$$

   a. What volume of 1.00 mol/l sulphuric acid is required to react, according to the equation, with 100 cm³ of 1.00 mol/l ammonia solution?
   b. Explain what is meant by the expression 'saturated solution'.
   c. Explain why crystals form when the saturated solution is cooled.
   d. Calculate the mass of ammonium sulphate formed when 100 cm³ of 1.0 mol/l ammonia solution react with sulphuric acid according to the equation above. (Relative atomic masses: H = 1, N = 14, O = 16, S = 32.)

8. Copy this passage, filling in the missing words.
   Nitrogen is obtained by the _____ _____ of liquid air. Hydrogen is obtained from _____ _____ by mixing it with _____ and passing the mixture over a heated _____ catalyst. Nitrogen and hydrogen combine when they are passed at

_____ pressure and _____ temperature over a _____ catalyst. The product is _____.

9 Where are the great reserves of nitrogen in nature?
Which nitrogen reserves are the easiest for us to use?
How does the human race upset the natural distribution of nitrogen?
Are we likely to run out of nitrogen?
What is the problem concerning nitrogen compounds that modern agriculture has created? How can this problem be solved?

10 The concentrations of nitrogen monoxide and nitrogen dioxide in polluted air are not toxic (harmful) to humans. Why, then, are people worried about the presence of these oxides in the air?

11 There is a fertiliser factory at Immingham in South Humberside (see Fig. 14.19). Explain this choice of site for the plant. Consider what the needs of a fertiliser factory are. Show how they are met in this area.

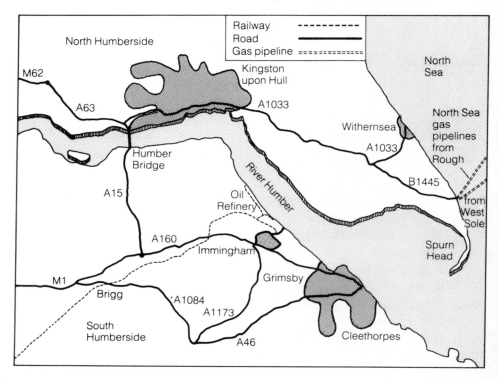

Fig. 14.19 Immingham and surroundings

12 Copy the wordsquare. (You may photocopy the wordsquare, but you may not photocopy other pages.) Ring the answers to the clues. Do not write in this book.

1 Foods of this kind contain nitrogen. (7)
2 A substance which replenishes the soil. (10)
3 Some of these organisms can 'fix' nitrogen. (8)
4 Litmus turns this colour with ammonia. (4)
5 A natural means of returning nitrogen to the soil. (6)
6 A type of fertiliser. (3)
7 An essential element in fertilisers. (9)
8 Plants want these, but we do not want them in our drinking water. (8)

F	I	X	B	L	U	E	N	P	K
M	E	P	A	A	E	R	U	N	Y
U	R	R	C	X	A	U	B	I	E
I	E	O	T	C	L	N	D	T	N
S	B	T	E	I	R	A	I	R	A
S	A	E	R	O	L	M	P	A	H
A	H	I	I	A	A	I	B	T	T
T	M	N	A	N	C	N	S	E	E
O	M	U	Y	T	R	E	Y	E	M
P	F	D	O	C	S	T	O	O	R

9 You find plenty of nitrogen in this. (3)
10 A nitrogen-containing compound which is used as a fertiliser. (4)
11 Plants use these to take in nutrients. (5)
12 He invented a method of making ammonia. (5)
13 Plants need _____ metals such as copper. (5)
14 To _____ nitrogen is to convert it into compounds. (3)

CHAPTER 15 Chemicals from salt

15.1 Sodium chloride

You contain about 0.25 kg (half a pound) of salt. This quantity is enough to fill several salt cellars (Fig. 15.1).

'Salt' is sodium chloride, a compound of the dangerously reactive metal, sodium, and the poisonous gas, chlorine. Salt is an example of a compound which differs greatly from the elements of which it is composed. Although the elements sodium and chlorine are poisonous, sodium chloride is essential for human life. Sodium ions must be present for our nerves to be able to send commands to our muscles. The presence of sodium ions is also essential for the contraction of muscles, including the largest and most important of muscles — the heart. Still another process which requires sodium ions is the digestion of proteins in our food. Sodium chloride regulates the flow of water into and out of our body cells. It controls the way in which food is carried into cells and waste materials are carried out. Without salt, our bodies become paralysed, and we die.

There is a limit to the amount of salt we need. The World Health Organisation recommends no more than 5 grams a day. There is strong evidence linking high salt intake with high blood pressure. High blood pressure often leads both to heart diseases and to strokes.

Through perspiration, the body loses water and salt. This salt must be replaced. Primitive man obtained enough salt from eating meat. When the human race turned to farming instead of hunting, they had to look for salt. A cereal diet does not provide enough salt.

Fortunately, all countries have supplies of salt. Sea water contains sodium chloride (26 g/l). Some countries obtain sodium chloride from sea water (see Section 2.3). Other countries mine salt. The salt deposits are the remains of the inland seas which covered large areas of the globe thousands of years ago. The Great Salt Lake in Utah, USA, was once a vast inland sea. It is saturated with sodium chloride. Its density makes the lake water so buoyant that no swimmer can sink in it (Fig. 15.2). The Dead Sea in Israel is another inland sea, so salty that no fish can live in it.

Salt keeps us alive. Salt keeps our industries alive too. It is the most important of the raw materials which our industries use. Figure 15.3 shows some of the ways we use salt and some of the useful substances which are made from it.

Figure 2.2 shows one way in which salt is mined: solution mining. This is the method used in Northwich, Cheshire. Large underground caverns are created. The ground above them would collapse if it were not supported. The cavities are therefore left full of saturated brine to support the ground above them. (A solution of sodium chloride is called **brine**.)

Fig. 15.1 This is how much salt is in your body

Fig. 15.2 The Great Salt Lake

Fig. 15.3 Salt (sodium chloride): some of its uses

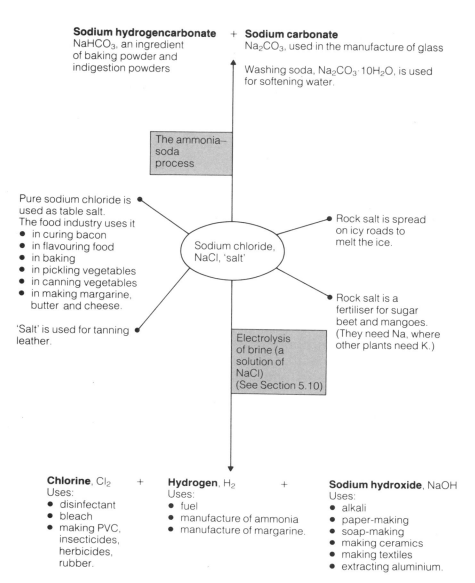

Sodium hydrogencarbonate + **Sodium carbonate**
NaHCO₃, an ingredient of baking powder and indigestion powders
Na₂CO₃, used in the manufacture of glass
Washing soda, Na₂CO₃·10H₂O, is used for softening water.

The ammonia–soda process

Pure sodium chloride is used as table salt.
The food industry uses it
- in curing bacon
- in flavouring food
- in baking
- in pickling vegetables
- in canning vegetables
- in making margarine, butter and cheese.

'Salt' is used for tanning leather.

Sodium chloride, NaCl, 'salt'

Rock salt is spread on icy roads to melt the ice.

Rock salt is a fertiliser for sugar beet and mangoes. (They need Na, where other plants need K.)

Electrolysis of brine (a solution of NaCl) (See Section 5.10)

Chlorine, Cl₂ + **Hydrogen**, H₂ + **Sodium hydroxide**, NaOH
Uses:
- disinfectant
- bleach
- making PVC, insecticides, herbicides, rubber.

Uses:
- fuel
- manufacture of ammonia
- manufacture of margarine.

Uses:
- alkali
- paper-making
- soap-making
- making ceramics
- making textiles
- extracting aluminium.

Fig. 15.4 A salt mine in Cheshire

Figure 15.4 shows a salt mine at Winsford, Cheshire. The operator is using the machine to drill holes in the rock face and insert charges of explosive. When detonated, a single charge can bring down 1 kilotonne of rock. Dumpers take the rock to be crushed.

In hot climates, **solar evaporation** is used. Sea water is evaporated by the sun until salt crystallises (Fig. 2.6).

SUMMARY NOTE

Sodium chloride, common salt, is essential for many of our body processes. Without it, we die. Every country has supplies of salt. It is obtained by solar evaporation, by underground mining and by solution mining. Salt is the basis of many industrial processes.

JUST TESTING 80

1 Why does the human body need salt?
2 What would you recommend to a person who does hard physical work and perspires heavily?
3 Roman soldiers were given a **salarium**, that is, money to buy salt. Salt seems to have been more highly valued in hot countries than in cold countries. Can you explain why? What does the word **salary** mean today?
4 Why is no country short of salt deposits? Where is salt mined in this country?

5 What advantages do swimmers in the Great Salt Lake have? What makes swimming in this lake unpleasant? (Hint: What happens easily in a saturated solution?)

6 Why is sodium chloride so different from the elements sodium and chlorine? (Remember Chapter 6.)

7
- When animals are fed diets with a high salt content, their blood pressure rises.
- When people are fed diets low in salt, their blood pressure drops.
- The average salt intake in the UK is 8 g/day.
- In a study of 3000 Scotsmen, aged 45–64, 40% had high blood pressure.
- In England and Wales in 1980, the number of deaths caused by high blood pressure = 7000, heart disease = 160 500, strokes = 73 500.

From this information, say
a How many deaths in England and Wales in 1980 were thought to be linked to high blood pressure?
b What is the connection between salt and blood pressure?
c What can be done about heart disease?
d What dietary factors other than salt may be involved?
e Imagine that someone asks you to find out whether fat intake or salt intake is more important in controlling blood pressure. You have a supply of laboratory animals and a machine for measuring blood pressure. Plan a set of experiments to answer the question.

SUMMARY NOTE

Hydrogen chloride is a fuming gas with a pungent smell. It shows typical acid reactions, reacting with bases and forming an acidic solution in water. A solution of this very soluble gas is made by passing the gas through an inverted funnel into water.

15.2 Hydrogen chloride

1 Hydrogen chloride is a colourless gas. It has an unpleasant, penetrating smell. When it comes into contact with air, it 'fumes', that is, it reacts with water vapour to form a mist.

2 Hydrogen chloride reacts with ammonia to form ammonium chloride (see Section 14.4).

3 Hydrogen chloride is a very soluble gas. Figure 15.5 shows how a solution of hydrogen chloride or any other very soluble gas must be made.

When hydrogen chloride dissolves in water, a solution of hydrochloric acid is formed:

hydrogen chloride(g) + water → hydrochloric acid(aq)

*A chemical reaction has taken place. Hydrogen chloride reacts with water to give hydrochloric acid, which contains **oxonium** ions, H_3O^+, and chloride ions. The oxonium ion can also be written as $H^+ \cdot H_2O$ or $H^+(aq)$, in which case it is called the **hydrogen** ion:

hydrogen chloride + water → oxonium ions + chloride ions

$HCl(g) + H_2O(l) \rightarrow H_3O^+(aq) + Cl^-(aq)$

Fig. 15.5 A method of making a solution of hydrogen chloride (and other very soluble gases)

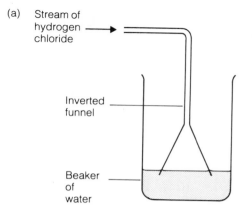

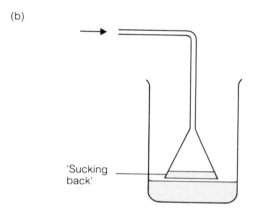

15.3 Reactions of hydrochloric acid

The reactions of acids were summarised in Section 7.2. You can study the reactions of hydrochloric acid in Experiment 7.1. They are summarised in Fig. 15.6.

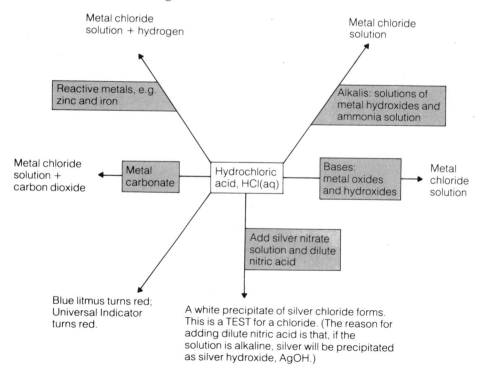

Fig. 15.6 Reactions of hydrochloric acid

*15.4 Solutions of hydrogen chloride in organic solvents

The reactions of hydrochloric acid shown in Fig. 15.6 depend on the ions present in the solution. These are the chloride ion, Cl^-, and the oxonium ion, H_3O^+ (the hydrogen ion, $H^+(aq)$).

A solution of hydrogen chloride in the covalent solvent methylbenzene behaves differently. It does not turn blue litmus red. It does not give a precipitate with silver nitrate solution. These tests show that there are no hydrogen ions or chloride ions present. When hydrogen chloride dissolves in an organic solvent, such as methylbenzene, these ions are not formed. Before hydrogen chloride can act as a proton donor, there must be a proton acceptor present. Water can act as a proton acceptor:

hydrogen chloride + water → oxonium (hydrogen) ions + chloride ions

$HCl(aq) + H_2O(l) \rightarrow H_3O^+(aq) + Cl^-(aq)$

15.5 Preparation of chlorides

All the reactions shown in Fig. 15.6 can be used for the preparation of chlorides (see Experiments 7.5 to 7.8). Some equations for typical reactions are:

magnesium + hydrochloric acid → magnesium chloride + hydrogen

$Mg(s) + 2HCl(aq) \rightarrow MgCl_2(aq) + H_2(g)$

copper(II) oxide + hydrochloric acid → copper(II) chloride + water

$CuO(s) + 2HCl(aq) \rightarrow CuCl_2(aq) + H_2O(l)$

potassium hydroxide + hydrochloric acid → potassium chloride + water

$KOH(aq) + HCl(aq) \rightarrow KCl(aq) + H_2O(l)$

calcium carbonate + hydrochloric acid → calcium chloride + water + carbon dioxide

$CaCO_3(s) + 2HCl(aq) \rightarrow CaCl_2(aq) + H_2O(l) + CO_2(g)$

silver nitrate + hydrochloric acid → silver chloride + nitric acid

$AgNO_3(aq) + HCl(aq) \rightarrow AgCl(s) + HNO_3(aq)$

SUMMARY NOTE

Hydrochloric acid reacts with many metals, with bases (metal oxides and hydroxides), with alkalis and with carbonates. These reactions are typical of acids. They result in the formation of chlorides, the salts of hydrochloric acid. The test for a chloride in solution is the formation of a white precipitate with silver nitrate solution and dilute nitric acid.

JUST TESTING 81

1 Nickel chloride, $NiCl_2$, is soluble. What would you need, in addition to hydrochloric acid, to make it? Describe briefly how you would make crystals of this salt.

2 Lead(II) chloride, $PbCl_2$, is insoluble. What would you need, in addition to hydrochloric acid, to make it? Briefly describe how you would make a dry sample of this salt.

3 Describe how you would make a solid specimen of ammonium chloride, starting from ammonia solution and hydrochloric acid.

4 a Name a substance which will react with hydrochloric acid to give hydrogen. Give the word equation for the reaction and a balanced chemical equation.

  b Name a substance which will react with hydrochloric acid to give carbon dioxide. Write a word equation for the reaction and a balanced chemical equation.

5 Say what you would need, in addition to hydrochloric acid, to make **a** copper(II) chloride, **b** potassium chloride, **c** zinc chloride, **d** silver chloride. In each case, describe what you would do with the starting materials to obtain a solid specimen of the product.
Write word equations for the reactions.
Write balanced chemical equations.

15.6 Chlorine in war

From 1914–1918, Germany was at war with Britain and France. In the spring of 1915, Germany prepared to perform a chemical experiment. Six thousand cylinders of chlorine were dug in over a $3\frac{1}{2}$ mile stretch of battlefield in Flanders. On 22nd April, the wind was blowing away from the German lines towards the British and French trenches. Masked technicians opened the cylinders and set the chlorine free. A green cloud rolled towards the Allied troops. It sank into the trenches, and stung the soldiers' eyes and noses. Minutes later, they were gasping for breath and coughing up blood. The Germans attacked successfully and made a big advance. The Allied casualties were 5000 killed, 6000 captured and 15 000 gassed.

The British were horrified. They had no defence whatsoever against this new menace. Fortunately for them, the German generals were not entirely satisfied with their new weapon. They had treated it as an experiment, and were not prepared to back it up. Its success depended on a favourable wind. For most of the time, winds in Flanders blew from west to east, away from the Allied lines, towards the German lines. Their deadly weapon could well be turned against them.

The choice of chlorine was made by Professor Fritz Haber (see Ammonia, Section 14.4). As well as being poisonous, a suitable gas must be dense; then it will remain for some time at ground level. It must also be easy to make in large quantities and to transport. Chlorine fits the bill. It can be made from rock salt; it is 2.5 times denser than air; it can be stored under pressure in cylinders. A disadvantage is that, having a colour and a smell, chlorine cannot take an enemy by surprise. Haber had planned to store the gas in shells and fire these into the Allied trenches. The shortage of shell cases forced him to settle for relying on the wind to transport the poisonous green cloud.

Chemical weapons have been used since 1915. In the war

against Vietnam, the USA used **defoliants** — chemicals which kill trees and plants. These are organic compounds of chlorine. The effects of these chemicals were more lethal than anyone expected (see Section 14.7). There is now an international agreement that no nation will use chemical weapons in war. The USA has signed the treaty banning chemical weapons.

15.7 Chlorine in peace

Chlorine is a non-metallic element. It is one of the halogens, the elements in Group 7 of the Periodic Table. Industrially, chlorine is made by the electrolysis of brine (sodium chloride solution; see Section 5.7). It has not been used to kill human beings since 1915. It is now used to kill bacteria and so to save human lives. It is used to sterilise drinking water (see Section 9.5). Some of the uses of chlorine are shown in Fig. 15.7. Some of the many useful organic compounds of chlorine are mentioned in Sections 14.7 and 16.3.

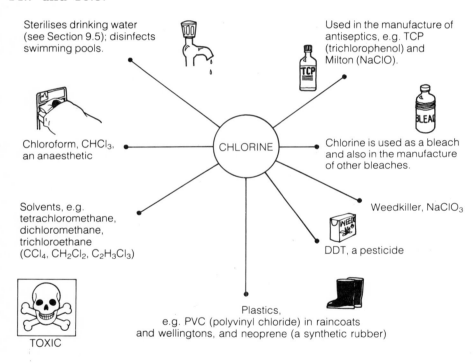

Fig. 15.7 Some uses of chlorine

15.8 Properties and reactions of chlorine

Table 15.1 Chlorine

1	Appearance	Dense green gas with a horrible, choking smell Very poisonous Moderately soluble in water
2	Damp, blue litmus paper	Turns red and then colourless: chlorine bleaches it. This is a **test for chlorine**.
3	Metals	Chlorine reacts with most metals. Sodium burns in chlorine to form sodium chloride (see Fig. 15.8):

Fig. 15.8 Sodium burns in chlorine

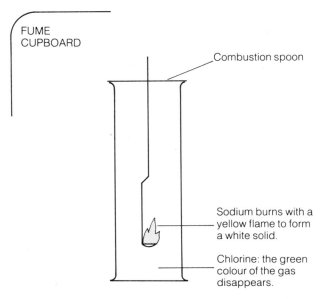

TOXIC

sodium + chlorine → sodium chloride
$2Na(s) + Cl_2(g) \rightarrow 2NaCl(s)$

Heated copper reacts with chlorine to form copper(II) chloride:

copper + chlorine → copper(II) chloride
$Cu(s) + Cl_2(g) \rightarrow CuCl_2(s)$

Heated iron reacts with chlorine to form iron(III) chloride (see Fig. 15.9):

iron + chlorine → iron(III) chloride
$2Fe(s) + 3Cl_2(g) \rightarrow 2FeCl_3(s)$

Fig. 15.9 Iron burns in chlorine

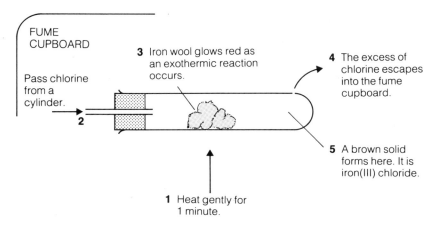

4 Metal ions   When a metal has a variable valency, chlorine will oxidise ions of the metal in its lower valency to ions of the metal in its higher valency. Iron(II) chloride is oxidised to iron(III) chloride, and iron(II) sulphate is oxidised to iron(III) sulphate:

chlorine + iron(II) chloride → iron(III) chloride
$Cl_2(g) + 2FeCl_2(aq) \rightarrow 2FeCl_3(aq)$

5 Non-metals   Hydrogen burns in chlorine to form hydrogen chloride (see Fig. 15.10):

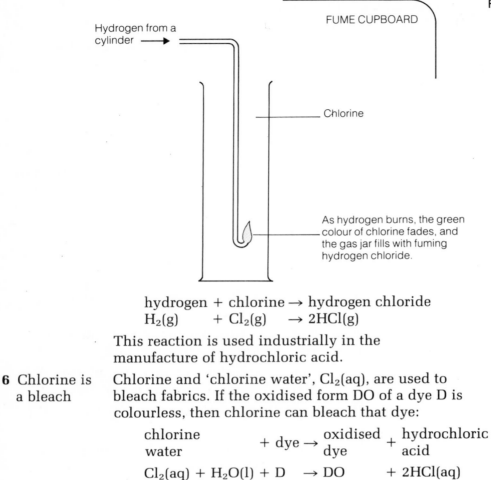

Fig. 15.10 Hydrogen burns in chlorine

hydrogen + chlorine → hydrogen chloride
$H_2(g) + Cl_2(g) → 2HCl(g)$

This reaction is used industrially in the manufacture of hydrochloric acid.

6 Chlorine is a bleach

Chlorine and 'chlorine water', $Cl_2(aq)$, are used to bleach fabrics. If the oxidised form DO of a dye D is colourless, then chlorine can bleach that dye:

chlorine water + dye → oxidised dye + hydrochloric acid

$Cl_2(aq) + H_2O(l) + D → DO + 2HCl(aq)$

7 Other halogens

Chlorine will displace bromine from bromides and iodine from iodides (see Experiment 15.1):

chlorine + bromide → chloride + bromine
$Cl_2(aq) + 2Br^-(aq) → 2Cl^-(aq) + Br_2(aq)$

chlorine + iodide → chloride + iodine
$Cl_2(aq) + 2I^-(aq) → 2Cl^-(aq) + I_2(aq)$

*15.9 Chlorine as an oxidising agent

The reactions of chlorine are dominated by its readiness to gain electrons and form chloride ions (see Chapter 6):

chlorine molecules + electrons → chloride ions
$Cl_2 + 2e^- → 2Cl^-$

Chlorine molecules are very reactive; chloride ions are very stable. A reagent which accepts electrons from another species (an atom or molecule or ion) is called an **oxidising agent**. Chlorine is an oxidising agent.

In the reaction between sodium and chlorine to form sodium chloride, sodium atoms give electrons, becoming sodium ions, Na^+. Chlorine molecules gain electrons to become chloride ions, Cl^-.

$$Na(s) \rightarrow Na^+(s) + e^-$$
$$Cl_2(g) + 2e^- \rightarrow 2Cl^-(s)$$

Sodium is oxidised by chlorine; chlorine is reduced by sodium. Sodium is a reducing agent; chlorine is an oxidising agent.

$$\underset{\text{Reducing agent}}{2Na(s)} + \underset{\text{Oxidising agent}}{Cl_2(g)} \xrightarrow[\text{REDUCTION}]{\text{OXIDATION}} 2Na^+(s) + 2Cl^-(s)$$

In the reaction with an iron(II) salt, the iron(II) salt is oxidised to an iron(III) salt, while chlorine is reduced:

$$Fe^{2+}(aq) \rightarrow Fe^{3+}(aq) + e^-$$
$$Cl_2(aq) + 2e^- \rightarrow 2Cl^-(aq)$$
$$2Fe^{2+}(aq) + Cl_2(aq) \rightarrow 2Fe^{3+}(aq) + 2Cl^-(aq)$$

The reactions of chlorine with bromides and iodides are oxidation–reduction reactions. You saw in Experiment 15.1 that chlorine displaces bromine from bromides. This happens because chlorine is a stronger oxidising agent than bromine is. Chlorine takes electrons away from bromide ions: bromide ions are oxidised

SUMMARY NOTE

Chlorine is a poisonous green gas. It is used to disinfect swimming pools, to sterilise drinking water, as a bleach and in the manufacture of a number of useful products. Chlorine is made industrially by the electrolysis of brine. Chlorine is a very reactive non-metallic element. It is an oxidising agent, which oxidises metals, metal ions, bromides and iodides.

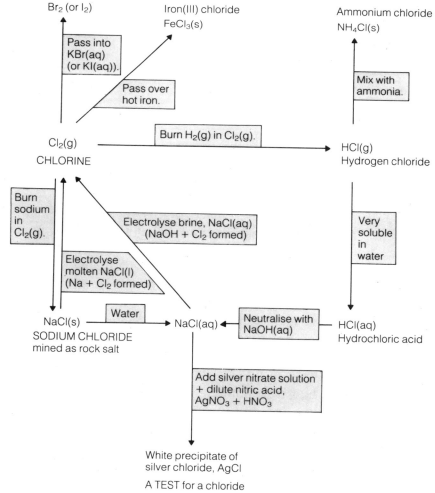

Fig. 15.11 A summary of chlorine and chlorides

to bromine molecules. Chlorine molecules are reduced to chloride ions:

$$\underset{\underset{\text{OXIDATION}}{\longleftarrow}}{\overset{\overset{\text{REDUCTION}}{\longrightarrow}}{Cl_2(aq) + 2Br^-(aq) \rightarrow 2Cl^-(aq) + Br_2(aq)}}$$

Similarly, chlorine displaces iodine from iodides because chlorine is a stronger oxidising agent than iodine is (see Exercise 15, Question 2).

Bromine displaces iodine from iodides because bromine is a stronger oxidising agent than iodine is (see Exercise 15, Question 3).

In order of oxidising power, these elements rank:

$Cl_2 > Br_2 > I_2$

> Bleaches which contain chlorine are powerful oxidising agents. For this reason, you should not use a chlorine bleach together with other chemical cleaning fluids. Detergents which contain ammonia and bleaches which contain hydrogen peroxide will react violently with chlorine. (Remember the story about chlorine in Chapter 3?)

Oxidation–reduction

Oxidation is
- the gain of oxygen or
- the loss of hydrogen or
- the loss of electrons.

An oxidising agent
- gives oxygen or
- takes hydrogen or
- takes electrons.

Reduction is
- the loss of oxygen or
- the gain of hydrogen or
- the gain of electrons.

A reducing agent
- takes oxygen or
- gives hydrogen or
- gives electrons.

JUST TESTING 82

1. Material which has been bleached by chlorine must be thoroughly rinsed. It is important to remove *two* substances which are present. What are they?

2. Write a word equation for the oxidation of an iodide by chlorine. Write a balanced chemical equation.

3. Write a word equation for the oxidation of an iodide by bromine. Write a balanced chemical equation.

4. You are given two gas jars. One contains hydrogen chloride, and one contains chlorine. What is the difference in the appearance and smell of the two gases? Describe two tests you could do to distinguish between them.

5. There are three bottles on the prep. room shelf. All contain crystalline white solids. Three labels lie on the floor. They read: *Potassium chloride*, *Potassium bromide* and *Potassium iodide*. How would you decide which label to stick on each of the three bottles if all you had to work with was a bottle of chlorine water and an organic solvent?

*6. Three identical half-litre bottles contain solutions of potassium chloride, potassium bromide and potassium iodide. The solutions are all colourless. All have a concentration of 2.0 mol/l. All have lost their labels. Can you think of a way of telling which solution is which without removing the stoppers? (See Section 11.11 for mol/l.)

15.10 The halogens

The elements fluorine, chlorine, bromine, iodine and astatine are called the **halogens**. (The name **halogen** is Greek for **salt-former**.) The halogens all combine with metals to form salts, called **halides**, in which the halogen is an anion, X^-. The halogens are all oxidising agents. They all have 7 electrons in their outer shells, and they form Group 7 of the Periodic Table. Their similarity led to their being regarded as a 'family' of elements, and was one of the factors which convinced Mendeleev that he was working along the right lines with the Periodic Table (see Section 4.6).

The first member of the group is fluorine, a poisonous yellow gas. It is so reactive that it is a dangerous element to work with.

Table 15.2 The halogens

Halogen, X_2	Chlorine, Cl_2	Bromine, Br_2	Iodine, I_2
1 State of the element at room temperature	Green gas	Reddish-brown liquid; easily vaporised	Shiny black solid; sublimes when heated
2 Vapour of X_2	Green	Orange-brown	Purple
3 Electron configuration of X	2.8.7	2.8.8.7	2.8.18.8.7
4 Halide ion, X^-	Chloride, Cl^-	Bromide, Br^-	Iodide, I^-
5 Sodium halide, NaX	Sodium chloride, NaCl, a white crystalline solid, soluble in water	Sodium bromide, NaBr, a white crystalline solid, soluble in water	Sodium iodide, NaI, a white crystalline solid, soluble in water
6 Oxidising agents	Cl_2 is $> Br_2 > I_2$	Br_2 is $> I_2$	I_2 is the least powerful.
7 Test for a halide ion in solution, X^-(aq): add $AgNO_3$(aq) + HNO_3(aq)	Precipitate of AgCl, a white solid, soluble in dilute ammonia solution	Precipitate of AgBr, a pale yellow solid, soluble in concentrated ammonia solution	Precipitate of AgI, a yellow solid, insoluble in ammonia solution
8 Hydrogen halide, HX	Hydrogen chloride HCl(g), a colourless gas, fumes in moist air, acidic (with water forms hydrochloric acid, HCl(aq))	Hydrogen bromide HBr(g), a colourless gas, fumes in moist air, acidic (with water forms hydrobromic acid, HBr(aq))	Hydrogen iodide HI(g), a colourless gas, fumes in moist air, acidic (with water forms hydriodic acid, HI(aq))
9 Reaction of X_2 with hydrogen	Hydrogen burns vigorously to form hydrogen chloride, HCl (see Fig. 15.10). Under some conditions, the reaction is explosive.	Hydrogen burns in bromine to form hydrogen bromide, HBr.	When hydrogen and iodine are heated together in a sealed tube, some hydrogen iodide is formed. The reaction does not go to completion.
10 Reaction of X_2 with iron	A vigorous, exothermic reaction gives $FeCl_3$ (see Fig. 15.9).	A less vigorous reaction gives $FeBr_3$ (see Fig. 15.11).	A gentle reaction gives FeI_2 (see Fig. 15.12).

Many of the chemists who did the early experiments on fluorine were injured in explosions. The last member of the group, astatine, is a short-lived radioactive element.

The elements show a gradual change in physical and chemical properties with increasing atomic mass (see Table 15.2).

SUMMARY NOTE

The elements of Group 7 of the Periodic Table are fluorine, chlorine, bromine, iodine and astatine. They are called the **halogens**. The halogens
- are non-metallic
- react readily with metals to form salts
- are oxidising agents
- combine with hydrogen to form acidic hydrogen halides.

There is a gradual change in physical and chemical properties down the group.

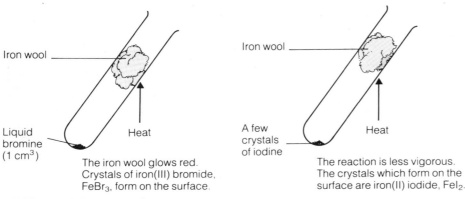

Fig. 15.12 Iodine reacts with heated iron

JUST TESTING 83

1. 'The halogens are a family of elements.' Explain why this statement is made.
2. Describe two experiments you could do to show that chlorine is more reactive than bromine.
   Which is the most reactive of all the halogens? Why do you not do experiments with this element?

15.11 Fluoride: friend or foe?

Fluorine is a poisonous yellow gas. It attacks glass, and its reactions are often explosive. Yet the British Government passed a law in 1985 to allow water authorities to put sodium fluoride into the drinking water! Research workers noticed that children in some cities had less tooth decay than average. They looked for a cause. They found that there was a high natural concentration of fluorides in the water in these cities. They recommended increasing the concentration of fluorides in other areas. Fluorides are most effective in preventing decay in teeth which are still forming and calcifying, that is, in children up to the age of 8 years. The fluoridation of drinking water must be carefully controlled. Enough sodium fluoride is added to bring the concentration of fluoride ions up to 1 p.p.m. (part per million). Excessive amounts of fluoride ion can damage teeth. Again, young children are more likely to be affected than other people by an excess of fluoride. Some authorities think that fluoridation of drinking water is too risky. They suggest using a toothpaste containing a fluoride instead.

How do fluorides work? The calcium compounds in teeth and bone are calcium carbonate, $CaCO_3$, and calcium hydroxide phosphate, $Ca_5(OH)(PO_4)_3$, 'hydroxyapatite'. Fluoride converts the

Fig. 15.13 Toothpaste with fluoride

latter compound into calcium fluoride phosphate, $Ca_5F(PO_4)_3$, 'fluoroapatite'. This is less likely to be attacked by food because it no longer contains the basic hydroxide ion, OH^-.

In large amounts, fluorides are pollutants. There have been cases of pollution arising from the use of fluorides in aluminium smelters (Section 10.12).

JUST TESTING 84

1. What word in the account you have just read means 'laying down calcium'? In which type of person is this process taking place the fastest? What should these people do to protect their teeth?
2. Emma is surprised to hear that fluorides in small quantities are good for teeth because she knows that fluorine is a poisonous gas. What would you say to her to explain the difference between fluorine and fluorides?

15.12 Iodides can be useful too

Scientists at the World Meteorological Organisation are doing a 7-year experiment. They are trying to find out whether they can alter the weather. The tool they are using is silver iodide. This salt has a crystal structure similar to ice. There is evidence that silver iodide will make it rain. Of course, there must be clouds; silver iodide cannot produce rain out of a clear blue sky! The salt has to be dropped into a cloud from a plane or fired into it from the ground. The crystals act as a 'seed' around which water vapour can condense.

You can imagine how the technique could be used to benefit people. It can also be used for evil. A treaty has been signed by members of the United Nations to stop countries from using the weather as a weapon. The USA used cloud seeding in Vietnam to produce floods, but the USA has signed the new treaty.

Scientists are doubtful about the ability of silver iodide to produce more than a 10% to 15% increase in the rainfall.

JUST TESTING 85

1. Cloud seeding is called 'weather modification'. Write a summary of ways in which weather modification could help farmers. In which countries would it help most?
2. Is silver iodide a magic 'rainmaker'? Explain your answer. If it is not magic, how does it work?
3. Are there any dangers connected with the research work? Do you think it should continue?

Exercise 15

1. **a** Why is chlorine added to the water in swimming pools?
   b What colour does chlorine turn damp litmus paper?
   c What colour does hydrogen chloride turn damp litmus paper?
   d What is formed when hydrogen chloride meets ammonia?
   e What is formed when hydrogen burns in chlorine?

2. What do you see and what new substances are formed when chlorine reacts with
   a burning sodium, **b** heated copper, **c** heated iron?
   Write (i) word equations for the reactions, (ii) balanced symbol equations.

3. With which one of the following pairs of reagents would a displacement reaction take place?
   a aqueous bromine and aqueous potassium chloride
   b aqueous bromine and aqueous sodium chloride
   c aqueous chlorine and aqueous potassium iodide
   d aqueous iodine and aqueous potassium bromide.

4. **a** Why do water boards add chlorine to water supplies?
   b Why is the element, chlorine, used, rather than a chloride?
   c Why do water boards add fluorides to water supplies?
   d Why are fluorides used, rather than the element, fluorine?

5. **a** Two chlorine atoms join by the formation of a covalent bond to give a molecule of chlorine. By means of a sketch, show what happens to the electrons in the outermost shells of the atoms.
   b A chlorine atom combines with a sodium atom by the formation of an electrovalent bond. Sketch what happens to the electrons in the outermost shells of the two atoms. What is formed in the reaction?
   c A chlorine atom combines with a hydrogen atom by the formation of a covalent bond. Sketch what happens to the electrons during bond formation. What is formed in the reaction?

6. The first four members of the halogens are
   $^{19}_{9}F$ $^{35}_{17}Cl$ $^{80}_{35}Br$ $^{127}_{53}I$
   a Write the electron arrangements in an atom of fluorine and an atom of chlorine.
   b State one characteristic which is common to the electron arrangements of all the halogen atoms.
   c State the number of protons in a chloride ion.
   d State the number of neutrons in an iodine atom.
   e Which contains more atoms, 1 gram of iodine or 1 gram of chlorine?

7. When dry chlorine is passed over heated iron, a black solid sublimes in the cooler part of the apparatus.
   a Name the black solid.
   b Explain what is meant by 'sublimes'.
   c Write a word equation and a chemical equation for the reaction.
   d Will the reaction between fluorine and iron be more or less vigorous than that of chlorine?

8. Hydrogen chloride reacts with water to form a solution which has a pH value less than 7.
   a State the type of bonding in hydrogen chloride.
   b Explain the reaction that takes place between hydrogen chloride and water.
   c Say what is formed when the solution is neutralised by potassium hydroxide solution.

9. Copy this table, and complete it.

Element	Hydrogen	Calcium	Carbon
Formula of the chloride			
State of the chloride at room temperature			

10. Choose your answers to the following questions from these compounds:
    $BaSO_4$ $ZnCO_3$ KOH $NaCl$ $AgNO_3$ ZnO $Ca(NO_3)_2$
    a Which compound dissolves in water to give a solution which will neutralise hydrochloric acid?
    b Which compound gives off a gas when dilute hydrochloric acid is added?
    c Solutions of two of the compounds form a white precipitate when they are mixed. Name the compounds.

d Which compound is an insoluble white solid which neutralises hydrochloric acid to form a salt?

11 An aqueous solution of chlorine was added to a colourless solution of a potassium salt. A brown solution formed. When this brown solution was shaken with the organic solvent 'trichlor', the organic liquid turned purple.
  a Name the substance causing the purple colour in the organic liquid.
  b Name the substance causing the brown colour in the aqueous solution.
  c Name the potassium salt.
  d Describe a test you could do to confirm that you are right.
  e Write a word equation for the reaction that took place.
  ***f** Write an ionic equation for the reaction.

Crossword on Chapter 15

Clues across
  3 See 14 across.
  6 This action of chlorine is useful in cleaning. (9)
  9 across, 7 down A fuming gas which contains chlorine. (8, 8)
  11 across, 8 down A description of chlorine. (5, 3)
  14 across, 3 across Used for preparing chlorides. (12, 4)
  17 Noble gas used in lights. (4)
  18 An essential element in teeth. (7)
  20 This type of compound always reacts with 14 across, 3 across. (6)

Clues down
  1 The colour of litmus in 14 across, 3 across. (3)
  2 down, 7 down The formation of this is a test for chlorides. (6, 8)
  4 A reddish-coloured metal with a green chloride. (6)
  5 A use for chlorine. (12)
  6 Evaporate on a steam ―――. (4)
  7 See 2 down.
  8 See 11 across.
  10 Evaporate a solution in this. (4)
  12 This gas gives a white solid when it reacts with 9 across, 7 down. (7)
  13 The chloride of this metal is 'common salt'. (6)
  15 To liquefy chlorine gas, you must make it much ―――― than room temperature. (6)
  16 Common salt is mined in this form. (4)
  19 Symbol for chlorine. (2)

CHAPTER 16 Hydrocarbon fuels

16.1 Biogas

India and China have millions of **biogas** generators. The biogas which they produce is a fuel used for cooking, heating and lighting. Biogas is the gas produced when organic matter (matter of plant and animal origin) decays in the absence of air. Figure 16.1 shows a biogas generator. The idea is to generate gas by the digestion of animal waste. With about three cattle to every person in India, there is a supply of cattle dung for digestion.

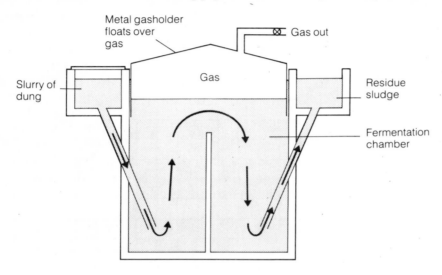

Fig. 16.1 Biogas plant

The gas burns much more efficiently than cakes of cattle dung. As another benefit, the residual sludge is a better fertiliser than raw dung. The output of a biogas digester provides twice as much heat and three times as much useful fertiliser as raw cattle dung. Human waste and some plant waste can also be processed. Both the gas and the residue are clean and odourless.

The biogas digester is simple and robust but, like any other chemical plant, it needs careful operation. Incorrect feed, the wrong temperature or poor mixing can bring production to a halt.

Biogas is about 50% methane. Methane is the gas we burn in Bunsen burners and gas cookers. It is the valuable fuel we call **natural gas** or **North Sea gas**. Methane is the gas which bubbles up through stagnant water. It forms slowly from the anaerobic (without air) decomposition of organic material in the water. It was methane that exploded in the Wyresdale water tunnel in Lancashire in May 1984, killing 11 people. It collects in coal mines, and has been the cause of many pit disasters.

Methane is the gas which forms in the landfill sites where cities store their rubbish (Fig. 16.2). Twenty-five million tonnes of organic waste go into landfills in the UK every year. While some landfill operators are burning methane to get rid of it, others are sinking pipes into the landfill and pumping out methane for sale.

SUMMARY NOTE

Methane is formed when plant and animal matter decays in the absence of air. It is found in coal mines, where it can cause explosions. The methane generated in landfill rubbish sites and in sewage works can be utilised as fuel. Some countries set up biogas generators to make biogas from human, animal and plant waste.

Fig. 16.2 Methane has to be burned off at this landfill in Liverpool

Fig. 16.3 Filling up with methane

Methane becomes a liquid fuel when stored under pressure; as such, it can be used to fuel vehicles. A town in California runs a fleet of cars on methane made from the municipal sewage. In Essex, methane from a sewage works is fuelling vans and road tankers (Fig. 16.3). The Anglian Water Authority expects the project to spread nationwide. Colchester sewage works is the site of the pilot project. Vehicles run on a mixture of 2 litres of diesel fuel to 1 litre of methane. The fuel costs £1 per gallon.

JUST TESTING 86

1 What are the advantages of biogas digesters **a** for developing countries, **b** for countries which have advanced technology?
2 Sometimes a flickering flame appears to dance over marshy ground. Country people used to think that the flame was a spirit called 'Will-o'-the-wisp', who lived in marshes. Can you explain what the flame really is?

16.2 Alkanes

Methane is a **hydrocarbon**, a compound of hydrogen and carbon. With formula CH_4, it is the simplest of hydrocarbons. Four covalent bonds join hydrogen atoms to a carbon atom (Fig. 16.4).

Hydrocarbons are **organic compounds**. At one time, the term organic compound meant a compound which was found in plant or animal material, e.g. sugars, fats and proteins. All these compounds contain carbon. Now the term organic compound is used for all carbon compounds, whether they come from plants and animals or were made in a laboratory. Simple compounds like carbon dioxide and carbonates are, however, not usually described as organic compounds. Most organic compounds are covalent. The organic acids form salts which have ionic bonds.

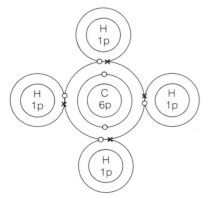
Fig. 16.4 Bonding in methane

 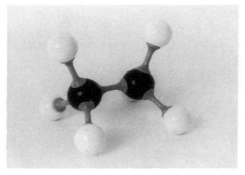 

Fig. 16.5 (a) Model of methane (b) Model of ethane (c) Model of propane

Methane is one of a series of hydrocarbons called the **alkanes** (Fig. 16.5). The next members of the series are ethane (C_2H_6), propane (C_3H_8) and butane (C_4H_{10}). Many hydrocarbons have much larger molecules.

As well as writing the molecular formulas, CH_4, C_2H_6 and C_3H_8, we write **structural formulas**. A structural formula shows the bonds between atoms.

```
      H              H  H            H  H  H
      |              |  |            |  |  |
  H—C—H          H—C—C—H         H—C—C—C—H
      |              |  |            |  |  |
      H              H  H            H  H  H

   Methane          Ethane           Propane
```

> **SUMMARY NOTE**
>
> The alkanes are hydrocarbons with the general formula C_nH_{2n+2}. They are a homologous series: each member differs from the next by a CH_2 group. Carbon compounds are called organic compounds.

The difference between one compound and the next is the group

A set of similar compounds in which one member of the series differs from the next by a CH_2 group is called a **homologous series**. The first members of the alkane series are shown in Table 16.1. The general formula C_nH_{2n+2} can be written for the alkanes. For methane, $n = 1$, giving the formula CH_4; for octane, $n = 8$, giving the formula C_8H_{18}. The alkanes are unreactive towards acids, bases, metals and many other chemicals. Their most important reaction is combustion. The alkanes contain only single bonds between carbon atoms. Such hydrocarbons are called **saturated** hydrocarbons. This is in contrast to the alkenes, which you will study in Section 16.5.

Table 16.1 The alkanes

Methane	CH_4
Ethane	C_2H_6
Propane	C_3H_8
Butane	C_4H_{10}
Pentane	C_5H_{12}
Hexane	C_6H_{14}
Heptane	C_7H_{16}
Octane	C_8H_{18}
and so on	
Alkanes	C_nH_{2n+2}

*Isomerism

Sometimes it is possible to write more than one structural formula for a molecular formula. For the molecular formula C_4H_{10}, there are two possible structures:

(a)
```
    H   H   H   H
    |   |   |   |
H — C — C — C — C — H
    |   |   |   |
    H   H   H   H
```
and

(b)
```
    H   H   H
    |   |   |
H — C — C — C — H
    |   |   |
    H   |   H
        |
    H — C — H
        |
        H
```

Butane   Methylpropane

The difference is that in (a) there is a straight, unbranched chain of carbon atoms, whereas in (b) there is a branched chain. The formulas belong to different compounds, which differ in boiling point and other physical properties. The compound with formula (a) is called butane; the compound with formula (b) is called methylpropane. These compounds are **isomers**. Isomers are compounds with the same molecular formula and different structural formulas.

> **SUMMARY NOTE**
>
> Isomers have the same molecular formula and different structural formulas.

> **JUST TESTING 87**
>
> 1 Explain what is meant by a **homologous series**.
> 2 What are alkanes?

16.3 Halogenoalkanes

Alkanes will react with halogens to form halogenoalkanes. For example,

$$\text{chlorine} + \text{methane} \xrightarrow{\text{in the presence of sunlight}} \text{chloromethane} + \text{hydrogen chloride}$$
$$Cl_2(g) + CH_4(g) \rightarrow CH_3Cl(g) + HCl(g)$$

This is a **substitution reaction**. A chlorine atom has been substituted for a hydrogen atom.

Many chloroalkanes are important solvents. Tetrachloromethane, CCl_4, was used for a long time by the dry-cleaning industry. 'Trichlor' (1,1,1-trichloroethane) is now widely used. Fluorohydrocarbons (hydrocarbons which contain fluorine) are very stable, unreactive compounds. They are called **freons** for short. The solvents in aerosol sprays and the liquids which circulate in refrigerators are **freons**.

Trichloromethane, $CHCl_3$, is better known as chloroform, the anaesthetic. Nowadays, there are better anaesthetics available, but chloroform (first used in 1846), ether (first used in 1847) and dinitrogen oxide (laughing gas) were the gases used by the pioneers. The use of anaesthetics brought about a revolution in surgery. Before anaesthesia, a surgeon had to make a quick job of an operation while the patient remained conscious. With the patient anaesthetised, a surgeon was able to explore the best way of dealing with an operation, rather than the quickest. Also, there was less chance of the patient dying of shock.

Chloroform is harmful if given in large quantities. Ether, being very flammable, caused some explosions in operating theatres. Dinitrogen oxide does not produce a very deep anaesthesia. Chemists searched for a better anaesthetic, one which would be safe for the patient, that is, non-toxic and non-flammable. Again, the stability of fluorohydrocarbons made them useful compounds. The fluorohydrocarbon chosen was

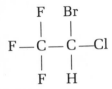

It is a colourless liquid with a pleasant smell. It is non-toxic and non-flammable. It is named Fluothane®. It went into use in 1956 and is the most widely used of anaesthetics.

16.4 Hydrogen

Do you remember how hydrogen is obtained from methane? If not, see Section 9.17.

16.5 Combustion

Hydrocarbons burn to form carbon dioxide and water (see Section 8.10). The reaction is **exothermic**: energy is released.

methane + oxygen → carbon dioxide + water   **Energy is released**

$CH_4(g) + 2O_2(g) \rightarrow CO_2(g) + 2H_2O(l)$

butane (in camping gaz) + oxygen → carbon dioxide + water   **Energy is released**

octane (in petrol) + oxygen → carbon dioxide + water   **Energy is released**

SUMMARY NOTE

The combustion of hydrocarbons is exothermic. The products of complete combustion are carbon dioxide and water. Incomplete combustion releases poisonous carbon monoxide. Halogenoalkanes are useful solvents and anaesthetics.

JUST TESTING 88

Do you remember what you read about combustion in Chapter 8?
Copy this passage, and fill in the blanks.

Hydrocarbons burn to form the harmless combustion products _____ and _____. If there is an insufficient supply of oxygen, however, combustion is _____, and the products _____ (a poisonous gas of formula _____) and _____ (a solid) are also formed. To avoid the formation of the poisonous gas, hydrocarbons must be burned _____.

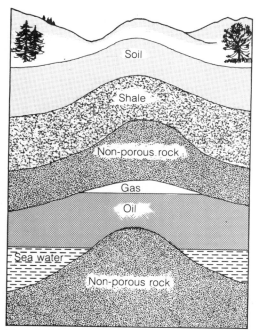

Fig. 16.6 Oil-bearing rock

16.6 Petroleum oil and natural gas

Alkanes are valuable fuels. We obtain them from petroleum oil (usually called simply **oil**; sometimes called **crude oil**) and natural gas. Oil and natural gas are **fossil fuels**: they are the remains of creatures which lived millions of years ago. Dead sea animals and plants sank to the ocean floor to become covered by silt, mud and rocks. Bacteria began to bring about the decay of the soft parts of the creatures' bodies. Decay was slow because there is little oxygen dissolved in the depths of the sea. The covering layer of mud and silt grew thicker over the years, and the pressure on the decaying organic matter increased. The combination of anaerobic bacterial decay, high pressure and heat from the interior of the Earth turned the organic matter into hydrocarbons. This mixture of hydrocarbons is petroleum oil. The sediment on top of the decaying matter became compressed to form rock. Oil is held in between rock grains in porous **oil-bearing rock** (Fig. 16.6). Natural gas is always found in the same deposits as oil.

Many parts of the world have deposits of oil and gas. Sometimes the deposits are buried deep below ground or below the sea. The USA has oil deposits, as have Iran, Nigeria, the USSR and the countries in the Arabian Gulf. The UK has piped ashore oil and gas from beneath the North Sea since 1972 (Fig. 16.7).

Fig. 16.7 Drilling for oil

Fig. 16.8 Pipeline showing laying and landscaping

Oil is transported from oil wells to refineries in pipelines and in oil tankers (Figs. 16.8 and 16.9). Modern tankers are huge, carrying up to 500 000 tonnes of crude oil. When a tanker has an accident at sea, oil is spilt. A huge oil slick floats on the surface of the ocean. It stays there for a long time. Air oxidises it very slowly. Bacteria decompose it very slowly. While the oil floats there, it poisons fish, and glues the feathers of seabirds together so that they cannot fly. When the oil slick washes ashore, it fouls beaches. In 1967 the oil tanker **Torrey Canyon** sank off the coast of the UK. A vast amount of oil escaped into the sea. People tried to disperse the oil by spraying it with huge amounts of detergent. This remedy did not work; it spread the oil over a larger surface, and the detergent killed seabirds and fish. When two tankers collided off San Francisco in 1971, the oil company used bales of straw to mop up the oil. Straw can

Fig. 16.9 An oil tanker

SUMMARY NOTE

Petroleum oil and natural gas were formed from animal and plant remains by slow bacterial decay in the absence of oxygen. They are found in many parts of the world. Drilling for oil and gas needs advanced technology.

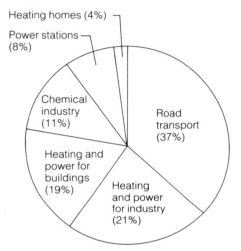

Fig. 16.10 This is how the fuels obtained from oil are used in the UK

SUMMARY NOTE

Crude oil is separated by fractional distillation into useful fuels. The use that is made of each fraction depends on its boiling point range, ignition temperature and other properties. The fuels obtained from crude oil are listed in Table 16.2. In addition, the petrochemicals industry makes useful chemicals from hydrocarbons. Cracking is used to make petrol and kerosene from heavy fuel oils.

soak up many times its own weight of oil. Recently at British Petroleum, chemists have discovered some chemicals which solidify oil. Sprayed on to a film of oil, they slowly turn it into a rubbery solid. Lumps of the solid can be picked up in fishing nets. This discovery has still to be tried out at sea. In addition to accidents to oil tankers at sea, spillage occurs at oil rigs and at oil terminals where tankers unload. There is also non-accidental discharge of oil by captains who want to save time by washing out their tanks at sea, rather than in port. This pollution is illegal.

There are two ways of getting oil ashore from an oil well in the sea. One way is to lay a pipeline along the sea bed and pump the oil through it. The other method is to use 'shuttle tankers' which pick up the oil from the oil field and transport it to a terminal on land. The choice of method depends on the size of the oil field and the distance between the field and the shore. Natural gas is almost always brought ashore by pipeline.

Crude oil is a black, viscous (syrupy) liquid which does not burn very easily. By fractional distillation (see Section 2.5 and Fig. 2.10) it can be separated into a number of important fuels. The fractions are not pure compounds. They are mixtures of alkanes which boil over a limited temperature range. Alkanes with small molecules boil at lower temperatures than those with large molecules. Alkanes with large molecules are more viscous than alkanes with small molecules. The fractions also differ in the ease with which they burn. When a fuel is heated, some of it vaporises. Eventually, there is enough vapour to be set alight by a flame. When the fuel reaches a temperature called the **ignition temperature**, there is enough vapour for a mixture of the fuel with air to be set alight and *continue to burn steadily*.

The use that is made of each fraction depends on its boiling point range, ignition temperature and viscosity (see Table 16.2). Our need for petrol, naphtha and kerosene is greater than our need for heavy fuel oils. Fortunately, chemists have found a way of converting the high boiling range fractions, of which we have more than enough, into the lower boiling range fractions, petrol and kerosene. The technique used is called **cracking**. Large hydrocarbon molecules are **cracked** (or split) into smaller hydrocarbon molecules. A heated catalyst (aluminium oxide or silicon(IV) oxide) helps the reaction to take place:

Vapour of hydrocarbon with large molecules and high b.p. →[CRACKING Passed over a heated catalyst]→ Mixture of hydrocarbons with smaller molecules and low b.p., and hydrogen

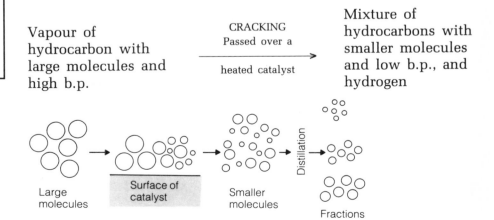

Fig. 16.11 Cracking

Table 16.2 Fractions and their uses

Fraction	Approximate boiling point range in °C	Approximate number of carbon atoms per molecule	Use
Petroleum gases	Below 25	1–4	These gases are liquefied under pressure, and sold in cylinders as 'bottled gas' for use in gas cookers and camping stoves. They burn easily at low temperatures. Sulphur compounds must be removed to make bottled gas clean to use and non-polluting.
Petrol	40–75	4–12	Petrol is liquid at room temperature, but vaporises easily at the temperature of the internal combustion engine.
Naphtha	75–150	7–14	Naphtha is used as a source of a huge number of useful chemicals. The petrochemicals industry manufactures plastics, drugs, medicines and fabrics (see Fig. 16.12).
Kerosene	150–240	9–16	Kerosene, another liquid fuel, needs a higher temperature for combustion. It is used in 'paraffin' stoves. The major use is as aviation fuel.
Diesel oil	220–250	15–25	Diesel fuel is more difficult to vaporise than petrol and kerosene. The special fuel injection system of the diesel engine allows this fuel to burn. It is used in buses, lorries and trains.
Lubricating oil	250–350	20–70	Lubricating oil is a viscous liquid. It does not vaporise sufficiently at the temperatures of vehicle engines to allow it to be used as a fuel. Instead, it is used as a lubricant to reduce wear.
Fuel oil	250–350	Above 10	Fuel oil is a viscous liquid with a high ignition temperature. It is used in ships, industrial machinery, heating plants and power stations. To help it ignite, fuel oil must be sprayed into the combustion chambers as a fine mist of small droplets.
Bitumen	Above 350	Above 70	Bitumen is the residue at the bottom of the distillation column. It has too high an ignition temperature to be used as a fuel. It is used to waterproof roofs and pipes and to tar roads.

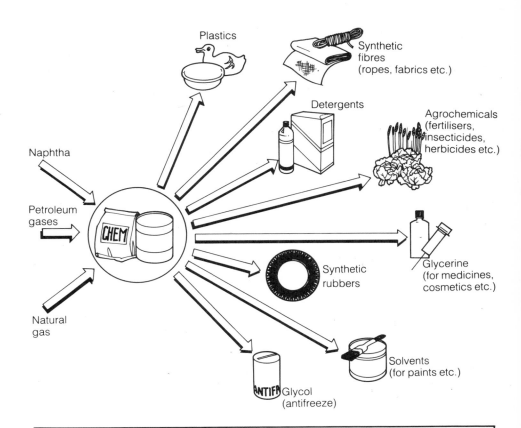

Fig. 16.12 Petrochemicals from naphtha

JUST TESTING 89

1. **a** Explain what is meant by a **homologous series**.
   b Explain what is meant by the term **isomerism**. Illustrate your answer by referring to pentane, C_5H_{12}.
2. How are halogenoalkanes made from alkanes? What is the name given to this type of reaction?
   What uses are made of halogenoalkanes?
3. How are petrol and kerosene made from high boiling point petroleum oil fractions? What name is given to this type of reaction? What are the economic reasons for carrying out this reaction?

16.7 Coal

Millions of years ago, the Earth was covered with trees and other plants. When the plants died, they started to decay. In the swampy conditions of that era, they formed peat. In time, layers of mud and sand covered the peat. The peat layer was compressed over millions of years by the deposits above it. Eventually, it turned into coal. The mud became shale and the sand became sandstone. Fuels that have been formed from the remains of living things are called **fossil fuels**. The energy of fossil fuels is derived from the energy of the sun. Coal is a fossil fuel derived from plants, and sunlight is needed for plants to grow. Oil and natural gas are fossil fuels derived from animals, and animals obtain their energy by eating plants or by eating animals which have eaten plants.

Many countries have coal deposits. The largest coal mining countries are the USSR, the USA, China, Poland and the UK. Coal is a complicated mixture of carbon, hydrocarbons and other compounds. When it burns, the main products are carbon dioxide and water:

carbon (in coal) + oxygen → carbon dioxide

hydrocarbons (in coal) + oxygen → carbon dioxide + water

Uses of coal

Three-quarters of the coal used in the UK is burned in power stations. The heat given out in burning raises steam. This steam drives turbines which generate electricity.

Coal can be **destructively distilled**. In this process, air is absent so that the coal does not burn. Four main fractions are obtained:

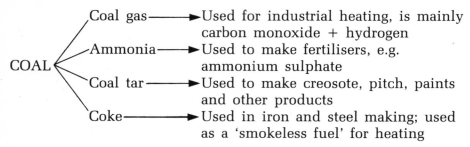

The importance of coal, oil and natural gas is discussed in Section 18.1.

SUMMARY NOTE

Coal is a fossil fuel derived from plant remains. It is burned in power stations. The distillation of coal gives useful products.

JUST TESTING 90

1 Study Fig. 16.13.
Why does the coal graph drop after 1960?
Why does the oil imports graph rise in 1960 and then fall after 1970?
Why is the oil production graph so recent an addition to the picture?
How long do you think the oil production graph will go on rising?

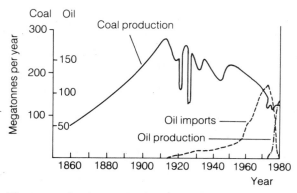

Fig. 16.13 Coal and oil in Britain, 1860–1980

Exercise 16

1. Explain what is meant by the terms
   a *organic* compound, **b** hydrocarbon, **c** alkane.

2. Where are alkanes found in nature? How were they formed? How are alkanes obtained from natural deposits? What is the chief use that we make of alkanes?

3. What important properties do halogenoalkanes possess?

4. The molecular formula for propane is C_3H_8.
   a Write the structural formula for propane.
   b What information does the structural formula give that the molecular formula does not tell you?
   c Write the molecular and structural formulas of two compounds which belong to the same **homologous series** as propane.
   d Explain what is meant by the term **homologous series**.

5. Coal and diamond are both minerals. Diamond is pure carbon; coal contains a high percentage of carbon. Both substances are valuable, but diamond is more expensive than coal. Can you explain why?
   What value does coal have which diamond does not share?
   Which of the two substances could we more easily do without? Explain your answer.
   How does diamond help in the search for fuels?

6. **a** Briefly describe how plant and animal material can be fermented to produce fuel gas.
   b What problems can you see in adapting the process to produce fuel gas for domestic use?
   c Can you think of any situations in which such a gas generator might be valuable both from an economic point of view and from an environmental point of view?

7. What is **crude oil**? Where is it found? What has to be done to crude oil before it is useful? What useful substances are obtained from crude oil?

CHAPTER 17 Alkenes, alcohols and acids

17.1 Plastic sand: what next?

Many Third World countries are plagued by drought. In long periods without rain, plants die and the soil becomes eroded. A British chemist called Allan Cooke has invented a plastic grain that could help these arid countries. He believes that his invention, Agrosoke®, will enable deserts to be turned into arable land. Agrosoke is a **polymer**. You will find out what a polymer is later in this chapter. Allan Cooke discovered something special about this polymer: it can absorb 40 times its own mass of water. It has the chemical name poly(propenamide). Cooke believes that a mixture of the polymer with sand should help plants to grow in arid regions. The idea is that the polymer soaks up water and then releases it gradually to the roots of plants. Field trials have been made on

- sunflower plants in Egypt, grown for the edible oil from the seeds
- acacia trees in the Sudan, grown for the gum arabic they produce
- eucalyptus trees in India, grown for shade
- seed germination in India.

In all these cases, plants grown with polymer have survived while control specimens have died (Fig. 17.1).

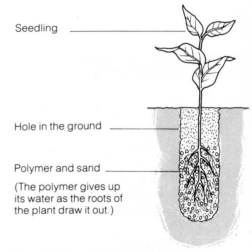

Fig. 17.1 Growing well thanks to a polymer

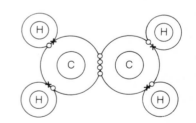

Fig. 17.2 A model of ethene

17.2 Alkenes

Ethene is a hydrocarbon of formula C_2H_4 (see Fig. 17.2). It is formed when alkanes are cracked (see Section 16.6). The bonding in ethene is shown in Fig. 17.3. The carbon atoms share two pairs of electrons: there is a double bond between them (see Section 6.3). Hydrocarbons, such as ethene, which contain double bonds between carbon atoms are described as **unsaturated** hydrocarbons. They will react with hydrogen to form saturated hydrocarbons:

ethene + hydrogen → ethane

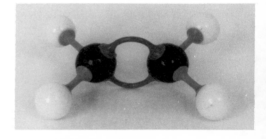

Fig. 17.3 The bonding in ethene

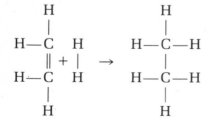

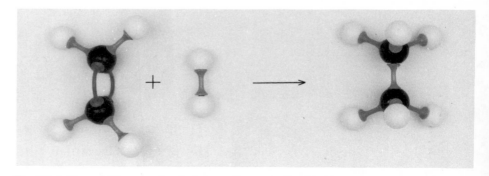

Reactions of this kind are called **addition reactions** (see Fig. 17.4).

Fig. 17.4 The addition reaction between ethene and hydrogen

Table 17.1 The alkenes

Ethene	C_2H_4
Propene	C_3H_6
Butene	C_4H_8
Pentene	C_5H_{10}
Alkenes	C_nH_{2n}

> **SUMMARY NOTE**
>
> Alkenes are a homologous series of unsaturated hydrocarbons. They possess a double bond between carbon atoms. The general formula is C_nH_{2n}.

Ethene is the first member of a homologous series called the **alkenes**. Propene (see Fig. 17.5) is the next member. Table 18.1 lists some members of the series. The general formula of alkenes is C_nH_{2n}.

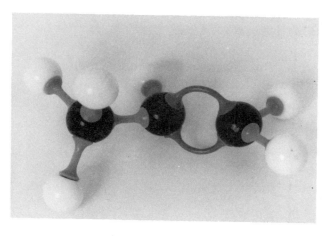

Fig. 17.5 A model of

17.3 Reactions of alkenes

Combustion

Alkenes burn in the same way as alkanes:

ethene + oxygen → carbon dioxide + water

$$C_2H_4(g) + 3O_2(g) \rightarrow 2CO_2(g) + 2H_2O(l)$$

Alkenes are not used as fuels because they are an important source of other compounds. Their unsaturated nature makes them chemically reactive. The petrochemicals industry uses alkenes as starting materials in the manufacture of plastics, solvents and other chemicals.

Addition reactions

The double bond enables alkenes to take part in **addition reactions**. Bromine adds to alkenes. A solution of bromine in an organic solvent or in water is brown. If an alkene is bubbled through a solution of bromine, the solution loses its colour (see Experiment 17.1). Bromine has added to the alkene to form a colourless compound. The reaction can be shown as

```
    H                          H
    |                          |
H—C        Br             H—C—Br
    ‖   +   |      →          |
H—C        Br             H—C—Br
    |                          |
    H                          H
```

ethene + bromine   1,2-dibromoethane

The product has single bonds: it is a saturated compound. With two carbon atoms in the molecule, it is named after ethane. With two bromine atoms in the molecule, it is a dibromo-compound,

1,2-dibromoethane. The numbers 1,2- tell you that one bromine atom is bonded to one carbon atom and the second bromine atom is bonded to the second carbon atom. The decolourisation of a bromine solution is used to distinguish between an alkene and an alkane. Chlorine adds to alkenes in a similar way.

Hydration

A molecule of water also will add across the double bond. Combination with water is called **hydration**:

```
   H                        H
   |                        |
H—C      H              H—C—H
   ‖   + |      →          |
H—C      O—H            H—C—O—H
   |                        |
   H                        H
```

The product is ethanol, C_2H_5OH. This is the compound we commonly call **alcohol**. It is an important industrial solvent. Ethanol is made by passing ethene and steam over a heated catalyst at high pressure. Only about 10% of the ethene is converted, and the unreacted gases are recycled.

ethene + steam $\xrightarrow[\text{(phosphoric acid), under pressure}]{\text{Pass over a heated catalyst}}$ ethanol

↑_____ Recycle unreacted gases over the catalyst ←_____|

Hydrogenation

The reaction between alkenes and hydrogen is used in the production of margarine. Animal fats, such as butter, are solid. Vegetable oils, such as olive oil and sunflower seed oil, are liquid. Insufficient butter is produced to satisfy our demand for solid fats, but more vegetable oil is produced than we need for cooking. Manufacturers therefore convert liquid oils into solid fats. They make use of the fact that solid fats are saturated, while liquid oils are unsaturated. Hydrogenation (the addition of hydrogen) will convert an unsaturated compound into a saturated compound. The vapour of a liquid oil is passed with hydrogen over a nickel catalyst:

vegetable oil (unsaturated) + hydrogen $\xrightarrow[\text{heated nickel catalyst}]{\text{Pass over}}$ solid fat (saturated)

The solid fat produced is sold as margarine. If some of the double bonds are left intact by partial hydrogenation, soft margarine is obtained.

SUMMARY NOTE

Alkenes burn to form carbon dioxide and water. The double bond makes alkenes reactive. They take part in addition reactions with bromine and chlorine, with water (in the presence of a catalyst) and with hydrogen. This reaction is used to turn vegetable oils into saturated fats.

Polymerisation

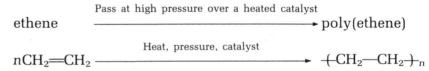

This reaction is called **polymerisation**. Many molecules of the **monomer**, ethene, join together (**polymerise**) to form the **polymer**, poly(ethene). In poly(ethene), 30 000–40 000 molecules of the monomer form one molecule of polymer. The conditions needed for polymerisation are high pressure, a moderate temperature and a catalyst:

$$\text{ethene} \xrightarrow{\text{Pass at high pressure over a heated catalyst}} \text{poly(ethene)}$$

$$n\text{CH}_2{=}\text{CH}_2 \xrightarrow{\text{Heat, pressure, catalyst}} {-}(\text{CH}_2{-}\text{CH}_2)_n{-}$$

You have seen poly(ethene) many times, although you may know it better by its trade name of **polythene**. It is the material used for making plastic bags, kitchenware (buckets, bowls etc.), laboratory tubing and toys. It is flexible and difficult to break.

17.4 Polymers

Polymers such as poly(ethene) and other polyalkenes are plastics. Plastics are materials which soften on heating and harden on cooling. Objects can be moulded easily from plastics. There are two subsets of plastics.

Thermosoftening and thermosetting plastics

Thermosoftening plastics can be softened by heating, cooled and resoftened many times:

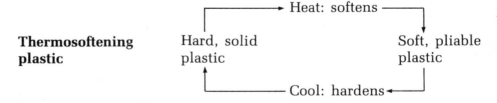

a thermosetting plastic can be softened by heat only once.

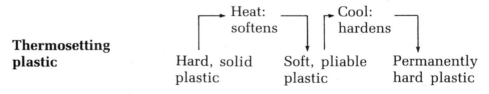

The reason for the difference in behaviour is a difference in structure. **Thermosoftening plastics** consist of long polymer chains. The forces of attraction between the chains are weak. Manufacturers find thermosoftening plastics convenient to work with. They can buy tonnes of thermosoftening plastic in the form of granules, melt the material and mould it into the shape of the object they want to make. Plastics of this sort can be moulded several times during the manufacture of an article. Another advantage is colour. A pigment can be added to the molten plastic and thoroughly mixed. Then the moulded objects will be coloured all through. This is a big advantage over a coat of paint which can be chipped.

Thermosetting plastics have a different structure. When a thermosetting plastic is softened and moulded, the chains react with one another. Cross-links are formed, and a huge three-dimensional structure is built up. This is why thermosetting plastics can be formed only once (Fig. 17.6).

The chief thermosetting polymer families are not poly(alkenes). They are epoxy resins (used as glues), polyester resins (used in glass-reinforced plastics), polyurethanes (used in varnishes), melamine (used in kitchen surfaces and on bench tops) and bakelite (used in electrical fittings). Both types of plastics have their advantages. The materials used for electrical fittings and counter tops must be able to withstand high temperatures without softening. For these purposes, 'thermosets' are used.

Different methods are used for moulding the two kinds of plastics. Three methods of moulding thermosoftening plastics are shown in Figs. 17.7, 17.8 and 17.9. Figure 17.10 shows a method for shaping thermosetting plastics.

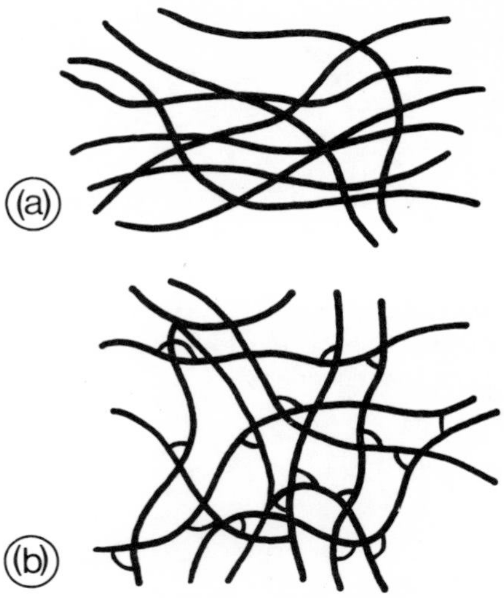

Fig. 17.6 (a) Thermosoftening plastic; (b) thermosetting plastic

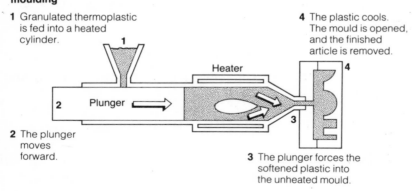

Fig. 17.7 Injection moulding is used for objects such as milk bottle crates, television set cases and construction kits

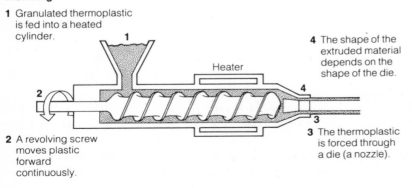

Fig. 17.8 The extrusion method is used for making pipes, threads of fabric and insulation for electrical cable

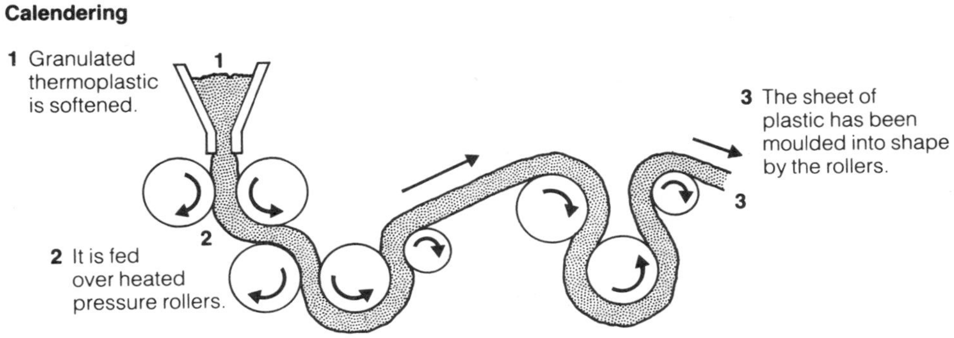

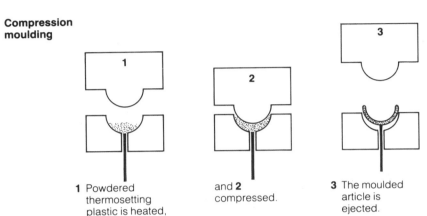

Calendering

1 Granulated thermoplastic is softened.
2 It is fed over heated pressure rollers.
3 The sheet of plastic has been moulded into shape by the rollers.

Fig. 17.9 Calendering is used for making large sheets of plastic, e.g. for car seat covers and floor covering

Compression moulding

1 Powdered thermosetting plastic is heated,
and 2 compressed.
3 The moulded article is ejected.

Fig. 17.10 Most thermosetting plastics are shaped by compression moulding

Sometimes gases are mixed with plastics during shaping to make low density plastic foams. Such foams are used in car seats, for thermal insulation of buildings, for insulation against sound and for packaging. Plastics can also be strengthened by the addition of other materials. Glass-fibre-reinforced plastic is strong enough to be used for the manufacture of boat hulls and car bodies (see Fig. 17.11).

SUMMARY NOTE

Plastics are either **thermosoftening**: can be softened by heat many times, or **thermosetting**: can be softened by heat only once before setting permanently.

Fig. 17.11 GRP (glass-reinforced plastic) boat

Poly(alkenes)

Table 17.2 Some poly(alkenes) and their uses

Poly(ethene); trade name polythene

Monomer

$$\begin{array}{c} H \quad H \\ | \quad | \\ C = C \\ | \quad | \\ H \quad H \end{array}$$

Polymer

$$\left(\begin{array}{c} H \quad H \\ | \quad | \\ C - C \\ | \quad | \\ H \quad H \end{array}\right)_n$$

Uses:
Polythene is used to make plastic bags. High density polythene is used to make kitchenware, laboratory tubing and toys.

Poly(chloroethene); trade name PVC

Monomer

$$\begin{array}{c} Cl \quad H \\ | \quad | \\ C = C \\ | \quad | \\ H \quad H \end{array}$$

Polymer

$$\left(\begin{array}{c} Cl \quad H \\ | \quad | \\ C - C \\ | \quad | \\ H \quad H \end{array}\right)_n$$

Uses:
PVC is used to make plastic bottles, wellingtons and raincoats, floor tiles, insulation for electrical wiring, gutters and drainpipes.

Poly(propene)

Monomer

$$\begin{array}{c} H \quad CH_3 \\ | \quad | \\ C = C \\ | \quad | \\ H \quad H \end{array}$$

Polymer

$$\left(\begin{array}{c} H \quad CH_3 \\ | \quad | \\ C - C \\ | \quad | \\ H \quad H \end{array}\right)_n$$

Uses:
Poly(propene) is resistant to attack by chemicals and does not soften in boiling water. It can be used to make hospital equipment which must be sterilised.
Poly(propene) is drawn into fibres and used to make ropes and fishing nets.

Poly(tetrafluoroethene); trade names PTFE and teflon

Monomer

$$\begin{array}{c} F \quad F \\ | \quad | \\ C = C \\ | \quad | \\ F \quad F \end{array}$$

Polymer

$$\left(\begin{array}{c} F \quad F \\ | \quad | \\ C - C \\ | \quad | \\ F \quad F \end{array}\right)_n$$

Uses:
PTFE is a hard, waxy plastic which is not attacked by most chemicals. Few substances can stick to its surface. It is used to coat non-stick pans and skis.

Perspex

Monomer

Uses:
Perspex is an important plastic because it is transparent and can be used instead of glass. It is more easily moulded than glass and less easily shattered.

Polystyrene

Monomer

$$\begin{array}{c} H \quad H \\ | \quad | \\ C = C \\ | \quad | \\ H \quad C_6H_5 \end{array}$$

Uses:
Polystyrene is a hard, brittle plastic used for making food containers (e.g. yoghurt cartons) and construction kits. Polystyrene foam is made by blowing air into the softened plastic. It is used for making ceiling tiles and packaging fragile goods, e.g. cameras.

SUMMARY NOTE

Polymerisation is an important reaction of alkenes. Many molecules of monomer join to form a huge molecule of polymer. Many poly(alkenes) are plastics with important uses. Poly(ethene), poly(chloroethene), poly(propene), poly(tetrafluoroethene), perspex and polystyrene are plastics with a huge number of uses.

There are some disadvantages in the use of plastics:
- they are non-biodegradable
- they ignite easily
- some of them have toxic combustion products.

SUMMARY NOTE

Plastics are petrochemicals. The more oil we use as fuel, the less oil we shall have in the future to give us petrochemicals.

Some drawbacks

You must at some time have drunk coffee and soft drinks out of disposable cups. These are made of polyurethane or polystyrene. Like other plastics, these polymers are **non-biodegradable**. They cannot be decomposed by natural biological processes. All the plastic cups, spoons and food containers which people use and throw away have to be burned or dumped in ever-increasing rubbish tips. Chemists are now trying to make new plastics which will be easier to dispose of.

Summerland Amusement Park in the Isle of Man used to have an enormous amusement hall covered with plastic panels. In 1973, some boys playing with matches set the whole structure on fire. Within ten minutes, more than 50 people had died. Many were not burned to death. Most of the casualties were overcome by poisonous gases that had been produced when the plastic burned. The tragedy made manufacturers of building materials stop and investigate their products. They needed to check whether, if plastics should happen to burn, they would produce harmful substances.

Many buildings are insulated with plastic foam or furnished with plastic materials. Plastics have lower ignition temperatures than materials like wood, metal, brick and glass. Fires can spread very rapidly when plastics burn.

Oil: a fuel and a source of petrochemicals

The number of plastics and the uses found for them are constantly growing. The raw materials used in their manufacture come from oil. The Earth's resources of oil will not last for ever. It seems wasteful to burn oil as fuel when we need it to make plastics and other petrochemicals.

JUST TESTING 91

1. What advantage has a plastic doll over a china doll?
2. Toy farmyard animals and soldiers used to be made out of lead. Why do you think lead was the metal chosen for the purpose? What advantages do plastic toys have over lead toys?
3. Which plastic would be used to make a motorbike windscreen? What advantages does it have over glass?
*4. Agrosoke® (see Section 17.1) is poly(propenamide). Propenamide is

   H—C—CONH$_2$
   ‖
   H—C—H

   Draw the formula of poly(propenamide).

   Poly(ethenol) can absorb hundreds of times its own mass of water. The monomer is CH_2=CHOH. Draw the formula of the polymer.

5 a What does the word **plastic** mean?
  b There are two big classes of plastics, which behave differently when heated. Name the two classes. Describe the difference in behaviour. Say how this difference is related to the molecular nature of the plastics.

17.5 Alcohols

Ethanol and party-goers

Angela thoroughly enjoyed the party (Fig. 17.12). She felt relaxed and happy and had a great time. Next morning, however, she woke up with a splitting headache and a feeling of sickness. The substance responsible for these effects on her body was **ethanol**, the liquid we usually call **alcohol** (Fig. 17.13).

Ethanol is a drug. It depresses the central nervous system. That does not mean that it makes people feel depressed. In small quantities, it makes people relaxed by suppressing feelings of fear and tension. Ethanol dissolves completely in water. When ethanol is swallowed, it is absorbed through the stomach and intestines. It can take up to 6 hours for the ethanol in a single drink to be absorbed when the stomach is full, but only about 1 hour when the stomach is empty. This is why people feel the effects of a drink faster on an empty stomach than on a full one. When ethanol is in the bloodstream, it moves rapidly into the tissues until the concentration of ethanol in the tissues equals that in the blood. The concentration of ethanol in the breath or in urine can be used to indicate the level of ethanol in the blood. As the concentration of ethanol in the blood increases, speech becomes slurred, vision becomes blurred and reaction times increase. This is why it is so dangerous to drive 'under the influence' of alcohol. Drinking large amounts of ethanol regularly causes damage to the liver, kidneys, arteries and brain. Many people do not **use** alcohol properly, that is, in moderation: they **abuse** alcohol. Such people become addicted to alcohol, and their health suffers.

Fig. 17.12 Angela enjoying the party

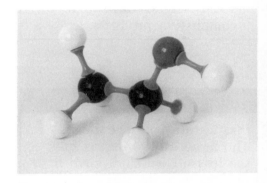

Fig. 17.13 A model of ethanol

Ethanol and the alcohols

Ethanol is a member of a homologous series of compounds called **alcohols**. Alcohols possess the group

and have the general formula $C_nH_{2n+1}OH$. Some formulas are given in Table 17.3. The formulas are written as CH_3OH etc. rather than CH_4O to show the —OH group and to make it clear that they are alcohols. The members of the series have similar physical properties and chemical reactions. Ethanol is the only alcohol that can be safely drunk. Methanol, CH_3OH, is very toxic. Drinking only small amounts of methanol can lead to blindness and death. The structural formulas of some alcohols are shown below.

Table 17.3 The alcohols

Methanol	CH_3OH
Ethanol	C_2H_5OH
Propanol	C_3H_7OH
Alcohols	$C_nH_{2n+1}OH$

SUMMARY NOTE

Alcohols are a homologous series of formula $C_nH_{2n+1}OH$. Ethanol is the alcohol which people drink. When the concentration of ethanol in the blood rises above a low safe level, it damages the body's co-ordination. Regular abuse of alcohol ruins your health.

SUMMARY NOTE

Ethanol is made from sugars by fermentation, a reaction which is catalysed by an enzyme in yeast. Starches can be hydrolysed to sugars and then fermented to give ethanol.

Women who drink heavily in pregnancy have babies with 'fetal alcohol syndrome'. The babies are very small at birth and have facial deformities such as a flattened nose, misshapen ears and a long lower jaw. Many have heart murmurs, hearing defects and defective hip joints. The babies do not grow out of these birth defects; they get worse. No-one knows how much alcohol, if any, a pregnant woman can safely drink.

Formulas of alcohols

Methanol

$$H-\overset{\overset{H}{|}}{\underset{\underset{H}{|}}{C}}-O-H$$

Ethanol

$$H-\overset{\overset{H}{|}}{\underset{\underset{H}{|}}{C}}-\overset{\overset{H}{|}}{\underset{\underset{H}{|}}{C}}-O-H$$

Propanol

$$H-\overset{\overset{H}{|}}{\underset{\underset{H}{|}}{C}}-\overset{\overset{H}{|}}{\underset{\underset{H}{|}}{C}}-\overset{\overset{H}{|}}{\underset{\underset{H}{|}}{C}}-O-H$$

Ethanol has been made for centuries by the fermentation of sugars and starches. These substances are **carbohydrates**. They are compounds of carbon, hydrogen and oxygen which have 2 atoms of hydrogen for every atom of oxygen, e.g. the sugar glucose, $C_6H_{12}O_6$, and starch, $(C_6H_{10}O_5)_n$. Glucose can be converted into ethanol by an enzyme called zymase, which is found in yeast. An enzyme is a catalyst which is found in a plant or animal. Yeast is a living plant. If you make ethanol in Experiment 17.3, you will see that carbon dioxide is evolved. The reaction is called **fermentation**:

$$\text{glucose} \xrightarrow{\text{enzyme in yeast}} \text{ethanol} + \text{carbon dioxide}$$
$$C_6H_{12}O_6(aq) \longrightarrow 2C_2H_5OH(aq) + 2CO_2(g)$$

Fruit juices contain sugars. When yeast is added, the juice ferments to give ethanol. When the ethanol content reaches 14%, it kills the yeast. A more concentrated solution of ethanol (96% ethanol) can be obtained by distillation.

Ethanol can be made from starchy foods, such as potatoes, rice, malt, barley, hops and others. Starch is hydrolysed by an enzyme in malt (germinated barley) to a mixture of sugars. Then yeast is added to ferment the sugars:

$$\text{starch} + \text{water} \xrightarrow{\text{enzyme in malt}} \text{sugar}$$
$$(C_6H_{10}O_5)_n(aq) + nH_2O(l) \longrightarrow nC_6H_{12}O_6(aq)$$

Ethanol is sold in four main forms:
- absolute alcohol: 96% ethanol, 4% water
- industrial alcohol or methylated spirit: 85% ethanol, 10% water, 5% methanol (this is added to make the liquid unfit for drinking)
- spirits: whisky, brandy, rum, gin, which contain about 35% ethanol
- fermented liquors: beer, wine, cider, etc. These contain flavourings, colouring matter and fragrant oils. Beers and ciders contain 3%–7% ethanol; wines contain 12%–14%.

Ethanol for industrial use is not made by fermentation. It is made by the catalytic hydration of ethene (see Section 17.3). When ethanol is dehydrated, it gives ethene. Figure 17.14 shows a method of passing the vapour of ethanol over a dehydrating agent to give ethene.

Ethanol is an important solvent. It is a volatile liquid, with b.p. 78 °C. It is used in cosmetics and toiletries, in thinners for lacquers and in printing inks. Being volatile, the solvent evaporates and leaves the solute behind. Other alcohols, e.g.

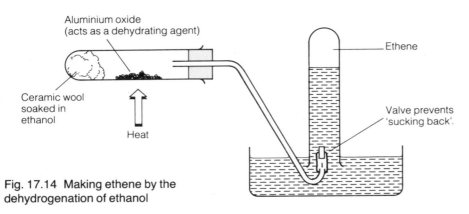

Fig. 17.14 Making ethene by the dehydrogenation of ethanol

methanol, propanol and butanol, also are used as solvents for shellacs, lacquers, paints and industrial detergents. Alcohols have a big advantage as solvents in that they are miscible with both water and a number of organic liquids.

Ethanol is oxidised by air if the right micro-organisms are present. The reason why wine goes sour if it is open to the air is that the ethanol in it is oxidised to ethanoic acid. Vinegar is 3% ethanoic acid.

$$\text{ethanol} + \text{oxygen} \xrightarrow{\text{certain micro-organisms}} \text{ethanoic acid} + \text{water}$$
$$C_2H_5OH(aq) + O_2(g) \longrightarrow CH_3CO_2H(aq) + H_2O(l)$$

A faster method of oxidising ethanol is to use acidified potassium dichromate(VI), $K_2Cr_2O_7$. This powerful oxidising agent is orange. Dichromate(VI) ions, $Cr_2O_7^{2-}$, are reduced by ethanol and other reducing agents to chromium(III) ions, which are blue. As the reaction proceeds, the colour changes from orange through green to blue. This colour change is the basis of the first 'breathalyser' test. A motorist suspected of having too much ethanol in his or her blood has to breathe out through a tube containing some orange potassium dichromate(VI) crystals. If they turn green, he or she is 'over the limit' (Fig. 17.15).

SUMMARY NOTE

Ethanol for industrial use is made from ethene by catalytic hydration. It is an important solvent. Ethanol is oxidised to ethanoic acid.

Fig. 17.15 The breathalyser test

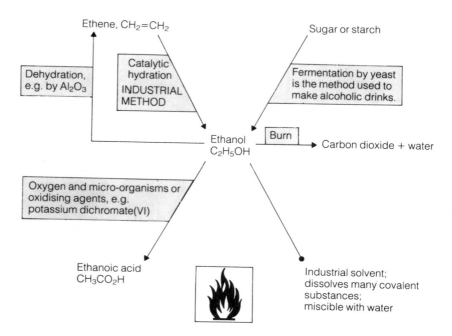

Fig. 17.16 Some reactions of ethanol

JUST TESTING 92

1. Ethanol is made from petroleum in three stages. Say what they are and how they are carried out.
2. a Explain what is meant by **fermentation**.
   b What commercially important substances are made by this method?
3. a Why does wine turn sour on standing?
   b What is the difference between drinking ethanol and drinking methanol?

Gasohol

During World War II, many German vehicles ran on ethanol which had been made from potatoes. Petrol engines are designed to operate at a temperature at which petrol will vaporise. Any substitute fuel must vaporise at the operating temperature and it must dissolve in petrol. Ethanol can be used in a petrol engine because it boils at the same temperature as heptane, it does not freeze and it dissolves in alkanes. The combustion of ethanol produces 70% as much heat per litre as petrol combustion. Ethanol burns well in vehicle engines, and the atmospheric pollution produced is small. A petrol engine will take 10% ethanol in the petrol without any adjustments to the carburettor (which controls the ratio of air to fuel in the cylinders).

Ethanol has made a come-back in the car. The costs of petroleum and gas and the energy spent in cracking oil fractions have increased. Countries which have to import oil want to find alternatives. The fermentation method of making ethanol needs sunlight to grow crops. Countries with land available for growing crops to supply sugar or starch for fermentation and sunlight to ripen the crops are interested in using ethanol as a fuel.

Brazil has already started producing ethanol for use as a vehicle fuel. Brazil has very little oil, but plenty of land and sunshine. Most of the petrol there now contains 10% ethanol, which has been made by the fermentation of cane sugar. Of the land area, 0.5% is devoted to growing sugar cane for fermentation. By the year 2000, Brazil hopes to provide for 75% of its motor fuel needs by using 2% of its land for growing crops for fermentation (Fig. 17.17).

Fig. 17.17 Alcohol in Brazil's tanks

SUMMARY NOTE

Ethanol is a clean fuel: it burns to form carbon dioxide and water. It can be used as a fuel in petrol engines. Brazil has no oil and has plenty of arable land to grow crops which can be fermented. Brazil is using ethanol mixed with petrol in vehicle engines. Other countries are following suit.

JUST TESTING 93

1. What is the source of energy in the cracking process?
2. What is the source of energy in fermentation?
3. Europe has a surplus of grain. Cereals can be turned into ethanol. This can replace tetraethyl lead as an octane-booster in lead-free petrol. You know that people in Asia and Africa are short of food. Should the surplus grain that Europe produces be used to fuel cars?

What's 'gasohol'? Do you drink it or burn it?

Better put it in the tank! It's the unleaded petrol with 10% ethanol that American 'gas' stations sell. The ethanol in it is made from corn.

17.6 Ethanoic acid

Ethanoic acid has the structural formula

$$\begin{array}{c} H O \\ | \| \\ H-C-C \\ | \diagdown \\ H O-H \end{array}$$

Ethanoic acid is a member of a homologous series called **carboxylic acids**. It is a weak acid (see Section 7.7). Some of the reactions of ethanoic acid are shown in Fig. 17.19.

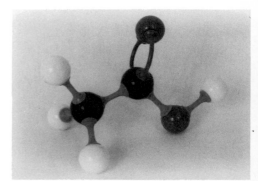

Fig. 17.18 A model of ethanoic acid

Fig. 17.19 Reactions of ethanoic acid

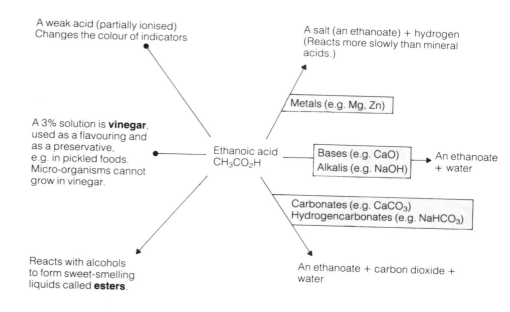

17.7 Esters

Esters with small molecules are liquids with fruity smells. They occur naturally in fruits. They are used as food additives to give flavour and aroma to processed foods. Esters are used as solvents. Many glues contain esters as solvents. Some people enjoy the effects of inhaling esters so much that they become 'glue sniffers'. It is really the solvent, which may be a hydrocarbon or an ester, that they are sniffing. This dangerous pastime is called **solvent abuse**. Solvent abuse produces the same symptoms as ethanol abuse. In addition, sniffers who are 'high' on solvents may believe that they will come to no harm if they jump out of windows or walk through traffic. Most of the deaths which occur through solvent abuse are caused by such disoriented behaviour or by sniffers passing out and suffocating on their own vomit.

Animal fats and vegetable oils are esters. Esters are liquids or solids, depending on the size of their molecules.

> **SUMMARY NOTE**
>
> Ethanoic acid is a carboxylic acid. It is a weak acid. It has the same reactions as mineral acids, but reacts more slowly. Carboxylic acids react with alcohols to form **esters**. These compounds are used as food additives and as solvents.

> **JUST TESTING 94**
>
> 1 Name three substances which will react with ethanoic acid.
>   Name the products of the reactions.
> *2 Explain why ethanoic acid is less reactive than hydrochloric acid. (See Section 7.5 if you need help.)

17.8 The chemical industry

Many of the chemicals we use are **petrochemicals**, derived from petroleum oil. The petrochemical industry is not the whole of the chemical industry. The industry can be divided roughly into ten sections.

1. Heavy chemical industry: oils, fuels etc. (see Chapter 16).
2. Agriculture: fertilisers, pesticides etc. (see Chapter 14).
3. Plastics: poly(ethene), poly(styrene), PVC etc. (see Chapter 17).
4. Dyes (see Chapter 14).
5. Fibres: nylon, rayon, Courtelle etc. (see Chapters 13 and 17).
6. Paints, varnishes etc. (see Chapters 13 and 18).
7. Pharmaceuticals: medicines, drugs, cosmetics (see Chapter 17).
8. Metals: iron, aluminium, alloys etc. (see Chapter 10).
9. Explosives: dynamite, TNT etc. (see Chapter 14).
10. Chemicals from salt: sodium hydroxide, chlorine, hydrogen, hydrochloric acid (see Chapter 15).

Figure 17.20 shows some of the petrochemicals which are obtained by the route

petroleum oil → naphtha → ethene → petrochemical

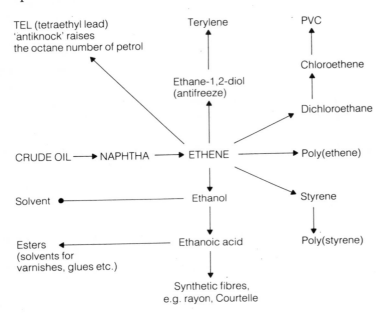

NOTE Propene reacts similarly. It is a source of rubber, detergents, paints and glues.

Fig. 17.20 Petrochemicals

17.9 Organic compounds in medicine

Many organic compounds are used in **chemotherapy**, the treatment of illness by chemical means. Chemotherapy makes our lives safer, longer and more free from pain. Some of the chemicals which relieve pain and cure diseases are mentioned in this section. There are many more.

Pain-killers

For the relief of moderate pain, aspirin is the all-time favourite. Aspirin is manufactured on a large scale. We in the UK each swallow an average of 200 aspirins a year. Many other drugs contain aspirin. When you swallow aspirin, there is slight bleeding of the stomach wall. In most people, this is not serious. In some people, severe loss of blood occurs. You should always swallow plenty of water with aspirin to reduce irritation of the stomach wall.

Codeine is a stronger pain-killer, used in headache tablets and in cough medicines. Morphine is used to relieve intense pain, and many a soldier wounded in battle has been glad of its almost magical quality. Morphine is *addictive*. People who use it become unable to do without it. Morphine can only be used under a doctor's supervision.

Heroin is even more potent and more addictive than morphine. The ease with which people become addicted to heroin explains why the medical profession has rejected it as a pain-killer. It also explains why it is so popular among drug pushers, who want to keep their customers eager to buy again and again.

The ideal pain-killer would be one which relieves even intense pain without causing addiction. A massive effort is being made by the pharmaceutical industry to find such a compound. New compounds are made and tested. In the search, a new compound 10 000 times more powerful and quick-acting than morphine has been made; this is the M-99 compound manufactured by Reckitt and Colman. They call it the 'rhino-dropper' and sell it to vets to use for anaesthetising large animals. The vet fires a pellet of M-99 into the animal; the animal drops unconscious. The new compound is, however, addictive, and the search for a new pain-killer continues.

Sedatives and tranquillisers

Barbiturates are drugs which have important medical uses, e.g. the treatment of high blood pressure and mental illness. They are also used as sedatives (sleeping tablets). They are habit-forming. Many people have accidentally killed themselves by taking an overdose of barbiturates.

Tranquillisers are substances which relieve tension and anxiety without producing sleep. They are sold in vast quantities by the drug industry. About one person in twenty takes tranquillisers every day. Medical researchers are beginning to look into the long-term effects of taking tranquillisers in this way.

During the 1950s, a tranquilliser called **thalidomide** was put on the market before it had been thoroughly tested. Terribly deformed babies were born to women who had taken the drug. The thalidomide scandal led to the passing of a new law. Since that time, drug companies have had to carry out a huge battery of tests on new drugs before they are sold.

Nicotine is a compound which occurs naturally in the tobacco plant. It is a poisonous oily liquid with an unpleasant smell, a burning taste and an alkaline reaction. Horrible though it sounds, people become addicted to it! Nicotine increases the pulse rate and the blood pressure. It depresses breathing. It makes the stomach wall contract, causing nausea, vomiting and diarrhoea. People become able to tolerate nicotine when they take small quantities regularly. This is why, although people feel sick or vomit when they start to smoke, after a while these symptoms do not occur. People like smoking because it produces a calming effect. Smoking may result in worsening eyesight, headaches, irregular heartbeat and cancer. Nicotine is the habit-forming compound in tobacco smoke. There are more than 1000 compounds in tobacco smoke. The tar that condenses in the lungs of smokers is a mixture of carcinogenic (cancer-producing) hydrocarbons.

Stimulants

Stimulants are produced naturally in the body. **Adrenaline** is a substance which the body produces when it needs to prepare for strenuous activity: for 'fight or flight'. It stimulates the heart to beat faster, makes a person keyed up and ready for action. Some people take **amphetamines** (pep pills) because they have similar effects on the body. With repeated use, people can become addicted. People who *abuse* (misuse) amphetamines are excitable and talkative, with trembling hands, enlarged pupils and heavy perspiration.

Antiseptics

A century ago, a surgeon called James Lister could not understand why so many patients who lived through operations died later in the wards. Something seemed to be going wrong *after* the operation. He realised that their wounds were becoming infected. He sprayed the operating theatre and the wards with a mist of phenol and water. The results were spectacular. The death rate was drastically reduced. Phenol is an **antiseptic**. It prevented micro-organisms from infecting the surgical wounds. Phenol is not a pleasant antiseptic to use. Its vapour is toxic, and solid phenol will burn the skin. Research chemists have discovered other compounds which work as well as phenol and are safer to use. TCP® and Dettol® are antiseptics which contain trichlorophenol.

Antibiotics

Antibiotics are used to fight diseases carried by bacteria. At the beginning of this century, thousands of people died every year from infectious diseases such as tuberculosis and pneumonia. The discovery of the **sulphonamide** drugs in 1935 completely changed the picture. They brought about marvellous recoveries from diseases which would otherwise have been fatal. Gerhardt Domagh was given the Nobel prize in 1939 for his discovery of the power of sulphonamide drugs. The first patient on whom he tested the drug was his daughter. She was dying of an infection called 'child-bed fever', which used to affect many women just after the birth of a baby. The drug worked! It is largely because of the discovery of powerful antibiotics that having a baby is so much safer than it was at the beginning of this century.

Penicillin was discovered in 1928 by Sir Alexander Fleming, who was working in a London hospital. During World War I, Fleming had seen many wounded soldiers dying in field hospitals. Many men died not from the severity of their wounds but from infection that set in later. Contamination from dirty clothing and mud caused gangrene, and the men died a slow, painful death. None of the existing antiseptics did much good. After the war, Fleming went back to his work as a **bacteriologist**. His ambition was to discover a substance which would kill bacteria. One day, a mould called **Penicillium** appeared on a dish of bacteria which he was culturing. It killed the bacteria. His efforts to treat infected patients with it were unsuccessful. When **penicillin**, an extract of the mould, was injected into a patient, substances in the blood made the **bacteriocide** inactive.

In 1940, two chemists called Howard Florey and Ernst Chain took up the work on penicillin. World War II had started, and the need for a powerful bacteriocide was urgent. Working in Oxford, Florey and Chain were able to purify Fleming's penicillin and make a stable extract. They tested penicillin on mice and then on human patients. In 1941, mass-production of penicillin started in the USA. The following year, penicillin was used on battle casualties with spectacular results. Fleming was knighted, and in 1945, he shared the Nobel prize for medicine with Florey and Chain.

Penicillin has been widely used since 1942 to treat infections of various parts of the body. Penicillin is broken down by acids, e.g. stomach acid. This disadvantage is not shared by **tetracycline**. Tetracycline is a 'broad spectrum' antibiotic, used against a large number of bacteria. Between them sulphonamides, penicillin, tetracycline and other antibiotics changed the pattern of medicine. Infectious diseases are no longer the killers they were.

JUST TESTING 95

1 What benefits does the manufacture of aspirins give us? Why should you take water with aspirins?

2 Do you think people should take tranquillisers regularly? Is there anything else people can do to relieve tension?

3 If you were employed as a pharmaceutical chemist by a drug company, and you discovered a new pain-killer, would you be in a hurry to see your discovery on the shelves of a 'chemist's shop'? Would you pressure your group-leader to speed up trials of your new drug?

4 Someone you know feels so tired that she is thinking of taking pep pills before going into an examination. What would you say to her?

5 A friend tells you that you should smoke for relaxation. Describe to him the effects of nicotine on the body. Tell him what other harmful substances are present in cigarette smoke. Suggest to him another way of relaxing.

6 What difference has the discovery of antiseptics made to surgery?
What difference has the discovery of anaesthetics made to surgery? (See Section 16.3 if you need help.)

7 How long did it take for Florey and Chain to go from tests on animals to tests on humans? It would take them much longer in the 1980s than it did in the 1940s. What has happened to slow down the trials of new drugs on patients?

8 Morphine is a very effective pain reliever. Why do doctors not allow most patients to use it continuously? For what types of patients is morphine prescribed?

Exercise 17

1 What is a polymer? Give the structural formula of a named example. Say what use is made of the polymer you name.

2 PTFE is a polymer of

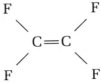

Draw the structure of the polymer. Give **a** the name of the monomer, **b** a use for the polymer.

3 List ten objects which are made from plastics. Say what material would have been used to make each object before the age of plastics. Explain what advantage plastics have over the other material.

4 Explain what is meant by 'cracking'. Why is it an important process?

5 What is the difference between thermosetting plastics and thermosoftening plastics? For which purposes are thermosetting plastics the better choice? Why are thermosoftening plastics preferred for other jobs?

6 PVC is a widely used plastic. Its proper name is poly(chloroethene). Draw the structural formula. What dangerous substances are formed if PVC burns?

7 **a** What class of chemical compound do soaps belong to?
 b Explain how soaps work.
 c What advantage do detergents have over soaps?
(See Section 9.11 if you need to revise.)

8 Alkenes can be made by passing the vapour of an alcohol over heated aluminium oxide.
 a Sketch an apparatus in which you could carry out this reaction. Show how you would collect the alkene.

 b Describe how you would test the alkene to show that it is **unsaturated**. Explain what the word **unsaturated** means.
 c Write a word equation and a symbol equation for the reaction of a named alcohol.

9 **a** Name three plastics, and give examples of the uses to which these plastics are put.
 b Point out why plastics are used for these purposes.
 c Give two reasons why plastic furniture is more dangerous than wood furniture in a fire.

10 'Ethanoic acid is a weak acid which can be made by the oxidation of ethanol.'
 a Write the structural formulas for ethanoic acid and ethanol.
 b Explain what is meant by **oxidation**.
 c Explain what is meant by **weak acid**.
 d Describe two experiments you could do to show that ethanoic acid is a weaker acid than hydrochloric acid.

11 The plastic containers we use and throw away are non-biodegradable. The mass of discarded plastic in our tips and landfills is steadily increasing. Do you think we should
 a burn discarded plastic in incinerators and use the heat generated or
 b pay for research to develop biodegradable plastics or
 c make it compulsory for the present almost indestructible plastics to be recycled? Explain the reasons for your answer.

12 In 1938, Dr Roy Plunkett was working for the Dupont chemical company. He needed a large quantity of tetrafluoroethene for his research. As no manufacturer made it, Plunkett built a small plant to make his own. He stored tetrafluoroethene in gas cylinders. In his experiments, he passed the

gas from a storage cylinder into a reaction chamber. One day, the flow of gas stopped only minutes after the cylinder had been opened. The weight of the cylinder showed that there were still many grams of gas inside it. The valve was open: a piece of wire could be passed through it. The gas must still be in the cylinder, but it would not come out. In desperation, Plunkett seized a hacksaw and cut the cylinder in two. Inside was a large quantity of a white powder. He did not know what it was, but he was soon busy doing experiments to find out.

  a Write the formula for tetrafluoroethene.
  b Name the white powder that formed in the cylinder.
  c Write its formula.
  d Say what use we make of this substance.
  e Do you think Dr Plunkett was just lucky to make this discovery? What else besides luck came into it?

13 Ethanol is made by using yeast to ferment a sugar solution.
  a What is the name given to the catalysts in yeast?
  b Why is the rate of fermentation not increased by raising the temperature above 25 °C?
  c When ethanol is to be used as a fuel, the solution that is obtained by fermentation is filtered and fractionally distilled. What is the reason for (i) filtration and (ii) fractional distillation?
  d Why is Brazil using ethanol made in this way to fuel motor vehicles? Why does the UK not do the same?
  e What is the advantage to the environment of burning ethanol instead of petrol?

14 On the one hand, in the UK,
  - The tax on alcoholic drinks brings in £4000 million a year.
  - The manufacture, distribution and sales of alcoholic drinks provides employment for $\frac{3}{4}$ million people.
  - Exports (mainly whisky) bring in £1000 million a year.

But, on the other hand, in the UK,
  - Alcoholism costs industry £1600 million a year.
  - One-third of the drivers killed in car accidents are drunk.
  - One-quarter of pedestrians killed in road accidents are killed by drunken drivers.
  - In many cases of murder, the murderer is drunk.

  a Can you add some more points for and against alcohol consumption? You may like to turn this question into a group discussion.
  b When does the *use* of alcohol turn into *abuse*?

CHAPTER 18 Energy

18.1 The energy crisis

You will have heard about the **world energy crisis**. It features constantly in newspaper headlines and television programmes. The problem is that we are consuming energy faster and faster. One day, unless we can find new sources of energy, we shall come to the end of the Earth's supplies of energy. The Earth's reserves of coal and oil and natural gas are limited. Coal, oil and natural gas are fossil fuels. They were made millions of years ago. Once we have used up the Earth's deposits, the supplies will not be renewed. Reserves of oil and natural gas are likely to run out in 20 to 30 years. Our present way of life is geared largely to oil and gas. Our agriculture depends on them: the petrochemicals industry provides farmers with pesticides to protect their crops from insects and weeds. The industries which manufacture most of our possessions depend on the oil industry. Our means of transport use oil: without oil, our cars, trains, boats and planes would be useless. Without the petrochemicals industry, we should have few modern drugs and medicines. Without oil and gas, our whole economy would collapse, leaving us shivering and starving in a primitive world. Unless scientists find alternatives to oil and gas, our future will be one of low technology.

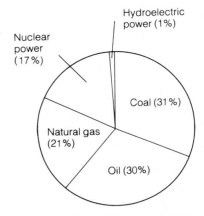

Fig. 18.1 Sources of energy in the UK

18.2 Some solutions

Nuclear power

The simplest solution is to obtain all the world's power needs from nuclear reactors. This solution is not free of difficulties. The dangers of accidents in nuclear reactors and the problems of storing long-lived radioactive material are discussed in Section 4.10.

Coal

Another solution is to make more use of coal. Already there are methods of making liquid fuels from coal:

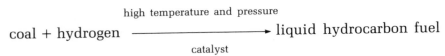

The fuel produced can be used to power diesel engines. German planes flew on it in 1945. South Africa has enormous plants for making transport fuel from coal, which make the country independent of oil imports. Coal 'liquefaction' plants are not healthy places to work because prolonged contact with many hydrocarbons can cause skin cancer.

Making gaseous fuels from coal is less risky and less expensive than making liquid fuels from coal. The production of gaseous fuels involves heating coal and then allowing the residue to react with steam. A mixture of gaseous fuels is formed:

coal $\xrightarrow[\text{2 React with steam}]{\text{1 Heat}}$ fuel gas (carbon monoxide + hydrogen + methane)

The UK has coal deposits that will last for 200–250 years. Coal mining on an increased scale, however, will blight our countryside. Open-cast mining devastates the landscape. Underground mining creates spoil heaps. A spoil heap is a great mountain of soil and rock that have been dug out of the mine. A spoil heap collapsed on the Welsh village of Aberfan in 1970, killing dozens of people. This has made people fearful of living under the shadow of a spoil heap.

Table 18.1 A comparison of coal, oil and gas

	Coal	Oil	Natural gas
Origin	Fossil fuel	Fossil fuel	Fossil fuel
UK reserves	200–250 years	20–30 years	20 years
World reserves	200–250 years	30–40 years	20 years
Effect on the environment	Open strip mining devastates the landscape. Spoil heaps are ugly and dangerous. Mining is still a dangerous job.	The sea is polluted by accidental spills from tankers and by ships illegally washing out their tanks at sea.	
Combustion	Air is polluted by sulphur dioxide, carbon monoxide, oxides of nitrogen, hydrocarbons, soot.	Air is polluted by sulphur dioxide, carbon monoxide, oxides of nitrogen, hydrocarbons, soot.	Natural gas is a clean fuel.

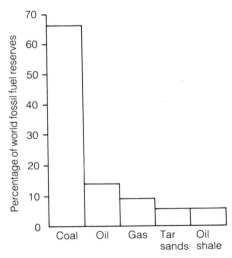

Fig. 18.2 Estimated reserves of fossil fuels in the world

Tar sands and oil shales

Some countries, e.g. Canada, have deposits of hydrocarbon tar mixed with sand. At present, the cost of obtaining a fuel from these deposits is too high to make the working of the deposits pay.

Oil shales are deposits of oil in porous shale rocks. Extracting the oil is too costly to be worthwhile at present.

Renewable energy sources

The idea of utilising **renewable energy sources** appeals to a lot of people. The sun, the wind and the tides are renewable energy sources. They derive their energy from the sun. As we use them, the sun renews them.

Solar energy. Solar energy is the name given to energy derived from sunlight. Sunlight is converted into electricity by the crystals in **photoelectric cells** (see Section 1.5). A solar power station is a vast area covered with rows of photoelectric cells. The cost of building a solar plant is high. The world's largest solar power plant is in southern California. California has plenty of sunshine and plenty of spare land to cover with solar panels. A big advantage of solar electric power plants is the complete absence of pollution.

Fig. 18.3 Solar panels on a house in Edinburgh

Windpower. Recently, the attention of scientists and engineers has turned to windmills and the attraction of their 'free' fuel. They have invented many new designs of wind machines (Fig. 18.4).

Waterpower. Hydro-electric power is obtained by damming a river at a high level and letting it flow under gravity to drive turbines lower down. The force of the tides can be used to drive turbines and generate electricity.

Geothermal energy. The rock layers thousands of metres below the surface of the Earth are hot. In some places, the hot rocks are only hundreds of metres down. Water can be pumped down over the hot rock and up to the surface. Water heated in this way is used to heat buildings in Paris. Steam generated in this way drives electric power generators in Italy.

Biomass. Material of biological origin can be converted into

Fig. 18.4 Wind machine

fuels. The use of methane from digesters of animal manure, toilet effluent and household waste has been described (see Section 16.1). Methane can be converted into methanol. This is a liquid fuel which can be used in the internal combustion engine. Some countries grow crops to be converted into fuel. Sugar cane and grain can be fermented to give ethanol (see Section 17.5). Biological material is called **biomass**.

18.3 Exothermic reactions

We need energy for many purposes. We obtain energy from the combustion of hydrocarbons, from the combustion of foods and from other chemical reactions. The chemical reactions and nuclear reactions which give out energy are **exothermic reactions** (ex = out; therm = heat). You have met many of them already, and only a summary is given in this chapter.

Combustion

Coal is a mixture of carbon and hydrocarbons. When it burns,

carbon (in coal) + oxygen → carbon dioxide   **Heat is given out**

hydrocarbons (in coal) + oxygen → carbon dioxide + water   **Heat is given out**

Natural gas is largely methane, CH_4. When it burns,

methane + oxygen → carbon dioxide + water   **Heat is given out**

$$CH_4(g) + 2O_2(g) \rightarrow CO_2(g) + 2H_2O(l)$$

To warm our homes and buildings, we use the combustion of methane and other hydrocarbons. To power our vehicles, we use the combustion of hydrocarbon fuels such as octane, C_8H_{18}.

Fig. 18.5 It gets its energy from combustion ▷

When octane burns,

octane + oxygen → carbon dioxide + water **Heat is released**

- An oxidation reaction in which heat is given out is a **combustion**.
- Combustion accompanied by a flame is **burning**.
- A substance which is oxidised with the release of energy is a **fuel**.

Respiration

We use a different kind of fuel to supply our bodies with energy. The 'energy foods' are sugars and starches. Sugars and starches are **carbohydrates**. These compounds contain carbon and hydrogen and oxygen in the ratio of 2H to 1O (as in H_2O). Glucose, $C_6H_{12}O_6$, is a sugar. It is oxidised to carbon dioxide and water with the release of energy:

glucose + oxygen → carbon dioxide + water **Energy is given out**

$C_6H_{12}O_6(aq) + 6O_2(g) \rightarrow 6CO_2(g) + 6H_2O(l)$

This reaction takes place inside our body tissues. Oxygen in the air we breathe oxidises sugars in our foods to provide us with energy. This reaction is called **respiration**. Respiration is an exothermic reaction.

A starch molecule consists of a string of $C_6H_{10}O_5$ units.

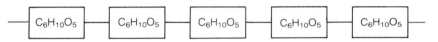

In the stomach, starch is hydrolysed (split up by water) to form sugars. Once formed, sugars are oxidised to give energy.

Neutralisation

When an acid neutralises an alkali (Experiment 18.2), heat is evolved. Neutralisation is an exothermic reaction.

hydrogen ion + hydroxide ion → water **Heat is given out**

$H^+(aq)$ + $OH^-(aq)$ → $H_2O(l)$

Hydration

When concentrated sulphuric acid reacts with water, heat is given out (see Section 13.8):

concentrated sulphuric acid + water → dilute sulphuric acid **Heat is given out**

When anhydrous copper(II) sulphate combines with water, it forms a **hydrate**. This is the name for a salt combined with water of crystallisation.

copper(II) sulphate + water → copper(II) sulphate-5-water **Heat is given out**

$CuSO_4(s) + 5H_2O(l) \rightarrow CuSO_4 \cdot 5H_2O(s)$

Fig. 18.6 Powered by combustion

SUMMARY NOTE

In **exothermic reactions**, energy is released. Examples are
- the combustion of hydrocarbons
- the combustion of sugars and starches in respiration
- neutralisation
- hydration
- chemical cells
- nuclear reactions.

Chemical cells

In chemical cells (see Section 5.11), chemical energy is transformed into electrical energy. Examples are the Daniell cell, dry cells, the lead–acid accumulator and fuel cells.

Nuclear reactions

You can read about nuclear reactions in Section 4.8.

18.4 Endothermic reactions

Some reactions take in energy; they are **endothermic reactions**.

Photosynthesis

Plants are able to convert the simple compounds carbon dioxide and water into sugars. They do this by converting the energy of sunlight into the energy of the chemical bonds in the sugar molecules. The reaction is called **photosynthesis**. It is an endothermic reaction. It takes place in the leaves of green plants, where the green pigment chlorophyll catalyses the reaction:

carbon dioxide + water → glucose + oxygen
(from the air) (from the soil) (sugar)

$6CO_2(g) + 6H_2O(l) \rightarrow C_6H_{12}O_6(aq) + 6O_2(g)$

Energy is taken in

Dissolving

Many salts dissolve in water exothermically; some dissolve endothermically. Ammonium chloride and potassium nitrate dissolve endothermically (see Experiment 18.1 and Section 18.5).

Thermal decomposition

Many substances decompose when they are heated. Examples are

zinc carbonate → zinc oxide + carbon dioxide **Heat is taken in**

$ZnCO_3(s) \rightarrow ZnO(s) + CO_2(g)$

copper(II) hydroxide → copper(II) oxide + water **Heat is taken in**

$Cu(OH)_2(s) \rightarrow CuO(s) + H_2O(l)$

copper(II) sulphate-5-water → copper(II) sulphate + water **Heat is taken in**

$CuSO_4 \cdot 5H_2O(s) \rightarrow CuSO_4(s) + 5H_2O(l)$

Compare this last reaction with **hydration**.

—Dehydration is endothermic→
copper(II) sulphate-5-water ⇌ anhydrous copper(II) sulphate + water
←Hydration is exothermic—

Thermal dissociation

Many compounds decompose when they are heated. If the products recombine on cooling, then the reaction is described as **thermal dissociation**. In the thermal dissociation of ammonium chloride,

⟶ The salt dissociates when heated ⟶
ammonium chloride ⇌ ammonia + hydrogen chloride
$NH_4Cl(s)$ ⇌ $NH_3(g)$ + $HCl(g)$
⟵ The products recombine when cooled ⟵

The reaction is **reversible**. It goes from left to right or from right to left, depending on the temperature. From left to right, it is endothermic; from right to left, it is exothermic.

You should revise the thermal dissociation of calcium carbonate (Section 12.2).

> **SUMMARY NOTE**
>
> In **endothermic reactions**, energy is taken in from the surroundings. Examples are
> - photosynthesis
> - the dissolving of *some* salts
> - thermal decomposition
> - thermal dissociation.

18.5 Heat of reaction

Why is energy (heat energy and other forms of energy) given out or taken in during a chemical reaction? Chemicals possess energy of two kinds. One is **kinetic energy** — energy of motion. The atom or ions or molecules of a substance are constantly moving (see Section 1.4). To do this, they must possess kinetic energy. The second kind of energy is the **energy of chemical bonds**. The atoms or ions or molecules in a substance are held together by chemical bonds (see Chapter 6). Energy must be supplied if these chemical bonds are to be broken. There are no chemical bonds that fly apart by magic: energy must always be supplied to break bonds. When bonds are created, energy is given out.

The reactants and the products will have roughly the same kinetic energy at the same temperature. They will have different bond energies because the bonds in the substances are different. The energy content of the products is different from the energy content of the reactants (see Figs. 18.7 and 18.8).

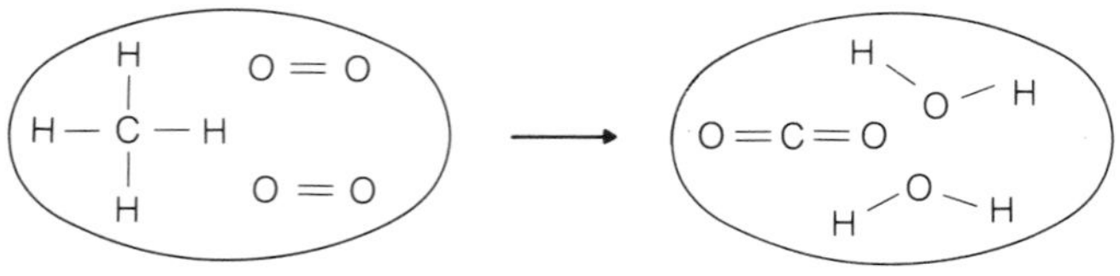

These bonds are broken. Energy must be taken in.

These new bonds are made. Energy is given out.

The energy taken in is less than the energy given out: this reaction is exothermic.

Fig. 18.7 Bonds broken and made when methane burns

The symbol H is used for the energy content of a substance. The symbol ΔH is used for the heat taken in or given out during a reaction. ΔH is called the **heat of reaction**. Figure 18.9 is an **energy diagram**. It shows the energy content of the reactants and

Fig. 18.8 Bonds broken and made when a crystal dissolves

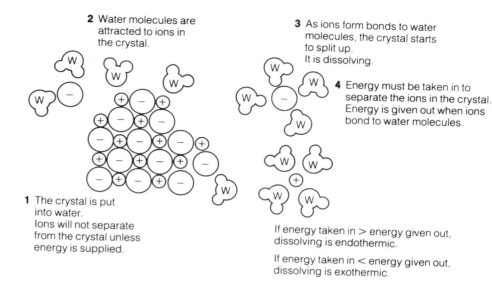

1 The crystal is put into water. Ions will not separate from the crystal unless energy is supplied.

2 Water molecules are attracted to ions in the crystal.

3 As ions form bonds to water molecules, the crystal starts to split up. It is dissolving.

4 Energy must be taken in to separate the ions in the crystal. Energy is given out when ions bond to water molecules.

If energy taken in > energy given out, dissolving is endothermic.

If energy taken in < energy given out, dissolving is exothermic.

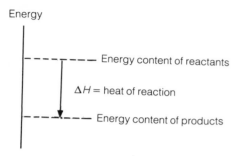

Fig. 18.9 An energy diagram for an exothermic reaction

SUMMARY NOTE

In a chemical reaction, bonds are broken and new bonds are made. Energy must be supplied to break bonds. Energy is given out when bonds are made. The difference between the bond energies of the products and the bond energies of the reactants is the **heat of reaction**. Energy diagrams are drawn to show the heat of reaction ΔH.
For an exothermic reaction, ΔH is negative.
For an endothermic reaction, ΔH is positive.

the products. The products of the reaction shown in Fig. 18.9 contain less energy than the reactants. When the reactants change into the products, they have to get rid of their extra energy. They give out heat to the surroundings: this is an exothermic reaction.

The difference ΔH between the energy of the products and the energy of the reactants is called the **heat of reaction**. By definition,

Heat of reaction ΔH = Energy of products − Energy of reactants

You can see that, because the products of the reaction shown in Fig. 18.9 have less energy than the reactants, ΔH is negative.

For an exothermic reaction, ΔH is negative.

Figure 18.10 is a second energy diagram. In this reaction, when the reactants change into the products, they have to ascend to a higher energy level. To do this, they must take energy from the surroundings: they cool the surroundings. This is an endothermic reaction. Since by definition,

Heat of reaction ΔH = Energy of products − Energy of reactants

in this reaction, ΔH is a positive quantity.

For an endothermic reaction, ΔH is positive.

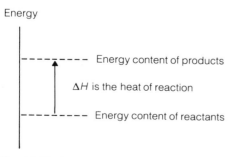

Fig. 18.10 An energy diagram for an endothermic reaction

JUST TESTING 96

1 Draw an energy diagram for the combination of hydrogen ions and hydroxide ions to form water. Re-read if you have forgotten whether it is exothermic or endothermic. Mark the heat of reaction on your diagram. State whether it is positive or negative.

2 Draw an energy diagram for the combustion of petrol to form carbon dioxide and water. Mark ΔH on your diagram, and state whether it is positive or negative.

3 Is the dissolving of ammonium chloride exothermic or endothermic? Illustrate your answer by an energy diagram.

18.6 Getting over the barrier

The combustion of hydrocarbon fuels is exothermic. There is no doubt about this: the energy given out drives our vehicles and warms our homes. When you mix petrol with air at room temperature, however, it does not immediately burst into flames. You have to warm it up before combustion takes place. The energy given out in combustion is greater than the energy you have to put in to start the reaction. The products are at a lower energy level than the reactants but there is a barrier to be surmounted before the reactants can roll down the energy hill to form the products (see Fig. 18.11).

SUMMARY NOTE

Chemical reactions need an input of energy to start them. The reactants will not be converted into the products until an energy barrier has been overcome.

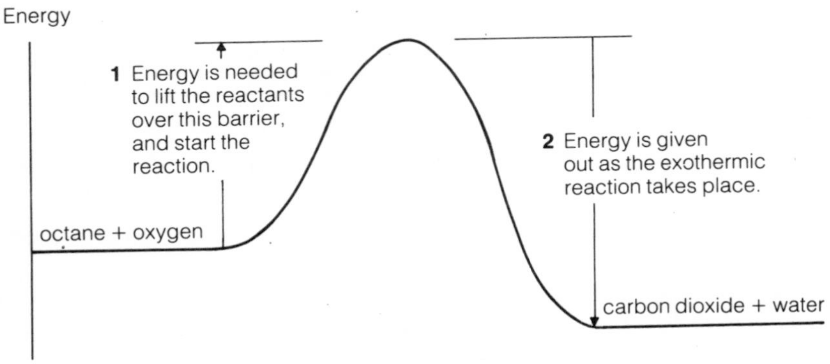

Energy

1 Energy is needed to lift the reactants over this barrier, and start the reaction.

2 Energy is given out as the exothermic reaction takes place.

octane + oxygen

carbon dioxide + water

Fig. 18.11 The barrier

JUST TESTING 97

1. Describe the reaction that takes place when you heat a mixture of iron and sulphur (see Chapter 3). What do you see to tell you that this reaction is exothermic?
   - Why do you have to heat to start the reaction?
   - Write an equation for the reaction.
   - Draw an energy diagram for the reaction.
2. Describe the reaction that takes place when you heat magnesium ribbon in air. What do you see to tell you that the reaction is exothermic? What forms of energy are released?
   - Why do you have to heat to start the reaction?
   - Draw an energy diagram.
*3. 'There is an energy barrier to be climbed before substances can react.'
   Explain this statement.
   Is this barrier a nuisance? If energy barriers were suddenly abolished, what would happen to **a** petrol and **b** magnesium ribbon?

*18.7 Calculations on heat of reaction

In stating the value of the heat change that occurs during a chemical reaction or a physical change, we give the value for 1 mole of substance. Heat is measured in joules (J) and kilojoules (kJ)

(1000 J = 1 kJ). Heats of reaction are expressed in kilojoules **per mole** (kJ/mol).

Molar heat of neutralisation is the heat taken in when 1 mole of hydrogen ions is neutralised:

hydrogen ions + hydroxide ions → water **Heat is given out**

$H^+(aq)$ + $OH^-(aq)$ → $H_2O(l)$ $\Delta H = -57$ kJ/mol

Molar heat of combustion is the heat taken in when 1 mole of a substance is completely burnt in oxygen:

methane + oxygen → carbon dioxide + water **Heat is given out**

$CH_4(g)$ + $2O_2(g)$ → $CO_2(g)$ + $2H_2O(l)$ $\Delta H = -890$ kJ/mol

Example 1 Camping gaz is butane, C_4H_{10}. The molar heat of combustion of butane is -2880 kJ/mol. Calculate the heat given out when 1 kg of camping gaz burns completely.

Method
Molar mass of butane, C_4H_{10} = (4 × 12) + (10 × 1) = 58 g/mol
1 mol of butane, that is, 58 g of butane, give out 2880 kJ
therefore 1000 g of butane give out
    2880 × 1000/58 kJ = 49 700 kJ

1 kg of camping gaz (butane) gives 49 700 kJ.

Example 2 50.0 cm³ of 1.00 mol/l sodium hydroxide solution are neutralised by an acid. The heat given out is 2.85 kJ. Calculate the molar heat of neutralisation.

Method
Amount (in moles) of sodium hydroxide
    = Volume (l) × Concentration (mol/l)
    = 50.0 × 10^{-3} × 1.00 mol
    = 5.0 × 10^{-2} mol

Heat evolved in the neutralisation of 5.0 × 10^{-2} mol = 2.85 kJ

Heat evolved in the neutralisation of 1.00 mol
    = 2.85 × 1.0/(5.0 × 10^{-2}) kJ
    = 57 kJ

The molar heat of neutralisation is -57 kJ/mol.

SUMMARY NOTE

Molar heat of combustion = heat taken in when 1 mole of substance is burned completely.

Molar heat of neutralisation = heat taken in when 1 mole of hydrogen ions are neutralised.

JUST TESTING 98

1 The molar heat of combustion of glucose ($C_6H_{12}O_6$) is -2800 kJ/mol. What mass of glucose must be combusted to release 400 kJ of energy?

2 The combustion of 0.80 g of methanol releases 17.9 kJ of energy. Calculate the molar heat of combustion of methanol, CH_3OH.

3 One-third of a litre of sodium hydroxide solution of concentration 0.1 mol/l was neutralised by dilute nitric acid. The heat given out was 1.90 kJ. Calculate the molar heat of neutralisation.

18.8 The nuclear debate

You may have heard of **the nuclear debate**. The burning question is nuclear power stations. One side of the debate is worried by the dangers of radioactive materials. The other side claims that we need nuclear power because coal and oil supplies will start to run out by the end of this century. A nuclear debate is in session at Newtown High School.

Jill I'm dead against nuclear reactors. After that terrible accident in Russia, I think we ought to scrap nuclear power stations.

Neil You mean the Chernobyl disaster in 1986? That was dreadful, when that nuclear reactor exploded and burned for days with no-one knowing whether they could put it out. It was two weeks before they brought it under control. At one time it got so hot they thought it might burn its way down into the Earth.

Beth You mean like that Jane Fonda film, The China Syndrome, where they thought the reactor might burn through the Earth to China?

Jill How did they put the fire out?

Neil Firemen put out the fire in the building which housed the reactor. Most of them died later because they got such a huge dose of radioactivity. The core of the reactor was white-hot graphite at 5000 °C. No-one could get near it. Helicopters dropped sand, lead and boron on the burning reactor to cool it.

Beth Besides the twenty people killed, there were hundreds taken to hospital with radiation sickness.

Mark People who got a smaller dose of radiation could get leukaemia or cancer in a year's time or in ten year's time.

Chris It wasn't only people in the plant who were affected. The explosion spread radioactive gases and particles over miles and miles of the Ukraine.

Jill The Ukraine is the farming area where most of the grain in Russia is grown, isn't it? I don't think they'll want to eat grain that has been contaminated with radioactivity.

Beth Does anyone know which radioactive isotopes are formed when uranium-235 splits?

Mark Strontium-90, iodine-131 and caesium-137.

Neil Strontium-90 is very dangerous. You see, strontium compounds behave like calcium compounds. They would be laid down in your bones and teeth.

Jill Iodine builds up in the thyroid gland in your throat. That makes iodine-131 a dangerous isotope too. Caesium compounds get into your body in the same way as sodium and potassium compounds do.

Chris Did you know that there was once a nuclear accident in Britain? Those radioisotopes escaped from Sellafield nuclear power station.

Mark You mean when the reactor caught fire in 1964. They called it Windscale nuclear power station then. The reactor overheated, and a load of graphite rods caught fire. They had to flood the reactor with water to put out the fire.

Neil The Lake District was showered with radioactivity. People were worried that cows would produce radioactive milk. When they tested the milk, they found there was only a very low level of iodine-131 in it. They dumped all the milk for a few weeks to be on the safe side.

Jill Iodine-131 has a half-life of one week. They were lucky it wasn't a long-lived radioisotope.

Chris Did the Russian accident happen in the same way?

Beth The Russians haven't explained what happened at Chernobyl. I expect it was similar to the accident at Three Mile Island in the USA in 1979. No-one was injured that time, but it was a big scare.

Mark That crisis started when the pumps which feed in cold water to cool the reactor stopped.

Chris It was mad not having a back-up system.

Mark There was one, but the emergency pumps couldn't come on because someone had shut off their water supply. Human error, they call it. Dozens of warning signals lit up. After a few minutes, they found out what was wrong and opened the valve.

Jill Was that the end of the panic?

Mark No. There was mechanical failure as well as human error. When the reactor started to heat up, a safety valve opened to release steam. It should have closed again, but it didn't. It let water and steam escape. Losing water sent up the temperature of the reactor. The hot metal reacted with steam to form hydrogen.

Neil Wow! I know hydrogen. It's an explosive. They must have been scared of a hydrogen explosion. I bet there was a hydrogen explosion at Chernobyl. Could there have been a nuclear explosion as well?

Chris No, that can't happen. The construction of a nuclear reactor is completely different from an atomic bomb. The danger is that the core might get hot enough to melt the metal and concrete shielding. Then *all* the radioactivity would escape. Also a steam explosion or a hydrogen explosion would let radioactive material escape.

Jill Well, what did happen at Three Mile Island?

Mark After two hours, the operators finally discovered that the steam valve was open. After they closed it, the temperature fell. It took them a week to get the temperature down. The plant is still closed because of the radioactive contamination.

Neil I still think that if a nuclear power station is operated properly, it's safe. You don't hear many people worrying about whether coal mining is safe, do you?

Beth Even when they are working properly, nuclear reactors produce plutonium. That's dangerous stuff, isn't it? I know it's the isotope they use to make atom bombs. I think terrorists could steal it and use it to make an atomic bomb. Then they could threaten to blow up a whole city if they didn't get what they wanted.

Chris The group of terrorists would have to include a pretty good physicist, a chemist and an engineer at least. They would need the right equipment. Making an atomic bomb is a really difficult and dangerous job.

Jill Perhaps you're right. I suppose terrorists can find easier ways of threatening people than making an atomic bomb.

Neil The biggest problem with nuclear reactors is the spent fuel that is taken out of them. One of the fission products it contains is plutonium-239. It is tremendously radioactive. It gives off heat as well. It needs to be cooled and stored in a safe place for a long time.

Jill A very long time. Strontium-90 and caesium-137 have half-lives of about 90 years. They need to be stored for 400 years. Plutonium-239 has a half-life of 24 000 years. It needs to be stored for a quarter of a million years before most of its activity has gone.

Beth West Germany stores radioactive waste in old salt mines. Britain pipes very low level radioactive waste into the sea and stores the rest until a proper storage place can be found for it. The USA packs radioactive waste in steel canisters and dumps them at sea.

Chris The French have worked out a method of turning liquid radioactive waste into glass. They pack the radioactive glass in steel drums. Then they bury the drums beneath concrete. This should store the waste safely until it isn't radioactive any longer.

Jill I hope you're right. Remember, that means a quarter of a million years for plutonium-239.

Chris What no-one has said yet is that we need nuclear power. Coal and oil won't last for ever. Well, not for more than 20 to 30 years, they say. Then what? I don't want the sort of life people used to have without electricity.

Beth Already we get 18 per cent of our electricity from nuclear power.

Neil The Americans get 17 per cent and the French 65 per cent.

Jill At least nuclear power stations are clean. They don't put the pollution in the air that coal and oil power stations do. You don't get sulphur dioxide and nitrogen oxides and soot from nuclear power stations.

Mark I think it's a terrible waste to burn oil when the petrochemical industry can make all those useful things from it.

Beth You mean plastics and paints and so on?

Chris I think there's a future for renewable energy sources. You know, wind and waves and solar power. People are building some modern windmills with great designs. Wind is free, and windmills don't make pollution.

Beth There are so many things to weigh up. I can't make up my mind about the best way to get energy.

Some questions for you to discuss

(The energy crisis and the nuclear debate are complicated topics. You may well like to form groups to discuss these questions. There are no short answers.)

1 Why is radioactive waste stored for a long time?
2 Why is it often stored underground? Why are old salt mines chosen?
3 Why is it often stored behind thick walls of concrete? What

is the advantage of vitrifying (making into glass) radioactive waste?
4. What is the danger of dumping it in steel cans at sea?
5. Someone suggested launching radioactive waste into space. Can you see anything wrong with this idea?
6. Why are people worried about earthquakes disturbing underground stores?
7. Why do strontium compounds behave as calcium compounds do?
   Why do caesium compounds behave as sodium compounds do?
8. Why are nuclear power stations so useful to us? What advantages do they have over coal-burning electric power stations?
9. Explain the advantages of making use of renewable energy sources.
   Describe some methods of utilising renewable energy sources which were not mentioned by the students.
   What difficulty is there in using solar power?
10. What is wrong with burning petrol and other hydrocarbon fuels?
    (There are at least two answers to this question.)
11. The accident at Chernobyl sent a radioactive cloud drifting towards Poland. The Russians did not warn the Poles of the danger that was approaching. Had the Poles known, they could have taken some precautions against radiation.
    a. What do you think should be the agreed code of behaviour for nuclear powers if an accident happens?
    b. If Poland had known that a cloud of radioactivity was approaching, what precautions could that country have taken?

In conclusion

I cannot offer you a solution to the problem. Beth, Jill, Chris, Neil and Mark have not covered all the points in the nuclear debate. You will see people discussing it on television. You will read about it in the papers. You would be wise not to make up your minds yet. Find out more about the problem. Keep thinking about it. Keep on discussing it.

Exercise 18

1. a. Explain the difference between **exothermic** and **endothermic** reactions.
   b. Give an example of an exothermic reaction of vital importance in everyday life. Explain why it is so important to you.
   c. Give an example of an endothermic reaction of vital importance. Explain why it is so important to you.
2. a. Why do people say there is an **energy crisis**?
   b. Do you think the energy crisis is going to affect your life?
   c. What are **renewable** energy sources? What advantages do they have over coal and oil?
   d. If someone asked you which renewable energy source Britain should invest in, what would you say? Explain your choice.
3. What advantages do nuclear power stations have over coal-fired and oil-fired electricity power stations?

4 a Name one radioactive element which is mined for use in nuclear power stations.
  b Name one radioactive element which is made in nuclear power stations.
  c Describe one of the risks involved in running a nuclear power station. Say what is done to avoid an accident.
  d Why do people worry more about the possibility of an accident in a nuclear power station than about accidents in coal mines?

5 The half-life of plutonium is 24 000 years.
  a What is meant by the term 'half-life'?
  b What problem is created by this long half-life of plutonium?
  c What attempts have been made to solve the problem?

6 a Name two sources of energy which do not depend on radioactive isotopes or fossil fuels.
  b Explain the advantages of utilising these sources of energy.
  c Outline the difficulties which have to be overcome before we can make good use of them.

7 What is **biomass**? Name two kinds of biomass from which fuels are obtained. Explain why some countries are exploiting this source of fuels more than the UK. In what way could the UK make more use of biomass as a source of fuel?

*8 a What is hydrocarbon?
  b One of the hydrocarbons in petrol is octane, C_8H_{18}. Write a word equation for the complete combustion of octane. Write a symbol equation.
  c The molar heat of combustion of octane is -5520 kJ/mol. Calculate the heat that is released by the complete combustion of 57 g of octane.
  d Why is the quantity of energy obtained when 57 g of octane burn in a car engine less than the calculated quantity?

9 'The reaction

anhydrous cobalt chloride + water → cobalt chloride-6-water

is exothermic.'
By means of a diagram, show an experiment by which you could find out whether this statement is true.

CHAPTER 19 The speeds of chemical reactions

19.1 Why reaction speeds are important

Making cheese takes time. When milk is allowed to age, bacteria in the milk feed on a sugar called **lactose**, which is present in milk. They turn it into **lactic acid**. Acids make some of the proteins in milk clot. When this happens, the sugar lactose and butterfat are trapped in the clotted milk, making a soft **curd** with a pleasant taste. The liquid part of the curdled milk is called **whey**.

As more proteins clot, curd turns gradually into cheese. Curd takes some days to age naturally. The process can be speeded up by adding a substance called **rennin**. Rennin is a **catalyst**. This is the name for a substance which makes a reaction take place faster, even though it is not itself used up in the reaction. Rennin **catalyses** the clotting of proteins. It is a great help to cheese-makers, who want to minimise the time for which they have to store ageing milk and maximise their production of cheese. Fortunately, rennin is in good supply. It is obtained from the gastric juices in calves' stomachs. A catalyst which is obtained from an animal or a plant is called an **enzyme**. You will learn more about catalysts and enzymes in Section 19.5.

Cheese manufacturers are interested in speeding up the chemical reactions which produce cheese. The more tonnes of cheese they can produce in a month, the more profit they will make. Other people are interested in slowing down chemical reactions. Car manufacturers want to slow down the rusting of iron. Butter manufacturers want to slow down the rate at which butter turns rancid. Conservationists want to slow down the rate at which acid rain attacks stone. Everyone prefers to store their milk in a refrigerator to slow down the rate at which milk turns sour.

In a chemical reaction, the starting materials are called the **reactants**, and the finishing materials are called the **products**. It takes time for the reactants to react to form the products of the reaction. If the reactants take only a short time to change into the products, that reaction is a **fast reaction**. The **speed** or **rate** of that reaction is high. If a reaction takes a long time before it is completely finished (before the reactants have changed into the products), it is a **slow reaction**. The speed or rate of that reaction is low.

Many people are interested in knowing how to alter the speeds of chemical reactions. A number of factors can be changed. These are investigated in Experiments 19.1–19.6. They are

- the size of the particles of a solid reactant
- the concentrations of reactants in solution
- the temperature
- the presence of light
- the addition of a catalyst.

Little Miss Muffet sat on a tuffet, eating the products of a slow chemical reaction.

SUMMARY NOTE

A fast chemical reaction is one in which the reactants change rapidly into the products. Such a reaction has a high **speed** or **rate**. A reaction with a low **speed** or **rate** takes a long time to change the reactants into the products.

A catalyst is a substance which can alter the rate of a chemical reaction without being used up in the reaction.

19.2 How does the size of solid particles affect the speed at which they react?

Remember the reaction you used for the preparation of carbon dioxide (see Section 12.5). One of the reactants, calcium carbonate (marble) is a solid:

calcium carbonate + hydrochloric acid → carbon dioxide + calcium chloride + water

$CaCO_3(s) + 2HCl(aq) \rightarrow CO_2(g) + CaCl_2(aq) + H_2O(l)$

You can use this reaction to find out whether large lumps of a solid react at the same speed as small lumps of the same solid. In Experiment 19.1, you compared the rate of this reaction for small marble chips and for large marble chips. In Fig. 19.1, you can see a different method for finding the rate of the reaction. The carbon dioxide formed, being a gas, escapes from the flask. As it escapes, the mass of the flask and contents decreases.

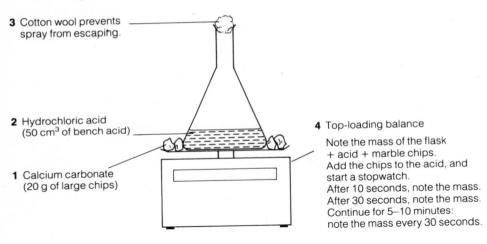

Fig. 19.1 Apparatus for following the loss in mass when a gas is evolved

The mass of the flask and contents is noted at various times after the start of the reaction. The mass can be plotted against time. Figure 19.2 shows typical results. These results show that the smaller the size of the particles of calcium carbonate, the faster the reaction takes place (see Fig. 19.3). There is a larger surface area in 20 g of small chips than in 20 g of large chips.

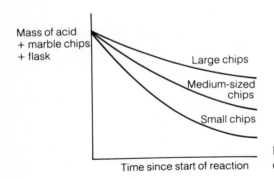

Fig. 19.2 Results obtained with different sizes of marble chips

SUMMARY NOTE

This section is about reactions in which one reactant is a solid. Such reactions take place faster when the solid is divided into small pieces. The reason is that a certain mass of small particles has a larger surface area than the same mass of large particles.

The acid attacks the surface of the marble. It can therefore react faster with small particles than with large particles.

> ## JUST TESTING 99
>
> 1 When potatoes are cooked, a chemical reaction occurs. Potatoes contain both starch and water. When the temperature is high enough, granules of starch react with water to form a gel. Two things decide how fast potatoes cook. One is the temperature of the oven. The other is the speed at which heat is transferred from the surface of the potato to its centre. What can you do to increase this speed?
>
> 2 There is a danger in coal mines that coal dust may catch fire and start an explosion. Explain why coal dust is more dangerous than coal.
>
> 3 Why does soluble aspirin act faster on a headache than aspirin tablets?
>
> 4 Your uncle is suffering from 'acid indigestion'. In the medicine cabinet are some *Fixit* indigestion tablets and some *Mendit* indigestion powder. He asks you which will work faster. Both these remedies are alkalis. Explain how you could test them in the laboratory with a bench acid before giving an opinion.

I'd rather peel 5 pounds of big potatoes than 5 pounds of littl'uns.

It's the ratio of surface area to mass that makes the difference!

Fig. 19.3 Large and small particles

19.3 Concentration and speed of reaction

Many chemical reactions take place in solution. In Experiment 19.2, the reaction studied is

sodium thiosulphate + hydrochloric acid →

sulphur + sodium chloride + sulphur dioxide + water

$Na_2S_2O_3(aq) + 2HCl(aq) \rightarrow S(s) + 2NaCl(aq) + SO_2(g) + H_2O(l)$

Sulphur appears in the form of very small particles of solid. The particles remain in suspension. Figure 19.4 shows how you can study the speed at which sulphur is formed.

Fig. 19.4 Experiment on reaction speed and concentration

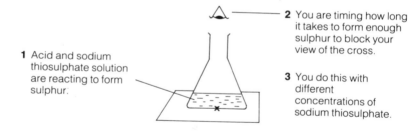

1 Acid and sodium thiosulphate solution are reacting to form sulphur.

2 You are timing how long it takes to form enough sulphur to block your view of the cross.

3 You do this with different concentrations of sodium thiosulphate.

(a) The measurement

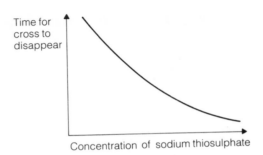

◁ Fig. 19.4

4 The results show that the cross disappears soonest when the solution is most concentrated.

(b) The results

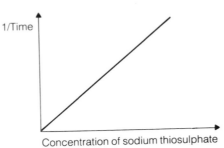

5 When 1/(time for cross to disappear) is plotted against concentration, a straight line is obtained. This shows that for this reaction,

1/time ∝ concentration

(c) A different way of plotting the results

The speed of the reaction is **inversely proportional** to the time taken for the reaction to finish:

Speed of reaction ∝ 1/time

Since, in **5**,

1/time ∝ concentration,
speed of reaction ∝ concentration.

In this experiment, only one concentration was altered. You may have extended Experiment 19.3 by keeping the concentration of sodium thiosulphate constant and altering the concentration of acid. You will have found that the speed of the reaction is *proportional to* the concentration of the acid. If the acid concentration is doubled, the speed doubles. This is because the ions are closer together in a concentrated solution. The closer together they are, the more often the ions collide. The more often they collide, the more chance they have of reacting.

SUMMARY NOTE

This section has covered reactions in which at least one of the reactants is in solution. For the reactions studied, the rate of the reaction is proportional to the concentration of the reactant (or reactants). (That is, the speed doubles when the concentration is doubled.)

JUST TESTING 100

1 Acid rain attacks stone buildings. Why does this reaction take place faster in some parts of the country than others?

2 Imagine you want to investigate the marble chips–acid reaction. You want to find out how changing the concentration of acid will affect the speed of the reaction. Explain how you could adapt the experiment shown in Fig. 19.1 for this purpose.

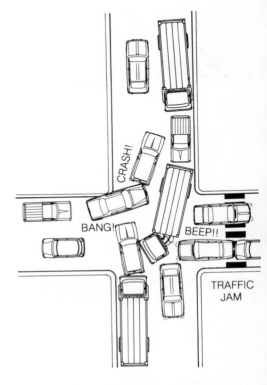

Fig. 19.5 Collisions are frequent in crowds

SUMMARY NOTE

Reactions between gases take place faster when the pressure is increased.

19.4 Pressure and speed of reaction

When gases react, the speed of the reaction can be increased by increasing the pressure. The result is to push the gas molecules closer together. This means that they collide more often and react faster.

19.5 Temperature and speed of reaction

The reaction between sodium thiosulphate and acid can also be used to study the effect of temperature on the speed of a chemical reaction (see Experiment 19.4). There is a steep increase in the speed of the reaction as the temperature is increased (see Fig. 19.6). This reaction goes approximately twice as fast at 30 °C as it does at 20 °C. It doubles in speed again between 30 °C and 40 °C and so on.

You raise the temperature of a solution by putting heat energy into it. At the higher temperature, the ions have more energy. They move through the solution faster. They collide more often and more vigorously. They have to collide before they can react. Once they have collided, there is a chance that they will react (Fig. 19.7).

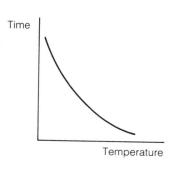
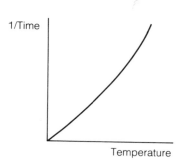

Fig. 19.6 The effect of temperature on the speed of reaction

Fig. 19.7 The faster they go, the more likely they are to collide

19.6 Light affects the speeds of some reactions

Heat is not the only form of energy that can speed up reactions. Some chemical reactions take place faster in the presence of light. The formation of silver from silver salts takes place when a photographic film is exposed to light (see Section 7.9). Photosynthesis, the reaction by which green plants synthesise sugars, takes place in sunlight (see Section 12.6).

SUMMARY NOTE

Reactions take place faster when the temperature is raised.
Some chemical reactions are speeded up by light.

> ## JUST TESTING 101
>
> 1 **Food storage**: The changes which occur when foods decay are chemical reactions. Like other chemical reactions, these changes take place quickly or slowly, depending on the temperature. Study this list of Australian cities and their average temperatures:
>
> | City | Average temperature in °C |
> |---|---|
> | Melbourne | 15 |
> | Sydney | 17 |
> | Brisbane | 19 |
> | Cairns | 25 |
> | Darwin | 28 |
>
> a In which of these cities will food keep longest?
> b In which of these cities will food go off fastest?
> c Look at the results you obtained from Experiment 19.4 on the reaction of sodium thiosulphate at different temperatures. What effect does a 10 °C rise in temperature have on the rate of this reaction?
> Suppose that a change in temperature has a similar effect on the rates of the reactions which take place when food decays. Then, how much more difficult is it to keep food fresh in Cairns than in Melbourne?
> How do Sydney and Darwin compare when it comes to keeping food fresh?
>
> 2 Someone asks you to do experiments to find out what effect temperature has on the rate of a chemical reaction. Choose a reaction. Say how you could find out how fast the reaction was going. Say what measurements you would make. Sketch the apparatus you would use.

19.7 Catalysis

Experiment 19.5 makes a good introduction to catalysis. The reaction studied is

hydrogen peroxide → oxygen + water
$2H_2O_2(aq) \rightarrow O_2(g) + 2H_2O(l)$

This is the reaction you used to prepare oxygen (see Section 8.8). Figure 19.8 shows how the oxygen evolved can be collected and measured in a gas syringe. The evolution of oxygen is very slow at room temperature. There are substances which will speed up the reaction. Manganese(IV) oxide is one of them. When manganese(IV) oxide is added to hydrogen peroxide, the evolution of oxygen takes place much more rapidly (see Fig. 19.9). Manganese(IV) oxide is not used up in the reaction. At the

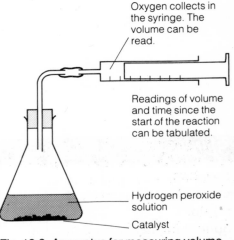

Fig. 19.8 Apparatus for measuring volume of gas evolved

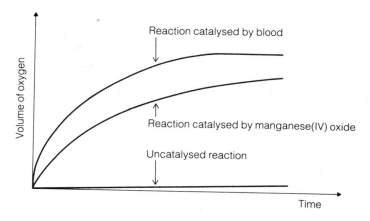

Fig. 19.9 Catalysis of the decomposition of hydrogen peroxide

end of the reaction, the manganese(IV) oxide is as good as new. It can be filtered off and used again. A substance which increases the speed of a chemical reaction without being used up in the reaction is called a **catalyst**. Manganese(IV) oxide will not catalyse all reactions. It catalyses a small number of reactions.

The chemical reactions that take place in animals and plants need catalysts. There are very powerful catalysts, called **enzymes**, in living things. Blood and potato contain enzymes. (You saw their effect in Experiment 19.6.) One of them is called **peroxidase**. It catalyses the decomposition of hydrogen peroxide.

Catalysts will only catalyse certain reactions. Platinum catalyses a number of oxidation reactions. Nickel catalyses hydrogenation reactions (see Table 19.1).

Table 19.1 Some catalysts

Catalyst	Reaction which is catalysed
Platinum	Ammonia + air → nitrogen monoxide Used in the oxidation of ammonia to give nitric acid
Vanadium(V) oxide	Sulphur dioxide + oxygen → sulphur trioxide Part of the contact process for making sulphuric acid
Nickel	Alkene + hydrogen → alkane Used in the hydrogenation of unsaturated oils to form fats
Iron	Nitrogen + hydrogen → ammonia The Haber process

Catalysts are very important in industry. Plastics are made under high pressure. Industrial plastics plants must be well built of strong materials to withstand pressure. A plastics manufacturer tries to find a catalyst which will enable the reaction to give a good yield of plastics at a lower pressure. Then the plant will not have to withstand high pressures. A less costly method of construction can be used. Industrial chemists are always doing research on new catalysts.

Some reactions need high temperatures in order to give good yields of product. This involves the manufacturer in high fuel costs. If an industrial chemist can find a powerful catalyst, the reaction will take place at a lower temperature. This will cut down running costs.

All the processes which take place in our bodies need enzymes. The digestion of food takes place in stages. Each stage needs the presence of a different enzyme. The conduction of nerve impulses along our nerves also needs enzymes.

There are substances which slow down reactions. They have a negative catalytic effect. They are called **inhibitors**. Nerve gases are inhibitors. They slow down the chemical reactions which take place when messages pass along nerve fibres. This leads to paralysis and then to death. Curare is an inhibitor of the same kind. It prevents the enzyme which enables the nerves to conduct messages to the muscles from doing its job. South American Indians used to tip their arrows with curare. What do you think happened to an animal or an enemy who was hit by one of these arrows?

SUMMARY NOTE

Sometimes the speed of a chemical reaction can be increased by adding a substance which is not one of the reactants. Such a substance is called a **catalyst**. A catalyst increases the rate of a chemical reaction without being used up in the reaction. A substance which decreases the rate of a chemical reaction is called an **inhibitor**.

JUST TESTING 102

1 Copy this graph into your workbook. Add a line to show the shape of the graph you would obtain for a catalysed reaction.

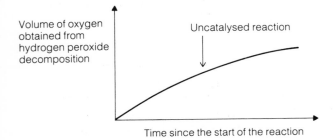

Does the rate of reaction increase or decrease as the time that has passed since the start of the reaction increases? Why?

Exercise 19

1 Zinc reacts with sulphuric acid to give hydrogen. Mention three ways in which you could speed up the reaction.

2 In a set of experiments, magnesium was allowed to react with hydrochloric acid. Each time, 0.5 g of magnesium was used. The volume of hydrochloric acid was different each time. The volume of hydrogen formed was measured each time. The results are shown in the table.

Volume of hydrochloric acid in cm³	Volume of hydrogen in cm³
5	100
15	300
25	500
35	500
45	500

a Plot, on graph paper, the volume of hydrogen produced against the volume of acid used.
b From the graph, find (i) what volume of hydrochloric acid will produce 400 cm³ of gas, (ii) what volume of gas is produced if 50 cm³ of hydrochloric acid are used, (iii) what volume of hydrochloric acid is just sufficient to react with 0.5 g of magnesium.

3 A few small marble chips were dropped into a large volume of hydrochloric acid. The gas evolved was collected and measured. The results obtained are shown in the table.

Time in minutes	Volume of gas in cm³
0.5	14
1.0	23
2.0	35
3.0	41
4.0	44
5.0	44

a What gas was formed during the reaction?
b On graph paper, plot the volume of gas against the time taken for it to be formed.
c From the graph, say what time was taken to collect (i) 20 cm³, (ii) 30 cm³ of gas.
d Suggest three ways in which you could make the reaction take place more slowly.

4 A solution of hydrogen peroxide decomposes slowly to give oxygen. Someone tells you that nickel oxide will catalyse the reaction. What do they mean by **catalyse**?
How could you find out whether your information is correct? Draw the apparatus you would use, and say what you would do.

5 What has keeping food in a refrigerator to do with chemical reactions?

6 What catalysts are there inside your body? What do they do?

*7 A question about deep freeze:
What is the **storage life** of a food? Trained food tasters are employed to find out how long it takes for the food to show the first signs of spoilage. The storage life of the food is judged to be six times this length of time. After its storage life has passed, the food will no longer be of high quality, but it will still be eatable.
Different foods have different storage lives. The storage life of any food depends on the temperature at which it is kept. Graph A shows how the storage life of frozen chickens depends on the temperature at which they are kept. Graph B shows the same for frozen chips and peas.

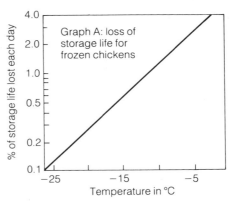

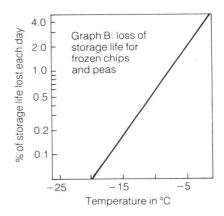

a How much faster does frozen chicken deteriorate (go off) at −5 °C than at −15 °C?
b How much faster do frozen chips and peas deteriorate at −5 °C than at −15 °C?
c Why do the foods keep better at the lower temperature?
d Compare chicken kept at −10 °C and peas kept at −10 °C.
(i) What is the rate at which chicken deteriorates?
(ii) What is the rate at which peas deteriorate?
Can you think of an explanation of the difference?

8 You are provided with
• marble chips in three sizes (small, medium and large)

- dilute hydrochloric acid
- a conical flask with a side-arm, rubber tubing and bung
- a measuring cylinder and a trough.

Design an experiment to find out which react fastest: small, medium or large chips. The experiment should be different from those in this book.

9 You are provided with
- magnesium ribbon
- dilute sulphuric acid
- a thermometer.

Describe how you would find out the effect of temperature on the rate of the reaction between magnesium and dilute sulphuric acid. Say what measurements you would make and what you would do with your results.

Crossword on Chapter 19

Clues across

2, 5 down This is what a catalyst does to a chemical reaction. (5, 2)
7 A catalyst for the decomposition of hydrogen peroxide. (6)
8 The size of these affects the speed of their reactions. (9)
9 The oxide of this metal catalyses the decomposition of hydrogen peroxide. (9)
10 As this increases, the rate of reaction increases. (11)
12, 18 across Needed for measuring reaction rates. (4, 5)
15 This is a negative catalyst. (9)
17 This is formed in the decomposition of hydrogen peroxide. (6)
18 See 12 across.

Clues down

1 This is what you often want chemicals to do. (5)
3 These react faster than lumps. (7)
4 A chemical word for 'split up'. (9)
5 See 2 across.
6 Same clue as 10 across. (13)
11 How to make 3 down from lumps. (5)
12 Symbol for antimony. (2)
13 Invert 'to'. (2)
14 Symbol for polonium. (2)
16 Symbol for platinum. (2)

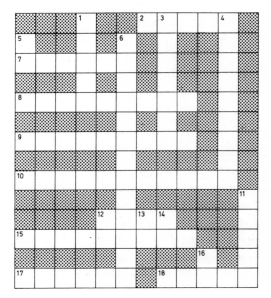

Career sketch: Pharmacist

Sarah is a pharmacist. She works in a 'chemist's' shop dispensing medicines prescribed by doctors. Sarah must understand the medicines she is dispensing. Her work is demanding as she must never make a mistake. She also supplies many medicines which people can buy without a prescription. Customers often ask her advice about what to take for their ailments. Sarah took A-levels, including chemistry and then went to university for a three-year course in pharmacy.

Some pharmacists work in hospitals, dispensing medicines and giving advice to doctors on the drugs available.

Career sketch: Hairdressing

Stuart is an apprentice hairdresser. He left school after GCSE to start work in a salon. He goes for one day a week to take a City and Guilds course at the College of Further Education. Chemistry is one of the subjects he studies. He must learn about the structure of hair and the effects of the chemicals in perming solutions, bleaches and colour rinses.

Career sketch: Synthetic organic chemist

Julie is one of a team of five organic chemists in a company which manufactures weed-killers. If the research department of the firm wants a certain compound, Julie and her group synthesise it (make it) for them. Julie may have to carry out a synthesis involving many stages. Her work is varied because she is always being asked for new compounds. She is looking forward to moving into the research department where she will be able to continue her work on synthesis and also test new compounds as herbicides.

Julie left school after A-levels and is taking Higher National Certificate by day release.

Career sketch: Technical writer

Bryn is a technical writer. His skill is to put scientific and technical information into everyday language which people can readily understand. To make sure his work is accurate, Bryn must spend a long time finding out about the area he is covering, say, lasers or liquid crystals. His writing must be clear and easy to follow. He must write good English and express complicated ideas in simple language. Bryn writes articles for a scientific magazine. Most technical writers work for industrial companies and Government departments.

Bryn has a degree in chemistry.

Table of symbols, atomic number and relative atomic masses

Element	Symbol	Atomic number	Relative atomic mass	Element	Symbol	Atomic number	Relative atomic mass
Actinium	Ac	89	227	Erbium	Er	68	167
Aluminium	Al	13	27	Europium	Eu	63	152
Americium	Am	95	243	Fermium	Fm	100	253
Antimony	Sb	51	122	Fluorine	F	9	19
Argon	Ar	18	40	Francium	Fr	87	223
Arsenic	As	38	75	Gadolinium	Gd	64	157
Astatine	At	85	210	Gallium	Ga	31	70
Barium	Ba	56	137	Germanium	Ge	32	72.5
Berkelium	Bk	97	247	Gold	Au	79	197
Beryllium	Be	4	9	Hafnium	Hf	72	178
Bismuth	Bi	83	209	Helium	He	2	4
Boron	B	5	11	Holmium	Ho	67	164
Bromine	Br	35	80	Hydrogen	H	1	1
Cadmium	Cd	48	112	Indium	In	49	115
Caesium	Cs	55	133	Iodine	I	53	127
Calcium	Ca	20	40	Iridium	Ir	77	192
Californium	Cf	98	251	Iron	Fe	26	56
Carbon	C	6	12	Krypton	Kr	36	84
Cerium	Ce	58	140	Lanthanum	La	57	139
Chlorine	Cl	17	35.5	Lawrencium	Lw	103	257
Chromium	Cr	24	52	Lead	Pb	82	207
Cobalt	Co	27	59	Lithium	Li	3	7
Copper	Cu	29	63.5	Lutecium	Lu	71	175
Curium	Cm	96	247	Magnesium	Mg	12	24
Dysprosium	Dy	66	162.5	Manganese	Mn	25	55
Einsteinium	Es	99	254	Mendelevium	Md	101	256

Element	Symbol	Atomic number	Relative atomic mass	Element	Symbol	Atomic number	Relative atomic mass
Mercury	Hg	80	201	Samarium	Sm	62	150
Molybdenum	Mo	42	96	Scandium	Sc	21	45
Neodymium	Nd	60	144	Selenium	Se	34	79
Neon	Ne	10	20	Silicon	Si	14	28
Neptunium	Np	93	237	Silver	Ag	47	108
Nickel	Ni	28	59	Sodium	Na	11	23
Niobium	Nb	41	93	Strontium	Sr	38	87
Nitrogen	N	7	14	Sulphur	S	16	32
Nobelium	No	102	254	Tantalum	Ta	73	181
Osmium	Os	76	190	Technetium	Tc	43	99
Oxygen	O	8	16	Tellurium	Te	52	127
Palladium	Pd	46	106	Terbium	Tb	65	159
Phosphorus	P	15	31	Thallium	Tl	81	204
Platinum	Pt	78	195	Thorium	Th	90	232
Plutonium	Pu	94	242	Thulium	Tm	69	169
Polonium	Po	84	210	Tin	Sn	50	119
Potassium	K	19	39	Titanium	Ti	22	48
Praesodymium	Pr	59	141	Tungsten	W	74	184
Promethium	Pm	61	147	Uranium	U	92	238
Protactinium	Pa	91	231	Vanadium	V	23	51
Radium	Ra	88	226	Xenon	Xe	54	131
Radon	Rn	86	222	Ytterbium	Yb	70	173
Rhenium	Re	75	186	Yttrium	Y	39	89
Rhodium	Rh	45	103	Zinc	Zn	30	65
Rubidium	Rb	37	85	Zirconium	Zr	40	91
Ruthenium	Ru	44	101				

Group	1	2							
Period 1					1 H hydrogen				
Period 2	3 Li lithium	4 Be beryllium							
Period 3	11 Na sodium	12 Mg magnesium							

Transition metals

			21 Sc scandium	22 Ti titanium	23 V vanadium	24 Cr chromium	25 Mn manganese	26 Fe iron	27 Co cobalt
Period 4	19 K potassium	20 Ca calcium							
Period 5	37 Rb rubidium	38 Sr strontium	39 Y yttrium	40 Zr zirconium	41 Nb niobium	42 Mo molybdenum	43 Tc technetium	44 Ru ruthenium	45 Rh rhodium
Period 6	55 Cs caesium	56 Ba barium	57 La lanthanum	72 Hf hafnium	73 Ta tantalum	74 W tungsten	75 Rh rhenium	76 Os osmium	77 Ir iridium
Period 7	87 Fr francium	88 Ra radium	89 Ac actinium	104* Ku kurchatovium	105* Ha hahnium				

Inner transition metals

58 Ce cerium	59 Pr praesodymium	60 Nd neodymium	61* Pm promethium	62 Sm samarium	63 Eu europium
90 Th thorium	91 Pa protactinium	92 U uranium	93* Np neptunium	94* Pu plutonium	95* Am americium

*Artificial element, does not occur naturally.

			3	4	5	6	7	0
								2 He helium
			5 B boron	6 C carbon	7 N nitrogen	8 O oxygen	9 F fluorine	10 Ne neon
			13 Al aluminium	14 Si silicon	15 P phosphorus	16 S sulphur	17 Cl chlorine	18 Ar argon
28 Ni nickel	29 Cu copper	30 Zn zinc	31 Ga gallium	32 Ge germanium	33 As arsenic	34 Se selenium	35 Br bromine	36 Kr krypton
46 Pd palladium	47 Ag silver	48 Cd cadmium	49 In indium	50 Sn tin	51 Sb antimony	52 Te tellurium	53 I iodine	54 Xe xenon
78 Pt platinum	79 Au gold	80 Hg mercury	81 Tl thallium	82 Pb lead	83 Bi bismuth	84 Po polonium	85 At astatine	86 Rn radon

64 Gd gadolinium	65 Tb terbium	66 Dy dysprosium	67 Ho holmium	68 Er erbium	69 Tm thulium	70 Yb ytterbium	71 Lu lutetium
96* Cm curium	97* Bk berkelium	98* Cf californium	99* Es einsteinium	100* Fm fermium	101* Md mendelevium	102* No nobelium	103* Lw lawrencium

Numerical answers

Just testing 44
5a 3×10^{10} tonne
  b 1.5×10^4 square miles

Exercise 8
  7a 500 g
  8 0.73 tonne
10a 300 tonne/day
  b £30/day

Exercise 9
  6 3×10^4 tonne
  8 0.1 tonne
  9a 10 g
  b 80 drops
10a 0.35 mg/day

Just testing 56
1a 137
  b 5
  c 4
  d 9

Just testing 59
1a 23 g
  b 480 g
  c 3.5 g
  d 7.0 g
  e 16 g
  f 128 g
2a 2 mol
  b 0.025 mol
  c 1 mol
  d 2 mol
3a 64 g
  b 980 g
  c 20 g
  d 200 g
4a (i) 27 g; (ii) 1350 g
  b (i) 54 g; (ii) 9 g

Just testing 60
1 159 g
2 50 g
3 88 g
4 18.6 g
5 142 g

Just testing 64
1 150 cm^3 O$_2$, 75 cm^3 CO$_2$
2 120 l
3 6 l
4 500 g

Just testing 65
1a 0.20 mol/l
  b 0.040 mol/l
  c 0.040 mol/l
  d 0.80 mol/l
2a 0.50 mol
  b 0.020 mol
  c 0.060 mol
  d 2.5×10^{-3} mol

Just testing 66
1 0.30 mol/l
2 0.24 mol/l
3 5.0 cm^3
4a 0.50 l
  b 0.125 l

Exercise 11
1 £30 million
2 2×10^{22}; 7.5×10^{-18} p
3a 0.20 mol
  b 0.40 mol
  c 1.20 mol
  d 0.20 mol
7 102 tonne
8 0.145 mol/l
10 1.6×10^{10} km
11 £120/day
12 19 kg

Exercise 12
5a 4.6 million tonnes
  b 8.2 million tonnes

Exercise 13
8d 3.9 kg

Just testing 98
1 26 g
2 −715 kJ/mol
3 −57 kJ/mol

Exercise 18
6c 2260 kJ

Exercise 19
2b (i) 20 cm^3; (ii) 500 cm^3; (iii) 25 cm^3
3c (i) 0.85 minute; (ii) 1.5 minute
7d (i) 1.2% per day; (ii) 0.55% per day

Index

α-particles 50
A_r: see Relative atomic mass
Acid 87, 92
  carbonic 92
  carboxylic 283
  citric 87, 92
  definition 93
  ethanoic 92, 94, 280, 283
  hydrochloric 92, 245
  mineral 87
  nitric 92
  organic 87, 270
  phosphoric 92
  properties 87
  rain 120–122, 131
  reactions 87
  snow 121
  soil 200
  strength 87, 91, 94
  strong 87, 91
  sulphuric 67, 92
  tartaric 87
  taste 87
  weak 87, 91, 94
Acidic oxides 113
Acidity 88, 93
Addition reaction 271
Adrenaline 286
Adsorb 21
Adsorbent 31
Agent Orange 237
Agriculture 218, 226, 228, 236–238, 284
Agrosoke^R 270
Air 105–128
  composition 115, 116
  liquid 106
  pollution 116
Airship 110
Alcohols 270, 278
Algae 140
Alkali metals 47, 154
Alkaline earths 154
Alkalis 89, 113
Alkanes 260
Alkenes 270
Allotropes, allotropy 30
Alloy steels 165, 166
Alloys 34, 147, 148
  uses 157
Alpha-particles 50
Aluminosilicate 209
Aluminium 170
  alloys 170
  extraction 161
  reactions 150, 157
  uses 170
Amount of substance 182
Amorphous solids 7
Ammonia 92, 194
  bonding 78
  manufacture 229, 268
  reactions 230, 232
  test for 231
Ammonium chloride 231, 232, 233
Ammonium salts 98, 228–232, 235
Ammonium sulphate 231
Amphetamines 286
Amphoteric hydroxides 95
  oxides 95
Anaesthetics 261
Analysis 22

Anhydrous 97
Anion 76
Anode 61
  sludge 163
Anodising 170
Antibiotics 286
Antiseptics 286
Arrhenius, Svante 88
Asbestos 209
Aspirin 284
Astronaut 107
Atomic bomb 55
  energy 55, 300
  fission 56
  fusion 57
  mass unit 43
  number 43, 315
  size 10
  theory 42
Atoms 10, 42–59
Avogadro 181
Avogadro constant 182

β-particles 50
Bacteria 131
Bacteriocide 74
Balancing equations 84
Barbiturates 285
Barium sulphate 98
Bases 87, 88, 92
  definition 93
  weak 94
Basic oxides 113
Basic oxygen furnace 165
Bauxite 161, 171
Bends 107
Beta-particles 50
Bilharzia 211
Biogas 258
Biomass 292
Bitumen 20, 266
Blast furnace 161
Bleach 216, 249, 250
Blue John 96
Bohr, Niels 45
Boiling point 3, 25
Bonding 74–81
  covalent 78
  double 79
  electrovalent 74
  ionic 74
  metallic 148
Bordeaux Mixture 223
Brass 157
Breathalyser 280
Breathing 107, 115
Brighteners 141
Brine 15, 242
Bromine 25, 271
Brönsted–Lowry definitions 93
Bronze 157
Brownian motion 6
Buchner funnel 15
Burning 114, 294

Calcium 150
Calcium carbonate 198, 199, 208, 306
Calcium chloride 98
Calcium hydrogencarbonate 135
Calcium hydroxide 92, 200
Calcium oxide 92, 199
Calcium sulphate 98, 197

Calculating
  equation for reaction 185
  formulas 185
  heat of reaction 299
  mass of product 184
  mass of reactant 184
  reactions in solution 191
  volumes of reacting gases 189
Cancer 1
Car 147, 281
Carbohydrates 279, 294
Carbon
  allotropes 30
  compounds 258–289
  cycle 203
  -14 dating 50
Carbon dioxide 109, 199, 203, 204
  fire extinguishers 206, 207
  preparation 201
  properties 202
Carbon monoxide 115, 118, 162
Carbonates 201
  effect of heat 199
  reaction with acids 201
  summary 202
Cast iron 164, 165
Catalysis 310
Catalytic converters 119, 124, 125
Cathode 61
Cation 76
Cells, chemical 61, 72, 295
Cement 197
Centrifugation/centrifuging 17
Chain, Ernst 287
Chain reaction 56
Chalk 197
Change of state 2, 4
Charcoal 30
Chemical
  bond 74–81
  calculations 179–196
  cells 61, 72, 295
  change 3, 74, 76
  industry 283
  reaction 3, 35–37
  reaction speed 305
Chemotherapy 284
Chernobyl 300
Chile nitre 229
Chlorides 246
Chlorine
  bonding 78
  manufacture 70
  properties 248
  reactions 248–252
  uses 74, 247, 248
Chloroform 248, 261
Chlorophyll 203
Cholera 132
Chromatogram 22
Chromatography 21
Chromium 166
  plating 60, 168
Clay 197
Clean air 126
Clean Air Act 123
Cloud seeding 254
Coal 267, 268, 290
  gas 268
  tar 268
Coastal waters 141
Cobalt 166
Codeine 285

Coke 268
Combustion 114, 262, 271, 268, 293
Common salt: see Sodium chloride
Composition, fixed 36
Compounds 9, 28, 35
   definition 35
   ionic and covalent 80, 81
   and mixtures 37
Computer 8
Concentration 191
   effect on speed of reaction 307
Concrete 197, 198
Condensation 3, 18
Conductors 61
Contact Process 219
Copper 157
   purification 163
Copper(II) oxide formula 187
Copper(II) sulphate 98, 109, 223
   crystals: formula 187
Corrosion 150–153, 157, 168
Cortez, General 213
Cosmetics 284
Courtelle 284
Covalent
   bond 78, 81
   compound 78, 81
   compound, formula of 84
   molecules 79
Cracking 265
Critical mass 56
Crude oil 20, 263
Cryolite 161
Crystallisation 11, 16
Crystals 7, 9, 16
Curie, Marie and Pierre 42

Dalton, John 42
Daniell cell 72
DDT 237, 248
Decant 17
Decomposition
   electrolytic 61–70
   thermal 37, 295
Dehydrating agent 222
Dehydration 222
Detergents 136, 264
Dettol 286
Deuterium 57
Diamond 28
   cutting 28
   structure 29, 79
   uses 29
Diaphragm cell 70
Diesel
   fuel 20
   oil 20, 266
Differences between
   compounds and mixtures 37
   ionic and covalent compounds 81
   metallic and non-metallic
      elements 155
Diffusion 5
Dinitrogen oxide 262
Dioxin 237
Direct combination 99
Discharge of ions 62, 65
Displacement reactions 155, 251, 253
Dissolution/dissolving 10, 15, 295, 297
Distillation 17
   destructive 268
   fractional 18
Divers 107
Domagh, Gerhard 286

Dot-and-cross diagram 78
Double covalent bond 79
Drugs 284
Drying agent 222
Duralumin 148, 157
Dyes 284
Dynamite 233

Electric
   current 65
   fire 174
   light filament 174
Electrical
   cell 60
   conductors and non-conductors 61
Electrochemical reactions 60–72
Electrode 61
   inert or reactive 68
Electrolysis 61
   aluminium oxide 161
   copper(II) chloride 61, 65
   copper(II) sulphate 66, 68
   lead(II) bromide 63
   molten salts 63
   sodium chloride 66, 70
   sulphuric acid 67
   water 134
Electrolyte 61
Electron 43
   arrangement 46
   configuration 46
   orbit 45
   shell 45
Electroplating 60, 69
Electrostatic precipitator 123
Electrovalent bond 76
Elements 9, 28, 34
   definition 29
   families of 47
   list of 315
   metallic and non-metallic 34, 35, 155
   repeating pattern of 47
Empirical formula 188
Emulsify 136
Emulsion 12, 136
Endoscope 208
Endothermic reactions 295, 297
Energy 290–304
   alternative sources 290
   atomic 290
   barrier 298
   of chemical bonds 296
   conservation 290
   consumption 290
   crisis 290
   diagrams 296
   kinetic 296
   nuclear 290
   renewable sources 292
   solar 292
Enzymes 311
Epsom salts 97
Equations
   balancing 84
   calculating 185
   chemical 40
   ionic 89, 102
Esters 283
Estuaries 141
Ethane 260
Ethanoic acid 92, 94, 280, 283
Ethanol 18, 278
Ethene 270, 284
Ether 262

Eutrophication 140
Evaporation 16
   solar 243
Exchange resin 138
Exhaust gases 123–126
Exothermic reactions 262, 293, 297
Explosives 233
Extraction of metals from ores 70, 143, 159, 160

Faraday, Michael 62
Fats 272, 283
Fermentation 279
Fertilisers 140, 143, 218, 226, 228
Fibres 284
Filter funnel and paper 15
Filtrate 15
Filtration 15
Fire
   extinguishers 206
   triangle 206
Fission 55
Fleming, Sir Alexander 286
Florey, Howard 287
Fluoride 172, 254
Fluorine 253
Fluorite 97
Fluothane 262
Foam 12
   extinguisher 207
Food
   chain 140, 238
   production 227, 236
Formulas 38
   calculating 185
   covalent compounds 84
   definition 38
   ionic compounds 82
   list 39
Fossil fuels 267, 290
Fractional distillation 18, 19, 20
Fractionating column 18, 19
Freezing 3
Freezing point 25
Freon 261
Fuel
   cells 72
   gas 291
   oil 20, 66
Fuels 114
   combustion 206
   fossil 263, 267
   hydrocarbon 258–269
   nuclear 55, 57
   pollution 118, 120, 123–126
Fungicides 238
Fusion 57

γ-rays 50
Galena 97
Galvanised iron 168
Gamma-rays 50
Gas pressure 4
Gases 1, 2, 4
Gasohol 281
Gasoline 20
Geiger–Müller counter 50
Geothermal energy 292
Giant molecules 30, 79
Glass
   ceramic 211
   fibre 208
   float 210
   optical 208

plate 208
Pyrex 209
soluble 211
Glucose 203, 222, 294, 295
Glue sniffing 283
Gold 14, 32, 157, 163
Goodyear, Charles 213
Graphite
    structure 30, 79
    uses 30
Greenhouse effect 204
Groups of Periodic Table 48
Gypsum 197

Haber, Fritz 229, 247
Haber process 229
Haematite 161
Haemoglobin 118
Hahn, Otto 55
Half-life 51
Hall–Héroult cell 161
Halogens 47, 253
    and the Periodic Table 47
    reactions 253
Hard water 137
Heat 206
    in chemical reactions 293, 295
    of combustion 299
    of reaction 296–298
    of neutralisation 299
Heavy metals 125
Herbicides 236
Heroin 285
Hiroshima 56
Homologous series 260
Humber Bridge 197
Hydration 272, 294
Hydrocarbons 258–269
    combustion 114, 262
    cracking 265
    in crude oil 18
    saturated 259
    in pollution 115, 124
    unsaturated 271
Hydrochloric acid 245
Hydroelectric power 171
Hydrogen 143
    -2 57
    bomb 57, 87
    ions 88
    manufacture 144, 262
    reactions 145
    reducing agent 143
    uses 143
Hydrogen chloride 244
    bonding 78
    solution 244, 246
Hydrogen ions 88, 90
Hydrogen peroxide 111
    decomposition 111
Hydrogenation 144, 272
Hydrogencarbonates 201
Hydroxide ions 89, 190
Hydroxides 95

Ignition temperature 206, 265
Immiscible liquids 21
Indicators 87, 88, 91
Inert gases 111
Infrared radiation 204
Inhibitors 312
Insecticides 237
Integrated circuit 9, 32
Iodide 254

Iodine 253
Ion exchange 138
Ionic bond 74, 76, 81
Ionic compounds 74, 76, 81
    formulas of 82
Ionic equations 89, 102
Ionic theory 62
Ions 62
    and atoms 63
    and bonds 74
    discharge of 62, 65, 67
    list of 64, 83
    spectator 89, 102
    valency 77
Irish Sea 142
Iron 34, 157, 164
    Age 34
    extraction 161
    pyrites 161
Iron(II) sulphate 98
Isomerism 260
Isotopes 48

Kerosene 20, 266
Kettle scale 138
Kinetic theory 4

Laser 8, 9
Latex 213
Layer structure 79
LCD 1
Lead 157
    in petrol 125
    pollution 125
Lead–acid accumulator 72
Liebig condenser 18
Lime kiln 199
Limestone 161, 162, 197, 199, 208
Limewater 109
Liquefaction 3
Liquid air 106
Liquid crystals 1
Liquids 1, 2, 4
Lister, James 286
Lithium 150
Litmus 91
Lubricants 20
Lubricating oil 20, 266
Lung disease 119
Lynemouth 173

M_r 179
Macromolecular structure 30, 79
Magnesium 150, 157
Magnesium fluoride 77
Magnesium oxide
    bonding 77, 92
    formula 186
Magnetite 161
Malaria 237
Manganese 166
Margarine 144, 272
Mass
    number 43
    of product 184
    of reactant 184
Matter 2
Medicines 284
Melting 3
Melting point 23
Mendeleev 48
Mercury 157
    pollution 125, 139

Mercury oxide 36
Mersey 142
Metal compounds, heat 159
Metal oxides 158
Metallic
    bond 148
    elements 155
Metals 147–178
    combustion 150
    competition for anions 156
    competition for oxygen 155
    extraction 70, 143, 159, 160
    in the Periodic Table 154
    properties 148
    reactions with acids 152
    reactions with air 149, 150
    reactions with water 150
    reactivity series 153, 154
    uses 157, 174
Methane 258
Methanol 278
Methyl orange 91
Mexico City 105
Microcomputer 8, 9
Microprocessor 8, 9
Minamata 139
Mist 12
Mixtures 7, 14
    and compounds 37
    heterogeneous 10, 12
    homogeneous 10
Molar mass 182
Mole 181
Molecular formula 188
Molecules 10
Molybdenum 166
Molluscicides 238
Monoclinic sulphur 214
Monomer 273
Morphine 285
Mortar 200
Moulding
    calendering 275
    compression 275
    extrusion 274
    injection 274

Nagasaki 56
Naphtha 20, 266
Natural gas 258, 263, 291
Neon 110
Neutralisation 88, 90, 192, 294
Neutron 43
    number 43
Nickel 166
Nicotine 285
Nitrates 141, 235
Nitric acid 233–235
Nitrifying bacteria 226
Nitrites 141
Nitrogen 106, 226–241
    atmospheric 106
    cycle 226
    fertilisers 140, 227
    liquid 109
    reactions 227
    uses 109
Nitrogen oxides 124
Nitrosoamines 141
Nobel, Alfred 234
Noble gases 47, 74, 110
Non-electrolytes 61, 62
Non-metallic elements 155
North Sea gas and oil 258
NPK fertilisers 229

Nuclear
　debate 300–302
　energy 55, 300
　fission 56, 57
　fusion 57
　power 290
　reaction 49, 295
　reactor 57
Nucleus 44
Nylon 284

Oil: see Petroleum oil
Optical fibre 208
Organic compounds 258–289
Ostwald 234
Oxidation 113
Oxides
　acidic 113
　amphoteric 95
　basic 88, 113
　neutral 113
　reduction of 158
Oxidising agent 250
Oxy-acetylene flame 107
Oxygen 105
　atmospheric 105
　dissolved 131
　preparation 111
　properties 112
　reactions 112
　reactions with metals 150, 155
　test 112
　uses 107

Pain-killers 284
Paints 284
Particle theory 4
Particles 4, 43, 123, 306
Peat 267
Penicillin 286
Percentage composition 180
Periodic Table 47
Permanent hardness 138
Permutit 138
Perspex 276
Pesticides 236
Petrochemicals 277, 283, 290
Petrol 20, 266
　engine 125, 281
Petroleum
　fractions 266
　gases 20, 263, 266
　oil 20, 263, 264, 277–291
pH 92
　meter 87
Pharmaceuticals 284
Phenol 286
Phenolphthalein 91
Phosphates 229
Phosphorus 229
Photoelectric cell 292
Photography 96
Photosynthesis 203, 295
Physical change 3
Pig iron 164
Plastics
　moulding 274
　thermosetting 273
　thermosoftening 273
　pollution 277
　uses 277
Plastic sand 270
Pollution
　agricultural 140, 228, 237

　atmospheric 116–127
　control 12, 107, 123, 228
　by detergents 141
　by fluorides 172
　by heavy metals 125
　by hydrocarbons 124
　land 171
　by lead 125
　by mercury 125
　by nitrogen oxides 124
　by oil 265
　by radioactivity 302
　of the sea 265
　by sulphur dioxide 120
　thermal 140
　of water 129, 139–142
Poly(chloroethene) 276
Poly(ethene) 273, 276
Polymers 273, 276
Polystyrene 276
Polythene 273
Poly(propene) 276
Port Talbot 177
Potassium 150, 229
Potassium dichromate(VI) 280
Potassium manganate(VII) 280
Potato blight fungus 223
Powder extinguisher 207
Precipitation 89, 101
　electrostatic 123
Preparation of salts 98
Propane 284
Proteins 226
Proton 43
　number 43
PTFE 276
Pure
　liquid 24
　solid 23
　substances 7, 14–27
Purity, tests for 23–25
PVC 248
Pyrex 209

Quartz 7
Quicklime 200

Radioactive
　decay 49
　isotopes 51
Radiation 50
Radioactivity 42, 49
　dangers 51
　protection from 53
　uses 50
Radiocarbon dating 50
Rayon 284
Reactions
　chain 55
　chemical 35
　in solution 191
　speed of 305–313
Reactivity series 153
Red mud pond 171
Reduction 143, 158
Relative atomic mass 43, 49, 179, 315
Relative formula mass 180
Relative molecular mass 179
Residue 15
Respiration 115, 204, 294
Rhombic sulphur 214
Roads 20
Rock phosphate 218

Roofing felt 20
Ruby 8, 9
Rust prevention 168
Rusting 116, 167
Rutherford, Lord 42, 44

Sacrificial protection 168
Salt: See Sodium chloride
Salts 87, 96
　electrolysis 61–70
　hydration 97
　list 97
　preparation 88, 89, 90, 99–102
　soluble 99
　insoluble 101
　uses 97
Sand 197, 208
Sapphire 8, 9
Satellite 8
Saturated hydrocarbons 260
Saturation/unsaturation 270
Schistosomiasis 211
Scrap metal 169
Scum 137
Second World War 56, 237, 281, 287
Selenium 8
Sellafield 300
Semiconductor 32, 155
Separating
　funnel 21
　immiscible liquids 21
　solid from liquid 14
　solids 14
　solute and solvent 17
　solutes in a solution 21
Sewage 132, 259
　works 132
Shale 197, 267
　oil 291
Shapes of molecules 79, 80
Shift reactor 144
Silica: see Silicon(IV) oxide
Silicon 8, 31, 209
Silicon chip 8, 32
Silicon(IV) oxide 208, 209
Silver 158, 163
Silver iodide 254
Slag 162, 165
Smelter 171
Smog 123
Smoke 12
Smokeless zone 123
Smoking 286
Sobrero, Professor 233
Soaps 136
Soda–acid extinguisher 206
Sodium 150, 158, 229
　extraction 160
Sodium carbonate 98
Sodium chloride 96, 242
　bonding 74
　chemicals from 242–257
　crystal 76
　electrolysis 66, 70
　mining 15
　uses 243
Sodium hexadecanoate 136
Sodium hydroxide 92
　manufacture 70
Solar
　cell 8
　energy 290
Solder 158
Solidification 3
Solids 1, 2, 4

Solubility 11
  curve 11
Solute 10
Solutions 10
Solvent 10
  abuse 283
Spectator ions 89, 102
Speed of reaction 305–313
  and catalysis 310
  and concentration 307
  effect of light 309
  and particle size 306
  effect of pressure 309
  effect of temperature 309
Stalactites 136
Stalagmites 136
Starch 279
State symbols 40
States of matter 1, 2
Steel 107, 158, 165, 166
  stainless 168
Stimulants 286
Stonehenge 50
Structural formula 260
Sub-atomic particles 43
Sublimation 3
Substitution reactions 261
Sugars 279
Sulphates 223
Sulphites 217
Sulphonamide drugs 286
Sulphur 213
  allotropes 214
  chemicals from 213–225
  plastic 214
  reactions 215
  uses 214
Sulphur dioxide 119, 120, 216, 218
Sulphur trioxide 220
Sulphuric acid 218
  concentrated 221
  manufacture 219
  reactions 221
  uses 219
Sulphurous acid 217
Supercooled liquid 208
Superphosphate 218
Suspension 12, 17
Symbols 37
  definition 38
  list 38, 315
Synthesis 36, 99

Tar 20, 266, 291
TCP 248, 286
Temperature
  effect on solubility 11
  effect on speed of reaction 309
  inversion 117
Temporary hardness 138
Tetracycline 287
Thalidomide 285
Thermal
  decomposition 37, 295
  dissociation 232, 295
  stability 159
Thermosetting plastics 273
Thermosoftening plastics 273
Thiosulphate 307
Three Mile Island 301

Tin 158
  plating 168
Titration 101, 193
Tobacco 285
Torrey Canyon 264
Trace elements 228
Transistors 31
Transition metals 154
Tranquillisers 285
Trichloroethane 261
Trichloromethane 261
Tungsten 166, 174

Universal indicator 91
Unsaturated hydrocarbons 270
Uranium-235 42, 55, 57
Uranium oxide 55
Urea 228

Vacuum tube 31
Valency 77, 79
Vanadium 166
Vaporisation 3, 4, 18
Vapour 2
Vegetable oil 272, 283
Volumes of reacting gases 189
Vulcanisation 213

Washing soda 138
Water 129–146
  in acids 93
  bonding 78
  coastal 141
  cycle 130
  drinking 132
  ground 132, 141
  hard 137
  pipes 175
  pollution 129, 135, 139–142
  power 292
  pure 135, 139
  purification 132, 135
  reactions 93, 150, 221, 231, 244
  recycling 132
  soft 137
  softening 138
  solvent 135
  tests 135
  treatment 132
  uses 134
  vapour 109, 204
Weak electrolytes 62
Weather balloons 143
Weedkillers 236
Windpower 292
Wine 87
Wrought iron 164

X-ray photographs 9

Yeast 279
Yellowcake 34

Zinc 150, 158